INTERNATIONAL TECHNOLOGICAL UNIVERSITY
This Book is Donated by:
PROF. WAI-KAI CHEN

Date:

1988
International Workshop on

NEW TRENDS IN STRONG COUPLING GAUGE THEORIES

1988 International Workshop on

NEW TRENDS IN STRONG COUPLING GAUGE THEORIES

Nagoya, Japan   24-27 August 1988

Editors

M. Bando
T. Muta
K. Yamawaki

World Scientific
Singapore • New Jersey • London • Hong Kong

Published by

World Scientific Publishing Co. Pte. Ltd.
P O Box 128, Farrer Road, Singapore 9128

USA office: World Scientific Publishing Co., Inc.
687 Hartwell Street, Teaneck, NJ 07666, USA

UK office: World Scientific Publishing Co. Pte. Ltd.
73 Lynton Mead, Totteridge, London N20 8DH, England

NEW TRENDS IN STRONG COUPLING GAUGE THEORIES

Copyright © 1989 by World Scientific Publishing Co. Pte. Ltd.

All rights reserved. This book, or parts thereof, may not be reproduced in any form or by any means, electronic or mechanical, including photocopying, recording or any information storage and retrieval system now known or to be invented, without written permission from the Publisher.

ISBN 9971-50-846-X

Printed in Singapore by JBW Printers & Binders Pte. Ltd.

PREFACE

Recently a considerable amount of attention has been paid to the strong-coupling regime of gauge field theories. The study in search of the strong-coupling phase of gauge field theories is by itself an exciting venture. At the same time it has been found that there exist important applications to the resolution of the long-standing problem of the flavor-changing neutral currents in technicolor theories and to the possible explanation of anomalous heavy-ion events.

In the summer of 1988 (August 24-27) more than a hundred physicists from around the world gathered in Nagoya for the workshop, New Trends in Strong Coupling Gauge Theories, held at Aichi University. The papers collected in this volume are written versions of all the talks given at the Workshop. It should be mentioned here that although Prof. V.A. Miransky and Prof. E. Dagotto were invited to the Workshop they could not participate in the Workshop because of some problems in the departure procedure (Eventually Prof. Miransky arrived in Nagoya right after the Workshop while Prof. Dagotto was not able to come). On account of an importance of their contribution to the Workshop we decided to include their papers in the Proceedings.

The Workshop was sponsored by Aichi University, Hiroshima University, Nagoya University and Physical Society of Japan. For having the Workshop we received financial supports through a number of organizations which are mentioned in the Opening and Closing Addresses. We would like to thank these organizations for their warm supports. We would also like to thank young physicists in Nagoya University and Hiroshima University for their devoted assistances in preparing the Workshop.

January 1989

Editors
M. Bando
T. Muta
K. Yamawaki

WORKSHOP ORGANIZATION

Workshop President   S. Ogawa (Nagoya)

Organizing Committee

    Chairman   Y. Ohnuki (Nagoya)

    K.-I. Aoki (RIFP, Kyoto)
    M. Bando (Aichi)
    R. Fukuda (Keio)
    K. Higashijima (KEK)
    T. Kugo (Kyoto)
    T. Maskawa (RIFP, Kyoto)
    K. Matumoto (Toyama)
    T. Muta (Hiroshima)
    M. Okawa (KEK)
    H. So (Niigata)
    K. Yamawaki (Nagoya)
    T. Yanagida (Tohoku)

Secretariat

    K.-I. Kondo (Nagoya)
    H. Mino (Nagoya)
    T. Nonoyama (Nagoya)

Sponsors

    Aichi University
    Hiroshima University
    Nagoya University
    Physical Society of Japan

viii

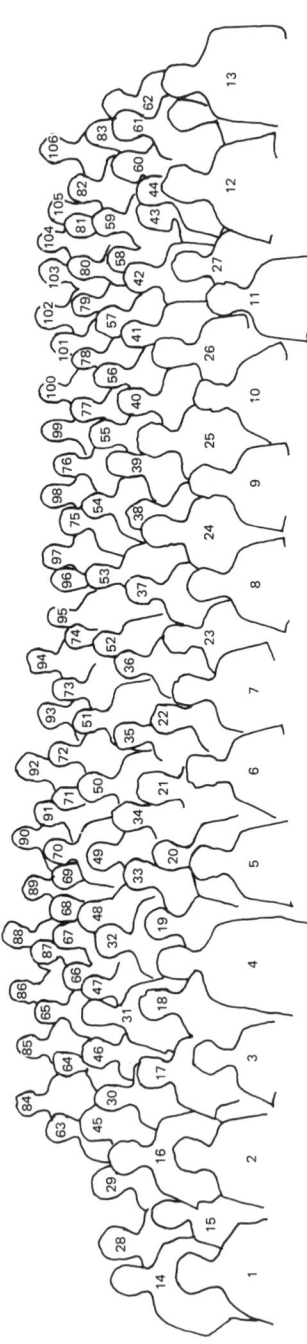

1	Kitamura,	I.		54	Aramaki,	S.
2	Appelquist,	T.W.		55	Kunihiro,	T.
3	Bando,	M.		56	Matsuda,	M.
4	Atkinson,	D.		57	Fujikawa,	K.
5	Ohnuki,	Y.		58	Kikugawa,	M.
6	Ogawa,	S.		59	Kanada,	H.
7	Maki,	Z.		60	Isogaya,	A.
8	Nambu,	Y.		61	Nonoyama,	T.
9	Nishijima,	K.		62	Suzuki,	T.B.
10	Casalbuoni,	R.		63	Minakata,	H.
11	Matumoto,	K.		64	Terasaki,	K.
12	Farhi,	E.		65	Midorikawa,	S.
13	Caldi,	D.G.		66	Fusaoka,	H.
14	Bardeen,	W.A.		67	Suwa,	M.
15	Hirata,	Y.S.		68	Inoue,	M.
16	Moshe,	M.		69	Furui,	S.
17	Muta,	T.		70	Saito,	J.
18	Yamawaki,	K.		71	Sakamoto,	M.
19	Ng,	Y.J.		72	Fukuda,	R.
20	Terazawa,	H.		73	Kikuchi,	H.
21	Nojiri,	S.		74	Yotsuyanagi,	I.
22	Kobayashi,	T.		75	Nakatani,	H.
23	Igi,	K.		76	Imai,	H.
24	Kondo,	K.-I.		77	Ninomiya,	K.
25	Kogut,	J.B.		78	Morii,	T.
26	Hasebe,	K.		79	Seo,	K.
27	Sawada,	S.		80	Kikukawa,	Y.
28	Maedan,	S.		81	Tanabashi,	M.
29	Kugo,	T.		82	Koretsune,	S.
30	Yu,	H.-L.		83	Mizutani,	K.
31	Holdom,	B.		84	Ohno,	S.
32	Takeuchi,	T.		85	Sakakibara,	K.
33	Hattori,	T.		86	Shiokawa,	K.
34	So,	H.		87	Morozumi,	T.
35	Matsunaga,	M.		88	Mitani,	N.
36	Mino,	H.		89	Katsube,	K.
37	Koide,	Y.		90	Kunitomo,	H.
38	Sugiyama,	Y.		91	Imachi,	M.
39	Kobayashi,	A.		92	Higashijima,	K.
40	Yanagida,	T.		93	Nakamura,	A.
41	Nishitani,	T.		94	Shiozaki,	T.
42	Suzuki,	C.		95	Hoshino,	Y.
43	Nagaya,	J.		96	Matsubara,	Y.
44	Meissner,	U.-G.		97	Suematsu,	D.
45	Nojiri,	M.		98	Asano,	H.
46	Aoki,	K.-I.		99	Inazawa,	H.
47	Kodaira,	J.		100	Yokota,	T.
48	Bernard,	V.		101	Saito,	S.
49	Kimura,	M.		102		
50	Suehiro,	K.		103	Morita,	K.
51	Wakaizumi,	S.		104	Ukita,	S.
52	Akiba,	T.		105	Otofuji,	T.
53	Abe,	Y.		106	Komachiya,	M.

CONTENTS

Preface v

Opening Address 1
   S. Ogawa

Quasi-Supersymmetry, Bootstrap Symmetry Breaking, and Fermion Masses 3
   Y. Nambu

Resurrecting Technicolor — Scale invariant Technicolor and a Technidilaton ... 12
   K. Yamawaki

Walking Technicolor 34
   T. Appelquist

Chiral Condensate in Technicolor Theories 44
   T. Nonoyama

Beyond the Ladder 54
   B. Holdom

Dynamics of the Nontrivial Fixed Point in Gauge Field Theories 69
   V. A. Miransky

The Application of Bifurcation Theory to Dynamical Symmetry Breaking 90
   D. Atkinson

Chiral Symmetry Breaking in QED Beyond the Ladder Approximation 95
   H. Mino

Collapse of the Wavefunction and Catalyzed Symmetry Breaking in QED4 ... 103
   J. B. Kogut

Dynamics of Symmetry Breaking in Strong Coupled QED 110
   W. A. Bardeen

Phase Structure of Quantum Electrodynamics in the Framework of the
Schwinger-Dyson Equation 127
   K. -I. Kondo

Renormalization Group Analysis of Strong Coupling Phase in QED 148
  K. -I. Aoki

A New Expansion for Quantum Field Theories 163
  M. Moshe

The Renormalization Group Study of the Effective Theory of Lattice QED ... 172
  Y. Sugiyama

Effective Theory of Strong Coupling QED 181
  H. So

On the Strong Coupling Phase of QED 189
  T. Morozumi

A New Phase of QED and Narrow Peaks in Heavy-Ion Collisions 198
  D. Caldi

Strong-Coupling QED and its Possible Relation to Anomalous Heavy-Ion
Events 206
  Y. J. Ng

Does QED4 Exist? 217
  J. B. Kogut

Tricritical Point in Compact QED 224
  M. Okawa

Phase Structure of Compact Lattice QED with Dynamical Fermions 227
  A. Nakamura

Chiral Symmetry Breaking in (2+1)–Dimensional QED 240
  E. Dagotto

Migdal-Kadanoff Renormalization Group Analysis of Lattice QED 246
  M. Imachi

Effective Potential for Bilocal Composite Fields and its Ambiguity 254
  T. Muta

Gauge Dependence of the Effective Potential for Bilocal
Composite Operators in Abelian Gauge Theories 266
   H. -L. Yu

On-Shell Expansion of the Effective Action 274
   R. Fukuda

Effective Action and the Ambiguity — Free Stability Criterion 285
   M. Komachiya

Hidden Local Symmetries 293
   M. Bando

Vector and Axial-Vector Resonances from a Strongly Interacting
Electroweak Sector 308
   R. Casalbuoni

The Strongly Coupled Standard Model 326
   E. Farhi

Chiral Gauge Theories on a Lattice 331
   T. Kashiwa

Concluding Remarks 339
   Y. Nambu

Closing Address 344
   Y. Ohnuki

List of Participants 346

OPENING ADDRESS

Shuzo Ogawa*

President of the Workshop

Department of Physics, Nagoya University
Nagoya 464, Japan

Ladies and Gentlemen,

It is my great pleasure to express our hearty welcome to all of you and to address a few words on behalf of the organizing committee of the workshop " New Trends in Strong Coupling Guage Theories ".
Since the introduction of the so called strong interaction as Yukawa's meson theory, our knowledge of it has got into a deeper and deeper level of matter, as the late Professor Sakata anticipated when he advocated his composite model of hadrons. In fact our cognition on the fundamental matter has surely attained to the quark

---

*) Professor Emeritus of Nagoya University.

level in cooperation of theory with experiment. For the time being, however, we have been lack of experimental impetus for further advance, and a cue is so much asked for the theoretical investigation. I hear that this workshop will persue to make clear the inner structure of matter behind quark and its interrelation with guage coupling, and the related problems, under the new theoretical phase. The verification of some of the expected results in those energy region would not be far from the present ability of experiment. So I am looking forward to the achievement of this workshop with keen interest, and hoping your success.

Before closing my address, I would like to express, on behalf of the organizing committee, my deep gratitude to Aichi, Hiroshima and Nagoya Universities, and also Japan Physical Society, for the sponsorship extended to us. For the financial aids, I also sincerely thank to Aichi University, Nagoya University(Kato Foundation), Miyoshi-cho, and the following Foundations : Casio, Daiko, Inoue, Ishida, Kashima, and Shimazu. Thank you.

QUASI-SUPERSYMMETRY, BOOTSTRAP SYMMETRY BREAKING, AND FERMION MASSES

Y. Nambu[†]
Enrico Fermi Institute and Department of Physics
University of Chicago, Chicago, IL 60637 USA

ABSTRACT

The characteristic features of the BCS mechanism, one mathematical and the other physical, are explored in relativistic field theories. The first is called quasi-supersymmetry among fermions and composite bosons, and various examples can be found at various energy scales. The other has to do with the notion of a hierarchical chain of symmetry breakings and that of a self-sustaining (bootstrap) mechanism. The latter is applied to electroweak unification, and some predictions are made.

1. INTRODUCTION

My talk will cover essentially the same ground as the one I gave at a recent conference in Poland.[1] For this reason the written version is only a slight variation on the article that will appear in the proceedings of that conference. There are, however, some results that have been obtained since then.

For the background and motivation behind the current work I refer the reader to the papers that have appeared in the last few years.[1-4] Basically it has developed in an attempt to analyze dynamical symmetry breaking mechanisms of the BCS type, with the eventual goal of understanding the origins of masses, especially the masses of quarks and leptons.

The BCS mechanism is defined here as a dynamical symmetry breaking due to a short range attraction between fermions, usually represented by point interactions, giving rise to low energy fermionic and bosonic excitations. They are the Bogoliubov-Valatin fermions (f), the Goldstone (pi) and the Higgs (sigma) bosons having universal mass ratios 1:0:2 (in the weak coupling limit). This situation suggests a kind of broken supersymmetry built into the BCS mechanism. When these low energy features are translated into an effective Ginzburg-

---

[†] Work supported in part by the National Science Foundation: grant no. PHY-85-21588.

Landau-Gell-Mann-Levy (GL) Hamiltonian H, one can in fact factorize the static part of it as a bilinear form in fermionic operators Q and $Q^\dagger$, which generate a deformation of a superalgebra. I call this quasi-supersymmetry.

My student M. Mukerjee and I have recently used the GL form of the BCS theory in a derivation of the Interacting Boson Model of nuclear excitations, in which the symmetry breaking patterns SU(4)-Sp(4) ($\approx$ O(6)-O(5)) and U(3)-O(3) can be explained by the BCS mechanism.[4] But our concern now is about the relevance of the BCS-GL theory in particle physics. There are two different aspects to it. One has to do with the mathematical possibilities of relativistic quasi-supersymmetry, and the other addresses more physical principles.

2. MODELS OF RELATIVISTIC QUASI-SUPERSYMMETRY

A prototype is obtained by emulating the BCS superconductivity. Let ψ and ψ^c be a two component Weyl spinor and its charge conjugate, Φ ($= \sigma + i\pi$) and Π be a complex scalar field and its canonical conjugate, and construct

$$Q = (\Pi^\dagger \sigma_0 \pm \sigma_i \nabla_i \Phi^\dagger)\psi + iW\psi^c, \; (\sigma_0 = 1; \sigma_{i\pm} = \pm \sigma_i, i = 1,2,3) \quad (1)$$

Then

$$\{\overline{Q}, \sigma_{\mu\pm} \overline{Q}^\dagger\}/2 = P_\mu, \mu = 0,..3,$$

gives the total energy and momentum, where bar means spatial integral. In particular,

$$P_0 = \overline{H}, H = \Pi^\dagger \Pi + |\nabla \Phi|^2 + W^2 + i\psi^\dagger \sigma \cdot \nabla \psi + G(\Phi^\dagger \psi^\dagger \psi + \Phi \psi^{\dagger c} \psi^c)/2.$$

is the proper Hamiltonian for a fermion coupled to a Higgs field that gives a Majorana mass with the standard BCS mass ratios mentioned above. In fact if one drops the $\nabla \Phi$ term in Q, this is nothing but the static part of a Ginzburg-Landau representation of BCS superconductivity, with ψ being the nonrelativistic electron field.

When kinetic terms are added, it is crucial in this construction that the bosonic and fermionic degrees of freedom are equal ($= 2$). The boson kinetic energy receives its weight from anticommutators of fermion fields, and the fermion kinetic energy receives its weight from commutators of boson fields, so the weights must be equal. Note also that Q and $Q^\dagger$ would be nilpotent if $W = 0$, giving an ordinary free supersymmetric system of spin 0 and 1/2. So one is dealing here with a deformation of supersymmetry.

This suggests that a (1/2, 1) supersymmetry may also be emulated. In fact

$$Q = \sigma \cdot F_\pm \psi, F_\pm = (E \pm iB)/\sqrt{2}, \quad (4)$$

will do, where B and E are the usual U(1) gauge field components. It leads (in the $A_0 = 0$ gauge) to

$$\{\overline{Q}, \overline{Q}^\dagger\}/2 = \overline{H}, H = (E^2 + B^2)/2 + i\psi^\dagger \sigma \cdot \nabla \psi. \tag{5}$$

The two cases can now be combined to handle a Dirac fermion. Thus let

$$Q_l = (\Pi^\dagger - \sigma \cdot D \Phi^\dagger)\psi_r + (\eta W + \sigma \cdot F_-)\psi_l,$$

$$Q_r = (\Pi^\dagger + \sigma \cdot D \Phi)\psi_l + (\eta W + \sigma \cdot F_+)\psi_r, \quad |\eta| = 1. \tag{6}$$

Here ∇ is relaced by a covariant derivative. One finds that either pair $(Q_l, Q_l^\dagger)$ or $(Q_r, Q_r^\dagger)$ gives a Hamiltonian for a Dirac fermion coupled to a U(1) gauge field and a U(1) × U(1) Higgs field. A few remarks are in order.

1) The suffix l or r implies left- or right-handed spinor (0,1/2) or (1/2,0), which goes with the + sign. Switching ψ_l and ψ_r in Eq.(8) would make Q and $Q^\dagger$ behave like components of (1,1/2) or (1/2,1), and reverse the sign of fermion kinetic terms. There seems nothing intrinsically wrong about this, and offers the possibility of assigning negative weights to some bosons in the degree matching. (See below.)

2) The anticommutator of $Q(x)$ and $Q^\dagger(x')$ must be constructed by inserting a Wilson line between them to get the correct covariant derivatives appearing in the kinetic terms for bosons and fermions.

3) The Yukawa terms get contributions from both the Higgs field [Π, W^2] and the gauge field [E, $D\phi$]. Their relative chiral phase (η) is arbitrary, and the BCS mass relation between m_f and m_σ is replaced by a more general one that also involves the vector meson mass m_V.

Putting aside the details for now, I will list some model theories that are compatible with quasi-supersymmetry just by the matching of the number n of degrees between bosons and fermions, allowing for the possibility of some bosons giving negative weights to fermions.

a) [U(1) × U(1)]$_G$ × U(1)$_g$: $n = 4$. (G and g stand respectively for global and local symmetries.) The above example Eq. (8). 1 Dirac fermion. The bosons are S(**1**) (scalar singlet), P(**1**) (pseudoscalar singlet), and V(**1**) (U(1) gauge vector with 2 helicities).

b) [U(1) × U(1)]$_G$ × SU(2)$_g$: $n = 8$ with 2 Dirac fermions, S(**1**), P(**1**), and V(**3**) (isotriplet).

c) $[SU(2) \times SU(2)]_g$: $n = 8$ with 2 Dirac fermions, (-S(**1**), -P(**3**), V(**3**), and A(**3**) (axial vector). (-S means weight -1.)

d) $[U(1) \times U(1)]_G \times [SU(2) \times SU(3)]_g$: $n = 24$. 2 flavors and 3 colors of quarks, a gauge octet V(**8**) added to the bosons in a).

e) $[SU(2) \times SU(2) \times SU(3)]_g$: $n = 24$, similarly obtained from c). Examples b) vs. d), and c) vs. e) show that a compatibility exists between chiral dynamics at the nucleon level and at the quark level, reminiscent of the 't Hooft consistency for chiral anomaly.[5] Furthermore, there seems to be a close relation between these models and the hidden symmetry schemes of Bando *et al.*[6]

f) $[SU(3) \times SU(2) \times U(1)]_g$ (color and electroweak): $n = 32$. 1 generation of fermions (quarks and leptons) vs. gauge bosons V(**8**), V(**3**), V(**1**), and 2 sets of Higgs H(**4**).

g) $SU(5)_g$: $n = 96$ with 3 generations of $16 = (1 + 5 + 10)$ chiral fermions vs. gauge bosons V(**24**), and 2 sets of adjoint Higgs H(**24**).

An important physical assumption implied here is that, at each energy scale, there should be a set of fields that satisfy the quasi-supersymmetry constraints, but some of them, especially the scalars, are composite fields so they need not exist at a higher energy scale. As one comes down in energy scale, only "massless" fields in the earlier stage will survive, and some new light composites will also appear. For example, g) does not contain the Higgs of f) that would have made the fermions superheavy. This is in line with the "naturalness" philosophy of 't Hooft.[5]

3. BOOTSTRAP SYMMETRY BREAKING

The quasi-supersymmetry explored above is so far of purely mathematical nature. I will now turn to more physical propositions inspired by the BCS theory. Suppose a symmetry breaking at a high energy constituent level gives rise, among other things, to a sigma boson at the low energy scale. The latter, being a scalar, will induce attraction between the quasi-fermions, which in turn may generate a second generation symmetry breaking, and so on. Such a scenario in particle model building has been called tumbling.[7] I would like to claim, however, that an example of tumbling already exist in nuclear physics. It is generally accepted that quantum chromodynamics causes chiral symmetry breaking and quark mass generation leading to hadron dynamics. In particular, the sigma meson supplies strong attraction between nucleons, which not only makes the formation of nuclei possible, but also is responsible for nuclear pairing. In fact one can estimate the pairing energy fairly well from basic nuclear parameters and the gap equation.[8]

A somewhat similar and perhaps more familiar example is the ordinary superconductivity, in the sense that the phonons responsible for the Cooper pairing are Goldstone modes for the breaking of translation (and rotation) invariance in solids.

Instead of a cascading hierarchy of symmetry breakings, one can also envisage a theoretical possibility of bootstrap: The sigma boson and the rest are responsible for their own existence. Even if there may well be a substratum at a higher energy scale, it may not be necessary to look beyond what one sees at the low energy scale; the dynamics may be self-consistent among the latter. Hadron dynamics may have this property[9] in view of the 't Hooft type consistency mentioned above. But how does one clearly formulate the bootstrap condition?

I will discuss this with the concrete example of the Weinberg-Salam electroweak unification with three generations of quarks and leptons. The mass m_i of the ith fermion is generated by the Yukawa coupling f_i of the Higgs as (Fig. 1)

$$m_i = f_i c \ . \tag{7}$$

But there are also tadpole diagrams (Fig. 2a) which represent a sigma-induced attractive interaction between fermions, i.e., a bootstrap mechanism just mentioned. The tadpole also couples to the Higgs and gauge boson (W and Z) loops, Figs. 2b and 2c. These are all quadratically divergent, going with a cut-off parameter Λ

$$\Lambda^2 - m^2 \ln \Lambda^2 \ . \tag{8}$$

If the cut-off is interpreted as a higher energy scale, the bootstrap must mean independence (at least insensitivity) of the result on Λ, i.e., cancellation of divergences among the three diagrams of Figs. 2. Two points of view are possible here: a weak condition (quadratic divergences only) and a strong one (both quadratic and logarithmic divergences). At any rate, the quadratic and logarithmic conditions respectively lead to the following two equations

$$\begin{aligned} I. \quad & \Sigma m_i^2 = 1/8 \ m_H^2 + 1/2 \ m_W^2 + 1/4 \ m_Z^2 \\ II. \quad & \Sigma m_i^4 = 1/8 \ m_H^4 + 1/2 \ m_W^4 + 1/4 \ m_Z^4 . \end{aligned} \tag{9}$$

The condition I is analogous to one that would follow from supersymmetry, since a weakening of the degree of divergences is a characteristic feature of supersymmetry. Here, however, there are no a priori constraints on the number of fields or the values of individual coupling constants and masses. An important consequence of I is that the sum of Yukawa couplings are bounded from below by the known gauge couplings. If the sum is dominated by the top quark, one finds $m_t \geq 70$ Gev.

In ordinary or quasi-supersymmetry that is pertinent to the present problem, however, one requires two Higgs. They must strongly mix so that each can couple to the top quark as the dominant term. In this case Eq. (9) has to be written down for each Higgs with proper modifications.

The weak condition I is a secure one to impose. It is gauge independent; the quadratic divergences in the Higgs self-energy is also eliminated by the same condition.

Let us now turn to the condition II that would be required by a strict adherence to bootstrap. But then there are the usual self-energy terms (Fig. 3) which are also logarithmically divergent, and add to the condition II cross terms in the masses. Keeping only the dominant top quark, I and II together should then determine m_H and m_t from the known m_W and m_Z. The following results emerge.

a. Eq. (9) reduces to a quadratic equation in $(m_H/m_W)^2$ which has no real solutions. (Basically, the sum of the coefficients in either equation must be $\sim > 1$ for the existence of solutions.) When the self-energies of Fig. 3 are added, however, II gets replaced by

$$II'. \quad m_t^4 + 1/16\, m_t^2 m_H^2 \sim 1/8\, m_H^4 + 1/4\, m_Z^4 + 1/2\, m_W^4 + 1/144\, (5m_Z^2 - 14m_W^2)m_H^2. \tag{10}$$

One finds now an essentially degenerate solution

$$m_t \sim m_W \sim 80 \text{ Gev}, \quad m_H \sim 100 \text{ Gev}. \tag{11}$$

(Although m_t is rather stable, the solutions for m_H are sensitive to changes in the parameters.)

b. The case of two Higgs doublets: In a model with the Peccei-Quinn symmetry, for example, the four degenerate (neutral and charged) Higgs and an axion modify Eq. (9) to

$$I. \quad m_t^2 \sim 13/24\, m_H^2 + 1/2\, m_W^2 + 1/4\, m_Z^2,$$

$$II. \quad m_t^4 \sim 1/2\, m_H^4 + 1/2\, m_W^4 + 1/4\, m_Z^4. \tag{12}$$

One finds two solutions

$$m_H \sim 180 \text{ Gev}, \quad m_t \sim 110 \text{ Gev};$$

$$m_H \sim 40 \text{ Gev}, \quad m_t \sim 80 \text{ Gev} \tag{13}$$

where minor contributions from Fig. 3 have been taken into account.

I close with a brief discussion of the entire mass matrix. The weak consistency condition is common to all fermions, but the strong one is not. The Fig. 3 diagrams are different for different quarks, and also depend on other quarks in the intermediate states. So the strong condition will lead to a set of coupled equations for the cancellation of logarithmic terms.

I have examined these self-energies Σ for the top and bottom quarks. The equality of cancellation conditions for top and bottom means $\Sigma_t/f_t = \Sigma_b/f_b$, which leads to the quadratic sum rule

$$m_u^2 - m_b^2 = k(m_Z^2 - m_W^2),$$

$$k = 2 \text{ for 1 Higgs,} \qquad 4 \text{ for 2 Higgs.} \tag{14}$$

Comparing this with Eqs.(11) and (13), one sees that the 1 Higgs model and the first solution of the 2 Higgs model gives $m_b^2/m_t^2 \sim 1/2$, which violates the original assumption of small m_b, The second solution of the 2 Higgs case yields $m_b^2/m_t^2 \sim 0$, albeit slightly negative, which may be acceptable considering the radiative corrections and other adjustable parameters in the model. So the strong bootstrap condition seems to make some meaningful predictions about the top and bottom quarks and the Higgs.

As for the rest of the mass matrix, one cannot expect the same consistency to persist miraculously. But one can make the ansatz

$$m_i = (f_i c + \delta_i), \tag{15}$$

where f_i is the Yukawa coupling, and d_i a correction term. By substituting the ansatz into the self-energies of individual fermions, one will get a set of coupled consistency conditions the structure of which is similar to the form considered by Fritzsch[10] and by Kaus and Meshkov.[11]

On the other hand, the strong bootstrap condition is also similar to the assumption that low energy theory is at a critical point of renormalization group equations since the latter implies absence of logarithmic dependences of physical parameters on the cut-off. Zimmermann and collaborators[12] have obtained, on the basis of renormalization group arguments, results which are not very different from Eq. (11).

I thank Professors G. E. Brown, J. L. Rosner, H. Fritzsch, S. Meshkov, C. Savoy, and Mr. R. Rosenfeld for useful discussions and comments.

REFERENCES

1. Y. Nambu, talk given at the Kazimierz Conference *New Theories in Physics*, 1988, U. Chicago preprint EFI 88-39.
2. Y. Nambu, Physica **15D** (1985) 147.
3. Y. Nambu, (1986), in *Rationale of Beings* Festschrift in honor of G. Takeda (eds. K. Ishikawa et al., World Scientific, Singapore, 1986), p. 3; EFI 85-56.
4. Y. Nambu and M. Mukerjee, Phys. Lett., to be published; EFI 87-57.
5. G. 't Hooft, in *Recent Developments in Gauge Theories* (Plenum Press, London), p. 135.
6. M. Bando, T. Kugo, and K. Yamawaki, Nucl. Phys. **B259** (1985) 493.
7. S. Raby, S. Dimopoulos, and L. Susskind, Nucl. Phys. **B169** (1980) 373.
8. Y. Nambu, in *Festi-Val - Festschrift for Val Telegdi*, (ed. Klaus Winter, publ. North Holland, 1988) p. 181.
9. G.E. Brown, private communication.
10. H. Fritzsch, talk at the Kazimierz Conference *New Theories in Physics*, 1988.
11. P. Kaus and S. Meshkov, Cal Tech preprint CALT-68-1492.
12. J. Kugo, K. Sibold, and W. Zimmermann, Nucl. Phys. **B259** (1985) 331.

FIG. 1.

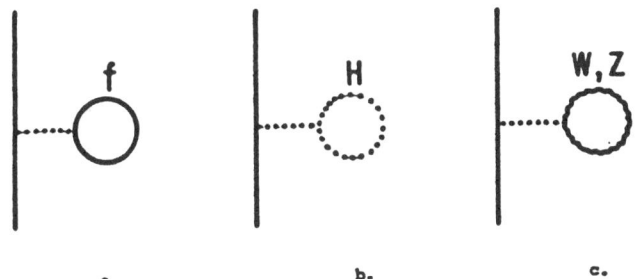

a.   b.   c.

FIG. 2

FIG. 3

Resurrecting Technicolor
—— Scale-invariant Technicolor and a Technidilaton

Koichi YAMAWAKI

Department of Physics
Nagoya University
Nagoya 464-01
Japan

ABSTRACT

I shall explain how the technicolor is resurrected in the light of NEW TRENDS IN STRONG COUPLING GAUGE THEORIES. Basic idea and ingredients of technicolor are critically reviewed. Then I shall describe the scale-invariant technicolor model characterized by the ladder Schwinger-Dyson equation. It is found that the theory has a non-trivial ultraviolet fixed point with large anomalous dimension, which naturally suppresses the flavor-changing neutral currents and at the same time raises the mass of the problematic light technipions. The model is expected to predict a dilaton (technidilaton), a light $J^{PC}=0^{++}$ pseudo Nambu-Goldstone bosons associated with the spontaneous breakdown of the scale invariance, whose dynamical issues are argued in detail.

1. Introduction

Since the pioneering work by Nambu and Jona-Lasinio the dynamical symmetry breaking has been an outstanding problem of particle theory. It is now well-established that the pions are composite Nambu-Goldstone (NG) bosons associated with the spontaneous breaking of the chiral symmetry in QCD, the underlying strong coupling gauge theory in hadron physics. It is the main purpose of this Workshop to discuss about the new possibility for such a strong coupling gauge theory other than QCD as to produce composite NG bosons like pions.

As it stands now, the standard model, QCD plus Glashow-Salam-Weinberg model, is a very successful framework for describing the elementary particles, quarks and leptons. However, the theory is far from fundamental: What is the origin of the spontaneous breaking of the electroweak gauge symmetry? How can one calculate the masses of quarks and leptons? These problems, the "naturalness" and "calculability" problems, are closely related to the characteristic features of the dynamics of the elementary Higgs bosons in the standard model.

Among others technicolor (TC)[1] is a very attractive idea to account for the origin of this symmetry breaking in a natural way, replacing the Higgs bosons by composite NG bosons produced in yet another strong coupling gauge theory with a scale parameter $\Lambda_{TC} \sim O(TeV)$. However, the theory has only been successful in explanation of the mass of W and Z bosons. As to the mass of quarks and leptons, the problem of excessive flavor-changing neutral currents (FCNC's), roughly 10^3 times the experimental limit, has long been a fatal disease to the TC theories.[2] Also the most of TC models predict light pseudo Nambu-Goldstone bosons (PNGB's) with $J^{PC}=0^{-+}$ in the region <10GeV, which are already excluded by the experiment. Because of these, there has been a widely spreading folklore that TC is dead.

In this talk[*),**)], I shall argue that this is no longer true in the modern version of TC, which, in contrast to the classical one, is not a naive scale-up of QCD. I shall explain a scale-invariant TC model[5] whose dynamics, being perturbatively an asymptotically non-free gauge theory like QED, is characterized by the ladder Schwinger-Dyson (SD) equation for the technifermion dynamical mass $\Sigma(p^2)$. It will be found[5] that this scale-invariant TC dynamics possesses a nontrivial ultraviolet fixed point and a large anomalous dimension near the fixed point,

---

*) For earlier reviews, see Refs.3 and 4.
**) The talk is based on the works done in collaborations[5),6),7),8),9),10),11)], with K-I. Aoki, M. Bando, K. Matumoto, K.-I. Kondo, H. Mino, T. Morozumi, T. Nonoyama, T.B. Suzuki and H. So.

$$\gamma^* = 1, \tag{1}$$

corresponding to an unusual asymptotic behavior of $\Sigma(p)$,

$$\Sigma(p^2) \sim 1/p \tag{2}$$

(up to logarithm), which automatically suppresses FCNC's <u>without fine tuning</u>, and at the same time remedies another syndrome, the problem of the light PNGB's with $J^{PC}=0^{-+}$, the technipions, by raising their mass to the order of the TC scale Λ_{TC} (~O(TeV)).

Moreover the model possesses a novel feature, namely, the possible existence of technidilaton[5] which is a new type of PNGB associated with the spontaneous breakdown of (approximate) scale invariance near the fixed point. As will be explained in details in the talk of Kondo[12], we have recently obtained a novel chiral symmetry breaking solution determing the full critical line in the scale-invariant TC with explicit four-technifermion interactions, based on which, I shall argue the dynamical issues of technidilaton: No technidilaton is found on the whole critical line in this approximation.[9],[10],[11]

After the proposal of the scale-invariant TC model[5], the above suppression mechanism of FCNC's, (1) and (2), has been attempted[13] to fit in with the asymptotically <u>free</u> TC theories, based on the "modified" ladder SD equation with the above fixed coupling constant simply replaced by the running coupling constant, especially by the slowly running ("walking") coupling constant[14]. However, as will be shown in the talk of Nonoyama[15], both the analytical[7] and numerical[8] studies of the above SD equation conclude that this "walking" coupling scenario as originally claimed does not yield an enough suppression factor of FCNC's in contrast to the scale-invariant TC model. We thus expect that TC will be the first place where the asymptotically non-free fixed point gauge theories really come into play.

2. Technicolor — Success and Failure

Let me briefly review the basic idea of the old TC developed extensively in the early 1980's.[2] The simplest TC is the "one-doublet model",[2] a simple scale-up of the massless two-flavored QCD in which the quark condensate $\langle \bar{q}q \rangle \sim (250\text{MeV})^3 \sim O(\Lambda_{QCD}^3)$ spontaneously breaks the chiral $SU(2)_L \times SU(2)_R$ symmetry down to $SU(2)_V$, yielding three massless NG bosons, the pions, with the decay constant $f_\pi \simeq 93\text{MeV}$. In much the same way, we assume technifermion condensate $\langle \bar{F}F \rangle \sim O(\Lambda_{TC}^3)$ which yields three composite NG bosons $\pi^\alpha \sim \bar{F} i \gamma_5 \tau^\alpha F$ ($\alpha = 1,2,3$) with the decay constant $F_\pi \simeq 250\text{GeV}$ ($\Lambda_{TC}/\Lambda_{QCD} = F_\pi/f_\pi \simeq 2600$).

In fact, as in the QCD case we can write down a low energy $(p < O(F_\pi))$ effective theory of this TC model in the form of an $SU(2)_L \times SU(2)_R$ linear sigma model

$$L = \frac{1}{2}[(\partial_\mu \vec{\pi})^2 + (\partial_\mu \sigma)^2] - \frac{\lambda}{4}(\vec{\pi}^2 + \sigma^2 - v^2)^2 \; , \tag{3}$$

where σ is the chiral partner of π^α, $\sigma \sim \bar{F}F$, and $F_{\pi^\alpha} = \langle \sigma \rangle = v$. Eq.(3) may be rewritten as

$$L = |\partial_\mu \phi|^2 - \lambda(|\phi|^2 - v^2/2)^2 \; , \tag{4}$$

with $\phi^t \equiv (i\pi^1 + \pi^2, \sigma - i\pi^3)/\sqrt{2}$ being an $SU(2)_L$ doublet, which precisely the standard model Higgs Lagrangian if we take $v \simeq 250\text{GeV}$. Switching on the $SU(2)_L \times U(1)_Y$ gauge interactions in this TC implies the gauging of (4); this immediately leads to the desired result $m_{W^\pm}^2 = (gF_{\pi^\pm}/2)^2$ and $m_{Z^0}^2 \cos^2\theta_W = (gF_{\pi^0}/2)^2$, where $\rho \equiv m_{W^\pm}^2/m_{Z^0}^2 \cos^2\theta_W = (F_{\pi^\pm}/F_{\pi^0})^2 = 1$ is guaranteed by the residual $SU(2)_V$ symmetry of the TC dynamics.

If (3) and (4) were the "fundamental" theory, $\langle \sigma \rangle$ could not be maintained to be $\simeq 250\text{GeV}$ without fine-tuning due to quadratic divergence of the radiative corrections for $p \gg 250\text{GeV}$. However in the composite (gauge) theory at hand, with the quadratic divergence changed into the logarithmic one for $p \gg F_\pi \simeq (\text{composite size})^{-1}$, we can naturally set the decay constant of composite NG bosons $F_\pi \simeq 250\text{GeV}$. Actually F_π is not an order parameter signaled by the vacuum expectation value of elementary scalar field but is the one given in terms of the (non-local) order parameter, $\Sigma(p^2)$, the dynamical mass of the fundamental fermion F generated by the spontaneous chiral symmetry breaking ($S_\chi SB$); more explicitly we have[16]*)

$$F_\pi^2 = N \int_0^\infty dp^2 \; \frac{p^2 \Sigma(p^2)(\Sigma(p^2) - p^2 \Sigma'(p^2)/2)}{(p^2 + \Sigma(p^2)^2)^2} \;, \qquad (5)$$

where we took an SU(N) technicolor for simplicity and set $S_F(p)^{-1} = \not{p} - \Sigma(p^2)$ (in Landau gauge) in Euclidean space.

It is well known from the operator product expansion and the renormalization group analysis[17] that $\Sigma(p^2)$ asymptotically behaves as

$$\Sigma(p^2) \stackrel{p\to\infty}{\sim} \frac{\langle \bar{F}F \rangle}{p^2} C(\alpha(t)) e^{\int^t \gamma_m(\alpha(t'))dt'} \;, \qquad (6)$$

where $t \equiv \log(p/\Lambda_{TC})$, $\alpha(t)$ is the running coupling constant of TC, $\gamma_m(\alpha)$ denotes the anomalous dimension of the composite operator $\bar{F}F$ and $C(\alpha)$ is a certain coefficient depending on the dynamics. In the QCD-like (asymptotically free) theory where $\alpha(t) = 1/bt$ ($p\frac{\partial\alpha}{\partial p} \equiv \beta(\alpha) = -b\alpha^2 + \cdots$), we have $C(\alpha(t)) \sim \alpha(t) = 1/bt$ and $\gamma_m(\alpha(t)) = (3c_2(F)/2\pi)\alpha = A/2t$ ($A \equiv 3c_2(F)/\pi b = 9(N-1)/N(33-2N_f)$ for N_f-flavored SU(N)TC; $c_2(F)$ is the quadratic Casimir of the TC-representation of F) which are substituted into (6) as[18],[19]

$$\Sigma(p^2) \sim \frac{\langle \bar{F}F \rangle}{p^2} (\log(p/\Lambda_{TC}))^{\frac{A}{2}-1} \;. \qquad (7)$$

Substituting (7) into (5), we have a convergent integral, namely, $\underline{F_\pi^2}$ is calculable in terms of the local order parameter $\langle \bar{F}F \rangle \sim O(\Lambda_{TC}^3)$, an infrared scale; it is then natural to keep $F_\pi \sim O(\Lambda_{TC})$.

The mass of W and Z bosons is thus "naturally" explained by the dynamical chiral-symmetry breaking in TC.

---

*) The formula, based on the crude (ladder-like) approximation, is generally valid as far as the convergence (ultraviolet) behavior is concerned, though the precise value of F_π^2 from this formula should not be taken seriously.

As to the mass, m_f, of quarks and leptons (f), however, one encounters a phenomenological disaster: One needs to introduce transitions between F and f to communicate $\langle \bar{F}F \rangle$ down to m_f, quite analogously to the Yukawa couplings in the standard model which communicates $\langle \phi^0 \rangle = v/\sqrt{2}$ down to m_f. This is typically done by the Extended Technicolor (ETC)[20] which in general yields the low energy effective theory below the ETC scale $\Lambda_S (\gg \Lambda_{TC})$;

$$L = L_0 + G_1 \bar{F}F\bar{F}F + G_2 \bar{F}F\bar{f}f + G_3 \bar{f}f\bar{f}f , \qquad (8)$$

where L_0 is the $U(1)_Y \times SU(2)_L \times SU(3)_C \times (TC)$ gauge dynamics for F, and the rest is the $U(1)_Y \times SU(2)_L \times SU(3)_C \times (TC)$-gauge-invariant four-fermion interactions[*] with $G_{1,2,3} \sim O(1/\Lambda_S^2)$. Eq.(8) with G_2 and G_3 terms is actually the cause of the notorious Flavor-Changing Neutral Currents (FCNC's) problem[2].[**]

In fact, m_f is written in terms of $\Sigma(p^2)$ as

$$m_f \sim G_2 \langle (\bar{F}F)_0 \rangle$$

$$\langle (\bar{F}F)_0 \rangle = \frac{N}{4\pi^2} \int_0^{\Lambda_S^2} dp^2 \frac{p^2 \Sigma(p^2)}{p^2 + \Sigma(p^2)^2} , \qquad (9)$$

which is estimated via (7) as $m_f \sim G_2 \langle \bar{F}F \rangle \sim \Lambda_{TC}^3 / \Lambda_S^2$ (up to $N/4\pi^2$ and logarithm). On the other hand, G_3 term contains FCNC's; the most stringent one is the $K^0 - \bar{K}^0$ mixing whose experimental upper bound reads $G_3 \theta_C^2 < 5 \times 10^{-7} TeV^{-2}$ (θ_C: Cabibbo angle, $\theta_C \sim 1/20$), implying $1/\Lambda_S^2 < 10^{-5} TeV^{-2}$ ($\Lambda_S > 350 TeV$). This value of Λ_S is responsible for the mass of s-quark relevant to the $K^0 - \bar{K}^0$ mixing, $m_s \lesssim \Lambda_{TC}^3 / \Lambda_S^2 \sim 10 \cdot (\Lambda_{TC}/TeV)^3 MeV$, which yields

---

[*] $SU(2)_R$ is generally violated in the four-fermion interactions. This violation of G_2 terms is necessary for the iso-spin breaking of m_f ($m_u \neq m_d$), while that of G_2 term could cause a considerable deviation $\rho \neq 1$ ($\rho \equiv m_{W^\pm}^2 / m_{Z^0}^2 \cos^2\theta_W = F_{\pi^\pm}^2 / F_{\pi^0}^2$).

[**] The same problem also applies to the technicolored preon model[21] which yields essentially the same effective Lagrangian as (8) but with more flexible values of $G_1 \sim G_3$ than in ETC models.

$m_s \lesssim 0.3$ MeV for Λ_{TC} 1/3 TeV*), roughly three orders of magnitude smaller than the realistic value $m_s \sim 10^2$ MeV (conversely, if one adjusted $m_s \sim 10^2$ MeV, one would obtain excessive FCNC's by three orders of magnitude).

Moreover, there is another syndrome in the old TC, namely the light PNGB's with $J^{PC}=0^{-+}$ (technipions) which are already excluded by the experiments.[2] Although this difficulty is a model-dependent one (e.g., absent in the "one-doublet model"), it is certainly common to many TC models having larger chiral symmetry than $SU(2)_L \times SU(2)_R$, quite independently of the mechanism responsible for m_f (ETC, TC preon model, etc.). Let us take the "one-family model", for example, which predicts very light colorless PNGB's $P^\pm$, P^3 and P^0.[2] $P^\pm$ acquire mass from electroweak interactions $m_{P^\pm}^2 \sim \alpha m_Z^2 < (10 \text{GeV})^2$ which contradicts the present experimental bounds from PETRA i.e., $>(20\text{GeV})^2$. One may consider another source of the mass of these PNGB's due to G_1 term in (8), which however is estimated to be

$$m_{PNGB}^2 \sim \frac{1}{F_\pi^2} G_1 <(\bar{Q}Q)_0><(\bar{L}L)_0> \, , \qquad (10)$$

where condensates of techniquark Q and technilepton L are given by the same integral as that for $<(\bar{F}F)_0>$ in (9): This yields through (7) $m_{PNGB}^2 \sim G_1 \cdot \Lambda_{TC}^6 / F_\pi^2$ (up to $N/4\pi^2$ and logarithm) $<(1\text{GeV})^2$, if $G_1 \sim 1/\Lambda_S^2 < 10^{-5}\text{TeV}^{-2}$.**)

These diseases, excessive FCNC's and light PNGB's, were actually fatal to the old TC.[2] However, Holdom[22] suggested an ingeneous mechanism for curing these diseases by simply assuming that TC is <u>not</u>

---

*) This is the value of the most popular "one-family model"[2] possessing four doublets, i.e. a techniquark Q and a technilepton L, with $F_\pi = 250\text{GeV}/\sqrt{4}$ (the "one doublet model" yields $m_s \sim 3$MeV).

**) There is actually no ETC interactions producing G_1 term responsible for (10), but one can introduce Pati-Salam type gauge interaction[2] or preonic interaction[21] connecting Q and L, with G_1 being much more stringent than ETC scale.

asymptotically free (below the ETC scale Λ_S) but has a nontrivial ultraviolet fixed point at $\alpha = \alpha^* \neq 0$, whose strong coupling ($\alpha > \alpha^*$) phase is identified with the confining and the SχSB phase having a large anomalous dimension γ_m of the operator $\bar{F}F$, $\gamma_m(\alpha(t)) \sim \gamma^* \equiv \gamma_m(\alpha^*) \gtrsim 1$, at the asymptotic region $t \gg 1$.[*)] The integral of γ_m in (6) now yields $\gamma^* t$, so that we have

$$\Sigma(p^2) \sim C(\alpha^*) \frac{\langle \bar{F}F \rangle}{p^2} \left(\frac{p}{\Lambda_{TC}}\right)^{\gamma^*}, \quad (11)$$

which is substituted into (9) to yield

$$\langle (\bar{F}F)_0 \rangle \sim \langle \bar{F}F \rangle (\Lambda_S/\Lambda_{TC})^{\gamma^*} \sim \Lambda_{TC}^3 (\Lambda_S/\Lambda_{TC})^{\gamma^*} \quad (12)$$

(up to numerical factor of O(1)) in comparison with the asymptotically free case, (7) and $\langle (\bar{F}F)_0 \rangle \gg \langle \bar{F}F \rangle$. The enhancement factor $(\Lambda_S/\Lambda_{TC})^{\gamma^*}$ in (12) <u>simultaneously</u> resolves the problems of FCNC's and PNGB's: From (9) and (10) we have

$$m_f \sim G_2 \langle \bar{F}F \rangle \left(\frac{\Lambda_S}{\Lambda_{TC}}\right)^{\gamma^*},$$

$$m_{PNGB}^2 \sim \frac{1}{F_\pi^2} G_1 \langle \bar{Q}Q \rangle \langle \bar{L}L \rangle \left(\frac{\Lambda_S}{\Lambda_{TC}}\right)^{2\gamma^*}. \quad (13)$$

For $\gamma^* \simeq 1$, we obtain $m_s \lesssim 300$ MeV and $m_{PNGB} \simeq 1$ TeV in the case that $\Lambda_S/\Lambda_{TC} \gtrsim 10^3$.[**)]

[*)] This mechanism was subsequently discussed in TC within a GUT scenario[23] and also in the technicolored preon model.[21),24]

[**)] $\gamma^* > 1$ implies a divergent m_{PNGB}^2 in the ETC (Pati-Salam gauge) scenario, while in the technicolored preon model[24] the form factor can be more rapidly damping than the massive vector meson propagator, thus yielding a finite (but enhanced) value $m_{PNGB} \sim (\Lambda_S/\Lambda_{TC})^{\gamma^*-1} \cdot (1\text{TeV})$. The same situation happens to the radiative mass of colored PNGB's[2] due to color forces. The radiative electroweak mass of $P^\pm$, on the other hand, is convergent in either model for $\gamma^* < 2$. (See, e.g., the discussion on $m_{\pi^\pm}^2 - m_{\pi^0}^2$ in Ref.25).

It is very important to note that (5) is still convergent and dominated by the infrared part as far as $\gamma^* < 2$, so that we can maintain the $F_\pi \sim O(\Lambda_{TC})$ without fine-tuning.

Unfortunately the idea, for all its attractive features, has not been paid much attention until recently, mainly because of the lack of explicit (field-theoretical) dynamical model realizing the mechanism, i.e., existence of a nontrivial ultraviolet fixed point and large anomalous dimension. Does such a theory really exist? The answer is yes. This is the "scale-invariant TC model"[5] based on the ladder SD equation for the asymptotically non-free gauge theory like QED, which indeed possesses a nontrivial fixed point and a large anomalous dimension $\gamma^* = 1$.

3. Scale-invariant technicolor model

Let me now explain the scale-invariant TC model.[5] The technifermion contents of this model are arranged in such a way that the TC gauge is asymptotically non-free; our basic assumption is that the dynamical feature of this model may be characterized by the ladder SD equation. We first consider only the L_0 part of (8) (four-fermion interactions will be discussed in the next section). As a typical asymptotically non-free theory, we shall first study the U(1) case, the (strong coupling) QED.

In Euclidean space, the ladder SD equation of QED in Landau gauge takes the form

$$\Sigma(p^2) = m_0 + \frac{3g^2}{(2\pi)^4} \int d^4q \, \frac{1}{(p-q)^2} \, \frac{\Sigma(q^2)}{q^2 + \Sigma(q^2)^2}, \qquad (14)$$

where the m_0 and g are the bare mass of the fermion and the gauge coupling constant, respectively. Eq.(14) has been extensively studied by many authors. In particular, it was clearly demonstrated in the cut-off version[26],[27] that there exists a critical point $\alpha(\equiv g^2/4\pi) = \alpha_c (= \pi/3)$ above which ($\alpha > \alpha_c$) the chiral symmetry is spontaneously broken, with $\Sigma(0)$ being non-zero for $m_0 = 0$ (SχSB solution).

This is easily seen by converting (14) (after angular integration) into a differential equation and the infrared and ultraviolet boundary conditions:[27]

$$(x\Sigma(x))'' + \frac{3\alpha}{4\pi} \frac{\Sigma(x)}{x+\Sigma(x)^2} = 0 \quad (x \equiv p^2), \tag{15}$$

$$\lim_{x \to 0} x^2 \Sigma'(x) = 0 \quad (\text{IRBC}), \tag{16}$$

$$(x\Sigma(x))'\big|_{x=\Lambda^2} = m_0(\Lambda) \quad (\text{UVBC}). \tag{17}$$

The asymptotic solution of (15) is given by

$$\Sigma(p^2) \sim \tilde{A} \frac{\Sigma(0)^2}{p} \times \begin{cases} \frac{1}{\sqrt{1-\frac{\alpha}{\alpha_c}}} \sinh(\sqrt{1-\frac{\alpha}{\alpha_c}} (\log(\frac{p}{\Sigma(0)})+\delta)) & (\alpha<\alpha_c), \quad (18) \\[6pt] \log\frac{p}{\Sigma(0)} + \delta & (\alpha=\alpha_c), \quad (19) \\[6pt] \frac{1}{\sqrt{\frac{\alpha}{\alpha_c}-1}} \sin(\sqrt{\frac{\alpha}{\alpha_c}-1} (\log(\frac{p}{\Sigma(0)})+\delta)), & (\alpha>\alpha_c). \quad (20) \end{cases}$$

where $\tilde{A}$ and δ are constants depending on α. It is evident that (18), behaving as $\Sigma(p^2) \sim p^{-1+\sqrt{1-\alpha/\alpha_c}}$, does not satisfy (17) for $m_0(\Lambda) \equiv 0$ and neither does (19) (these solutions are actually the explicit chiral-symmetry breaking solutions with $m_0(\Lambda) \neq 0$ and $m_0(\Lambda) \to 0$ ($\Lambda \to \infty$)). Only the oscillating solution for $\alpha > \alpha_c$ satisfies the UVBC (17) for $m_0(\Lambda) \equiv 0$, to be identified with a SχSB solution; (17) now reads

$$\sqrt{\frac{\alpha}{\alpha_c}-1} (\log\frac{\Lambda}{\Sigma(0)}+\delta) + \tan^{-1}\sqrt{\frac{\alpha}{\alpha_c}-1} = n\pi \quad (n \geq 1), \tag{21}$$

with n=1 being the ground state solution.

Eq.(21) requires a nontrivial dependence of α on the cut-off Λ if $\Sigma(0)$ is required to be kept finite for $\Lambda \to \infty$;[27],[28]

$$\frac{\alpha(\Lambda)}{\alpha_c} = 1 + \frac{\pi^2}{[\log(e^{\delta+1}\Lambda/\Sigma(0))]^2} \quad (\delta \to 0.714^{[8]}). \tag{22}$$

Recently the cut-off (Λ) dependence of this SχSB solution has been properly understood by Miransky[28],[29] in the sense of the continuum limit of lattice gauge theories, leading to a surprising field-theoretical insight; the critical point $\alpha=\alpha_c$ should be interpreted as a nontrivial ultraviolet fixed point, defining the continuum limit of the theory.[30] Actually the β function $\beta(\alpha)\equiv\Lambda\frac{\partial\alpha(\Lambda)}{\partial\Lambda}$ takes the form[29]; $\beta(\alpha)=0$ $(\alpha<\alpha_c)$ and $\beta(\alpha)=-2/3(\alpha/\alpha_c-1)^{3/2}$ $(\alpha>\alpha_c)$, indicating an evidence for the nontrivial ultraviolet fixed point.

Still another amazing fact is that the anomalous dimension of the fermion bilinear operator, $\gamma_m(\alpha)\equiv-\Lambda\frac{\partial}{\partial\Lambda}\log m_0(\Lambda)$, becomes unity at the fixed point; $\gamma_m(\alpha)=1-(1-\alpha/\alpha_c)^{1/2}$ $(\alpha<\alpha_c)$ and

$$\gamma_m(\alpha) = 1 \qquad (\alpha>\alpha_c) . \qquad (23)^{[5],[28],[29]}$$

As a consequence the asymptotic form of $\Sigma(p^2)$, "renormalized" à la Miransky, can be written in the form[5],[7]

$$\Sigma(p^2) \sim \frac{\Lambda_{QED}^2}{p} \cdot \log(\frac{p}{\Lambda_{QED}}) \qquad (\alpha>\alpha_c) , \qquad (24)^{*)}$$

where $\Lambda_{QED}\equiv\mu\cdot\exp(-\pi/\sqrt{\alpha(\mu)/\alpha_c-1})$ is a renormalization-group-invariant scale parameter of QED in the strong coupling phase. The appearance of Λ_{QED} is due to the dimensional transmutation similarly to Λ_{QCD} in QCD.

Now in the asymptotically <u>non-free</u> TC model, the ladder SD equation remains the same as (14) except that α is replaced by $c_2(F)\alpha$.**) Thus the nontrivial ultraviolet fixed point does exist at $\alpha=\alpha_c=\pi/3c_2(F)$, the β function being given as $\beta(\alpha)=-(2/3c_2(F))(\alpha/\alpha_c-1)^{3/2}$, with the anomalous dimension $\gamma^*\equiv\gamma_m(\alpha_c)=1$. Accordingly we have the same asymptotic behavior of the SχSB solution $\Sigma(p^2)$ as that in the U(1) case, (24);[5],[7]

---

*) (20) becomes $\Sigma(p^2)\sim\Sigma(0)^2/p\cdot(\log\frac{p}{\Sigma(0)}+\delta)$ in $\Lambda\to\infty$ $(\alpha/\alpha_c\to 1)$.
**) U(1)TC model[3] is not excluded, though not quite successful.

$$\Sigma(p^2) \sim \frac{\Lambda_{TC}^2}{p} \cdot \log(\frac{p}{\Lambda_{TC}}), \qquad (25)$$

where Λ_{TC}, like the Λ_{QED} in the above, stands for the scale parameter of our TC in the strong coupling phase. This is precisely the explicit dynamical model that realizes the mechanism described in Section 2, Eqs.(11)-(13), with a concrete value of the anomalous dimension $\gamma^*=1$, which naturally resolves the long standing problems of FCNC's and light PNGB's simultaneously.*⁾

4. Technidilaton

Let us next include four-fermion interactions, the G_1 term in (8) (G_2 and G_3 terms are irrelevant to our SD equation). We have been pretending that the G_1 term as well as G_2 and G_3 terms is merely a perturbation to the L_0 dynamics. This may not be true when the four-technifermion interaction becomes strong enough to trigger the $S_\chi SB$. One of the most relevant issues to this possibility would be a technidilaton[5],[6], a new type of PNGB associated with the spontaneous breakdown of the (approximate) scale invariance expected in our scale-invariant TC model. This technidilaton, having $J^{PC}=0^{++}$, would be an outstanding low energy signature of the TC of this kind, in sharp contrast with the asymptotically free TC models whose low energy signature is the existence of light 0^{-+} PNGB's, the technipions.

Actually our ladder SD equation (14), with $m_0=0$ and $\Lambda\to\infty$, possesses a scale-invariance under the transformation $\Sigma(p^2)\to\kappa\Sigma(p^2/\kappa^2)$, with κ being a constant. Then the ultraviolet fixed point $\alpha=\alpha_c$ is naively expected to be a scale-invariant point ($\Lambda\to\infty$). Since a dimensionful parameter $\Sigma(0)\neq 0$ is already acquired via the spontaneous breakdown of the chiral symmetry, we also expect a spontaneous breakdown of the scale-invariance as well, which would require the existence of dilaton.[29] Unfortunately the ladder approximation does not yield such a dilaton

---

*⁾ Similar mechanism on the line of Ref.14 has also been pointed out[31] within the cut-off version[26],[27], which, as it stands, is distinguished from the fixed point theory. Incidentally, Ref.14 as well as Ref.31 does not refer to Ref.22.

pole, which might endanger the fixed point identification[28] of $\alpha = \alpha_c$.

As a way out, Bardeen, Leung and Love[29] introduced a chiral-invariant four-fermion interaction, $(G_0/2)[(\bar\psi\psi)^2 - (\bar\psi\gamma_5\psi)^2]$, into the ladder QED; the SD equation (14) (after angular integration) now reads

$$\Sigma(p^2) = m_0 + \frac{g}{\Lambda^2}\int^{\Lambda^2} dq^2 \frac{q^2 \Sigma(q^2)}{q^2 + \Sigma(q^2)^2}$$

$$+ \lambda \int^{\Lambda^2} dq^2 \frac{\Sigma(q^2)}{q^2 + \Sigma(q^2)^2} [\frac{q^2}{p^2}\theta(p-q) + \theta(q-p)] , \qquad (26)$$

where we have defined dimensionless couplings $g \equiv G_0 \Lambda^2/4\pi^2$ and $\lambda \equiv 3\alpha/4\pi$. Then they looked for a "true" (scale-invariant) fixed point in the two-parameter space (λ, g) of the gauge and the four-fermion couplings. They actually solved (26) only in the "strong coupling region" ($\lambda > \lambda_c = 3\alpha_c/4\pi = 1/4$) and identified the point (1/4, 1/4) with a "fixed point", where again no dilaton pole was observed.

To draw a decisive conclusion on the dilaton pole, however, we need to know the structure of the chiral phase transition in the whole (λ, g) plane including $\lambda \lesssim \lambda_c$. Quite recently we have found[9],[10],[11]*) a complete set of SχSB solutions of (26) in the chiral symmetric limit $m_0(\Lambda) \equiv 0$ and a full critical line in the whole (λ, g) plane in both the analytical and numerical methods, the results being identical to each other as they should be. Eq.(26) may be converted into precisely the same differential equation as (15) with the same IRBC (16), only the UVBC (17) being changed into

$$(x\Sigma(x))' + \frac{g}{\lambda} x\Sigma'(x)\Big|_{x=\Lambda^2} = m_0(\Lambda) . \qquad (27)$$

It is important that the four-fermion coupling constant enters only through UVBC.

Analytical solutions take the same form as (18)-(20), with the UVBC (21) for $m_0(\Lambda) \equiv 0$ being replaced by

---

*) See also Appelquist et al.[32]

$$2\sigma(\log(\Lambda/\Sigma(0))+\delta) + \tanh^{-1}(2\sigma(\lambda+g)/(\lambda-g)) = 0 \quad (0<\lambda<\lambda_c), \quad (28)$$

$$\log(\Lambda/\Sigma(0)) + \delta + (\tfrac{1}{4}+g)/(\tfrac{1}{4}-g) = 0 \quad (\lambda=\lambda_c), \quad (29)$$

$$2\rho(\log(\Lambda/\Sigma(0))+\delta) + \tan^{-1}(2\rho(\lambda+g)/(\lambda-g)) = n\pi \quad (\lambda>\lambda_c), \quad (30)$$

where $2\sigma \equiv \sqrt{1-\lambda/\lambda_c}$, $2\rho \equiv \sqrt{\lambda/\lambda_c - 1}$, $0<\tan^{-1}(2\rho(\lambda+g)/(\lambda-g))<\pi$ and $n=1,2,\cdots$. Note that thanks to $g \neq 0$, the non-oscillating solution (18) ($\lambda<\lambda_c$) (and (19) ($\lambda=\lambda_c$) as well) satisfies the UVBC even in the chiral symmetry limit $m_0(\Lambda) \equiv 0$, thus yielding a SχSB solution with much slower damping $\Sigma(p^2) \sim p^{-1+\sqrt{1-\lambda/\lambda_c}}$ than the oscillating one $\Sigma(p^2) \sim p^{-1} \cdot (\log p)$ ($\lambda>\lambda_c$), (20) ((24) and (25)).

More explicitly we find bifurcation solutions[9], nontrivial (SχSB) solutions $\delta\Sigma(p^2) \neq 0$ bifurcated from the trivial one $\Sigma(p^2) \equiv 0$ (a chiral-symmetric solution) for $m_0(\Lambda) \equiv 0$, which take the same form as (18)-(20) (for all p^2, $\mu^2 \leq p^2 \leq \Lambda^2$), with replacement $p/\Sigma(0) \to p/\mu$, $\Sigma(0)^2 \to \delta\Sigma(\mu^2) \cdot \mu$, $\bar{\Lambda} \to \sqrt{\lambda/\lambda_c}$, and $\delta \to (1/2\sigma)\tanh^{-1}(2\sigma)$, 1 and $(1/2\rho)\tan^{-1}(2\rho)$ ($0<\tan(2\rho)<\pi$), respectively.[*] Eqs.(28)-(30) now read

$$2\sigma \log \frac{\Lambda}{\mu} = \tanh^{-1}\left(\frac{\sigma}{g-\tfrac{1}{4}-\sigma^2}\right) \quad (0<\lambda<\lambda_c), \quad (31)$$

$$2\log \frac{\Lambda}{\mu} = \frac{1}{g-\tfrac{1}{4}} \quad (\lambda=\lambda_c), \quad (32)$$

$$2\rho \log \frac{\Lambda}{\mu} = \tan^{-1}\left(\frac{\rho}{g-\tfrac{1}{4}+\rho^2}\right) + n\pi \quad (\lambda>\lambda_c), \quad (33)$$

where $-\pi < \tan^{-1}\left(\frac{\rho}{g-1/4+\rho^2}\right) < 0$ with $n \geq 1$.

Then we find a <u>critical line</u> (Fig.1)[9] in the continuum limit ($\Lambda/\mu \to \infty$);

$$g = (\tfrac{1}{2} + \sigma)^2 \quad (0<\lambda<\lambda_c),$$

---

[*] In the bifurcation method[33], Eqs.(15), (16) read $(x \cdot \delta\Sigma(x))'' + \lambda \cdot \delta\Sigma(x)/x = 0$ and $(\delta\Sigma(\mu^2))' = 0$, respectively, where infrared cut-off $\mu (>0)$ was also introduced.

$$g \to \frac{1}{4} \quad (\lambda=\lambda_c) , \tag{34}$$

$$\rho \to 0 \ (g < \frac{1}{4}) \quad (\lambda>\lambda_c) .$$

Note that the points $(\lambda,g)=(1/4,0)$ and $(0,1)$ represent the critical couplings of the pure QED and the pure Nambu-Jona-Lasinio model, respectively, and the point $(1/4,1/4)$[29] seems to be a special point in view of the full critical. The critical line implies that the critical gauge coupling λ can be small ($<1/4$) if we include enough ($g>1/4$) attractive forces due to four-fermion interactions.

The same critical line ($\Lambda/\Sigma(0)\to\infty$) as (34) can also be obtained by other approximation, with the analytical solutions for $\Lambda/\Sigma(0)<\infty$ being somewhat different from (31)-(33).[10]

Now that we have found a complete set of solutions and a full critical line, we can discuss the dilaton pole in a decisive manner. Following Bardeen et al.[29], we study the fermion-antifermion scattering amplitude with $J^{PC}=0^{++}$ at zero momentum; a possible massless pole only comes from the sum of bubble diagrams containing ladder-like radiative corrections which is essentially given by $1/D_S^R(0)$ where $D_S^R(0)$ is the renormalized scalar denominator

$$D_S^R(0) = [1 + G_0 B_S^0(0)] Z_S^2 / G_0 , \tag{35}$$

with B_S^0 and Z_S being the bare scalar bubble function at zero momentum and the renormalization constant of the scalar vertex, respectively. The dilaton pole would imply $D_S^R(0)=0$ at $\Lambda\to\infty$ (on the critical line).

It is rather straightforward to rewrite (35) into

$$D_S^R(0) = [\frac{2}{(2\pi\Sigma(0))^2} \cdot (1+ \frac{2\lambda}{g} \frac{\lambda+g}{1+\Sigma(\Lambda^2)^2/\Lambda^2})]$$
$$\cdot [(\frac{1}{1+\Sigma(\Lambda^2)^2/\Lambda^2} - \frac{g}{(\lambda+g)^2}) \cdot (\Lambda\Sigma(\Lambda^2))^2] . \tag{36}$$

The first [ ] yields trivially a finite result near the critical line but the second [ ] is rather delicate; the factor $[1+\Sigma(\Lambda^2)^2/\Lambda^2]^{-1}$

$-g/(\lambda+g)^2 \to 1-g/(\lambda+g)^2$ ($\Lambda\to\infty$) which vanishes on the critical line (34), while $(\Lambda\Sigma(\Lambda^2))^2$ yields a divergence for $\Lambda\to\infty$, both of which are precisely cancelled by each other, resulting in a finite expression for $D_S^R(0)$. In fact we obtain from (28)

$$S(\lambda,g) \equiv 1-g/(\lambda+g)^2 = (\lambda_c/\lambda)\cdot(4\sigma^2)/\sinh^2[2\sigma(\log(\Lambda/\Sigma(0))+\delta)] , \quad (37)$$

and from the solution (18), we have

$$(\Lambda\Sigma(\Lambda^2))^2 = \tilde{A}^2\Sigma(0)^4 \cdot \sinh^2[2\sigma(\log(\Lambda/\Sigma(0))+\delta)]/(4\sigma^2) , \quad (38)$$

which leads to $S(\lambda,g)\cdot(\Lambda\Sigma(\Lambda^2))^2 \to (\lambda_c/\lambda)\tilde{A}^2\Sigma(0)^4$ with $\tilde{A}\sim O(1)$. Then we obtain*)

$$D_S^R(0) \to (\lambda_c/\lambda)(\tilde{A}^2/2\pi^2)(1+2\lambda(\lambda/g+1))\Sigma(0)^2 \neq 0 \quad (0<\lambda<\lambda_c) , \quad (39)$$

i.e., <u>no dilaton exists on the critical line</u>. The same is also true for $\lambda>\lambda_c$ (and $\lambda=\lambda_c$ as well), where σ and sinh in (37) and (38) are simply replaced by ρ and sin, respectively (see (30) and also (29)), the result being identical to that of Bardeen et al.[29] for $\lambda\to\lambda_c=1/4$ with g=1/4 fixed.

The same conclusion can be reached by the effective potential approach.[10] First, vacuum energy must vanish in $\Lambda\to\infty$ limit if the dilaton existed.[34] However it takes the negative definite form[29),34)], or more explicitly[10]

$$\bar{V}(\Sigma) = -\frac{1}{16\pi^2}\cdot S(\lambda,g)\cdot(\Lambda\Sigma(\Lambda^2))^2 + O(\Sigma(\Lambda^2)^4) , \quad (40)$$

---

*) Of course the same conclusion is reached through the bifurcation solution[9], with a simple replacement, $(\lambda_c/\lambda)\tilde{A}^2\Sigma(0)^2 \to (\delta\Sigma(\epsilon))^2$. Only difference is the absence of $1/\lambda$ singularity for $D_S^R(0)$. This singularity is actually the $\log(\Lambda/\Sigma(0))^2$ divergence in the $\lambda\to 0$ limit in a more elaborate calculation[10], in accord with the Nambu-Jona-Lasinio model.

which yields a finite result for the same reason as mentioned above. Also, the dilaton (mass)^2 can be estimated through Dashen's formula by the second derivative of the effective potential with respect to the parameter κ of the scale transformation $\Sigma(p^2) \to \kappa \Sigma(p^2/\kappa^2)$ [34],[35]; we have [10]

$$F_D^2 m_D^2 = \frac{d^2 V}{d\kappa^2}\Big|_{\kappa=1} = -4(\Lambda^2 \frac{dV}{d\Lambda^2} + \Lambda^4 \frac{d^2 V}{d(\Lambda^2)^2})$$

$$= \frac{1}{\pi^2} (\frac{3}{2} - \frac{\lambda+g}{1+\Sigma(\Lambda^2)^2/\Lambda^2}) [(\frac{1}{1+\Sigma(\Lambda^2)^2/\Lambda^2} - \frac{g}{(\lambda+g)^2}) \cdot (\Lambda\Sigma(\Lambda^2))^2] ,$$

(41)

where F_D is the dilaton "decay constant" of order $O(\Sigma(0))$. Again the factor [ ] is the same as that in (36) and thus yields a finite m_D^2 in $\Lambda \to \infty$ limit.

Thus we found no signature of light dilaton in the whole parameter space of the two coupling QED model in the ladder approximation. This is rather peculiar, since $(\lambda,g)=(1/4,1/4)$ as well as the whole critical line $(\lambda<\lambda_c)$ seems to be an UV fixed point from the analysis of the renormalization group flow.[11] Something is yet to be understood about the dynamics near the fixed point. Maybe, a true (scale-invariant) fixed point could be obtained in the multi-coupling space with more than two couplings. Or maybe, a dilaton could be obtained in the framework beyond the ladder.

Although at this moment we have no explicit dynamical model for the light dilaton, we may speculate on a possible property of the technidilaton[5],[6] which may be an excitement of TeV physics. Actually the technidilaton is very similar to the standard neutral Higgs boson. The mass and Yukawa couplings of the technidilaton were estimated[6]; $m_{TD}^2 = -\pi\beta(\alpha)/\alpha^2 \cdot <\alpha/\pi \cdot (F_{\mu\nu}^a)^2>/F_{TD}^2$ and $g_Y = 2m_f/F_{TD}$, where F_{TD} denotes the "decay constant" of the technidilaton of order $O(\Lambda_{TC})$ and $<\alpha/\pi \cdot (F_{\mu\nu}^a)^2>$ is the technigluon condensate of order $O(\Lambda_{TC}^4)$. It is expected that $\beta(\alpha)/\alpha^2 \ll 1$ near the fixed point, so that the mass of technidilaton may be small, $m_{TD} \ll \Lambda_{TC}$. It is also noted that g_Y is different from that of Higgs $(=m_f/v; v=250\text{GeV})$, although on the same order of magnitude; it is possible to distinguish quantitatively among these two.[6],[36] It is extremely intriguing to find out such a Higgs-like object in future experiments.

In conclusion, the general TC model described by the Lagrangian (8) can survive both the FCNC's problem (G_2 and G_3 terms) and the light PNGB's problem (G_1 term) within the framework of the ladder SD equation, a scale-invariant TC model. The model possesses a nontrivial ultraviolet fixed point at $\lambda=\lambda_c=1/4$ (for $g_1 \equiv G_1 \Lambda_S^2 < 1/4$) and an anomalous dimension $\gamma^* = 1$ at the fixed point, corresponding to $\Sigma(p^2) \sim p^{-1} \cdot \log p$, which enhances both m_f and m_{PNGB}^2 so as to be consistent with the FCNC's and the PETRA experiments on $P^{\pm}$ mass.

We also investigated the possibility for light technidilaton in this model including the full dynamics of the G_1 term ($g_1 \geq 1/4$ as well as $g_1 < 1/4$). We conclude no light technidilaton on the full critical line. The critical line also implies that for $g_1 > 1/4$ the gauge coupling λ can have a critical coupling (a fixed point in this model) at lower values than $\lambda_c = 1/4$, with $\Sigma(p^2) \sim p^{-1+\sqrt{1-\lambda/\lambda_c}}$, which enhances m_f and m_{PNGB}^2 still more than the case for $g_1 < 1/4$. This may or may not be useful for constructing realistic TC theories.

Although the ladder approximation in our scale-invariant model might be subject to criticism, it is very much encouraging that quite recently Monte Carlo studies of lattice QED, both noncompact[37] and compact[38] versions, have suggested the existence of a nontrivial ultraviolet fixed point. It is extremely interesting to clarify the detailed dynamics that leads to such a fixed point and a possible existence of dilaton. Just as the asymptotically free gauge theory was in GeV Physics, the fixed point gauge theories might become a new paradigm in TeV physics. Time will tell.

I would like to thank B. Holdom, K.-I. Kondo, H. Mino, V.A. Miransky, T. Nonoyama and T.B. Suzuki for valuable discussions. This work is supported in part by a Grant-in-Aid for Scientific Research from the Ministry of Education, Science and Culture (#62540202).

[Note added]

The same conclusion as ours concerning the dilaton pole, Eq.(39), has also been obtained by Bardeen, Leung and Love[39] and by Gusynin and Miransky[40].

Quite recently the anomalous dimension of $\bar{\psi}\psi$ in the model (26) has been calculated on the critical line by Miransky, V.A. and Yamawaki, K., Nagoya University Preprint DPNU-88-39, to be published in Mod. Phys. Lett. $\underline{A4}$, No.2; $\gamma_m = 1+\sqrt{1-\lambda/\lambda_c}$ ($0<\lambda<\lambda_c$) in the $S_\chi SB$ phase. This is in accord with Eq.(11).

References

1) Weinberg, S., Phys. Rev. $\underline{D13}$, 974 (1976); $\underline{D19}$, 1277 (1979); Susskind, L., Phys. Rev. $\underline{D20}$, 2619 (1979).
2) For a review, Farhi, E. and Susskind, L., Phys. Rep. $\underline{74}$, 277 (1981).
3) Yamawaki, K., in Proc. Int. Workshop on "Superstrings, Cosmology and Composite Structures" eds. by Gates, S.J. and Mohapatra, R.N., University of Maryland (World Scientific Co., 1987).
4) Yamawaki, K., Nagoya University Preprint, DPNU-88-34, to be published in the Proceedengs of the 12th Johns Hopkins Workshop on Current Problems in Particle Theory ("TeV Physics"), Baltimore, Maryland, June 8-10, 1988.
5) Yamawaki, K., Bando, M. and Matumoto, K., Phys. Rev. Lett. $\underline{56}$, 1335 (1986).
6) Bando, M., Matumoto, K. and Yamawaki, K., Phys. Lett. $\underline{178B}$, 308 (1986).
7) Bando, M., Morozumi, T., So, H. and Yamawaki, K., Phys. Rev. Lett. $\underline{59}$, 389 (1987).
8) Aoki, K-I., Bando, M., Mino, H., Nonoyama, T., So, H. and Yamawaki, K., Nagoya University Preprint, DPNU-88-7 (May, 1988).
9) Kondo, K.-I., Mino, H. and Yamawaki, K., Nagoya University Preprint, DPNU-88-18 (June, 1988); DPNU-88-18-Rev (Oct., 1988).
10) Nonoyama, T., Suzuki, T.B. and Yamawaki, K., Nagoya University Preprint, DPNU-88-36.

11) Kondo, K.-I., Mino, H., Nonoyama T., Suzuki, T.B. and Yamawaki, K., in preparation.
12) Kondo, K.-I., in this Proceedings.
13) Appelquist, T., Karabali, D. and Wijewardhana, L.C.R., Phys. Rev. Lett. 57, 957 (1986); Appelquist, T. and Wijewardhana, L.C.R., Phys. Rev. D35, 774 (1987); D36, 568 (1987); Appelquist, T., Carrier, D., Wijewardhana, L.C.R. and Zheng, W., Phys. Rev. Lett. 60, 1114 (1988); Appelquist, T., in this Proceedings.
14) Holdom, B., Phys. Lett. 150B, 301 (1985).
15) Nonoyama, T., in this Proceedings.
16) Jackiw, R. and Johnson, K., Phys. Rev. D8, 2386 (1973); Pagels, H. and Stoker, S., Phys. Rev. D20, 2947 (1979).
17) Weinberg, S., Phys. Rev. D8, 3497 (1973).
18) Politzer, H.D., Nucl. Phys. B117, 397 (1976).
19) Lane, K., Phys. Rev. D10, 2605 (1974); Miransky, V.A., Sov. J. Nucl. Phys. 38, 280 (1983); Higashijima, K., Phys. Rev. D29, 1228 (1984).
20) Dimopoulos, S. and Susskind, L., Nucl. Phys. B155, 237 (1979); Eichten, E. and Lane, K., Phys. Lett. 90B, 125 (1980).
21) Yamawaki, K. and Yokota, T., Phys. Lett. 113B, 293 (1982).
22) Holdom, B., Phys. Rev. D24, 1441 (1981).
23) Georgi, H. and Glashow, S.L., Phys. Rev. Lett. 47, 1511 (1981).
24) Yamawaki, K. and Yokota, T., Nucl. Phys. B223, 144 (1983).
25) Yamawaki, K., Phys. Lett. 118B, 145 (1982).
26) Maskawa, T. and Nakajima, H., Prog. Theor. Phys. 52, 1326 (1974); 54, 860 (1975).
27) Fukuda, R. and Kugo, T., Nucl. Phys. B117, 250 (1976).
28) Formin, P.I., Gusynin, V.P., Miransky, V.A. and Sitenko Yu.A., Riv. Nuovo Cim. 6, 1 (1983); Miransky, V.A., Nuovo Cim. 90A, 149 (1985).
29) Bardeen, W.A., Leung, C.N. and Love, S.T., Phys. Rev. Lett. 56, 1230 (1986); Leung, C.N., Love, S.T. and Bardeen W.A., Nucl. Phys. B273, 649 (1986).
30) For a recent review, Miransky, V.A., in this Proceedings. See also Aoki, K-I. in this Proceedings; Kogut, J.B., in this Proceedings.

31) Akiba, T. and Yanagida, T., Phys. Lett. 169B, 432 (1986).
32) Appelquist, T., Soldate, M., Takeuchi, T. and Wijewardhana, L.C.R. Yale University Preprint, YCTP-P19-88 (Aug., 1988).
33) Atkinson, D. and Johnson, P.W., J. Math. Phys. 28, 2488 (1987); Kondo, K.-I. and Kikukawa, Y., Nagoya University Preprint, DPNU-88-20.
34) Gusynin, V.P. and Miransky, V.A., Phys. Lett. 198B, 79 (1987).
35) Holdom, B. and Terning, J., Phys. Lett. 187B, 357 (1987); 200B, 338 (1988).
36) Clark, T., Leung, C. and Love, S., Phys. Rev. D35, 997 (1987).
37) Kogut, J.B., Dagotto, E. and Kocic, A., Phys. Rev. Lett. 60, 772 (1988); Kogut, J.B., in this Proceedings.
38) Okawa, M., KEK Preprint KEK-TH-204; in this Proceedings.
39) Bardeen, W.A., in this Proceedings.
40) Gusynin, V.P. and Miransky, V.A., private communication.

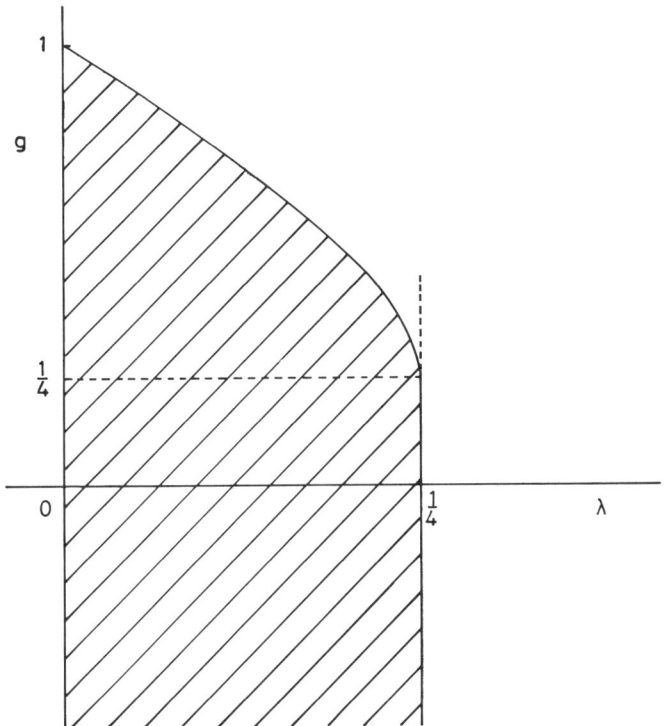

Fig.1

Critical line in (λ, g) plane is depicted by the solid line. The shaded region bounded by the critical line and the vertical axis is the purturbative phase where the chiral symmetry is not spontaneously broken.

WALKING TECHNICOLOR

Thomas Appelquist

Yale University, Department of Physics, New Haven, CT 06511

ABSTRACT

In this talk, I will briefly review the idea of walking technicolor theories and describe a few recent developments. I will then briefly comment on a related idea that attempts to address the same problem. Finally, I will describe a recent analysis of chiral symmetry breaking in Quantum Electrodynamics in three space-time dimensions. This model, when treated in a 1/N expansion, is similar in structure to walking gauge theories.

TECHNICOLOR

The well known difficulties with electroweak theories containing elementary Higgs scalars and the absence of any experimental evidence for their existence has led to a search for dynamical alternatives for electroweak symmetry breaking. An especially attractive idea is technicolor[1] in which a new set of massless fermions interacting by an asymptotically free gauge theory is imagined to be the agent of electroweak breaking. At its characteristic energy scale $\Lambda_{tc} \simeq \sqrt{2}\, G_F \simeq 250$ GeV, the technicolor (TC) interaction breaks its fermions' chiral symmetries. When the electroweak symmetry $SU(2) \times U(1)$ is properly embedded in these symmetries, it is broken down to electromagnetic $U(1)$.

To communicate this symmetry breaking to ordinary fermions, and so give them nonvanishing current masses, the TC gauge group was placed into a larger one, known as extended technicolor (ETC).[2] This couples the ordinary fermions to technifermions and generates quark and lepton masses m_f of order $m_f \sim g^2 \langle\bar{\Psi}\Psi\rangle_{etc}/\Lambda_{etc}^2$. Here, $\langle\bar{\Psi}\Psi\rangle_{etc}$ is the technifermion bilinear condensate cut off at the ETC energy scale Λ_{etc} at which the breakdown ETC → TC x ... takes place, and g is the ETC coupling strength at that scale.

An estimate of Λ_{etc}/g may be obtained if the asymptotic freedom of the TC interaction leads to a weak coupling setting in rapidly above Λ_{tc}. The anomalous dimension γ_m of $\bar{\Psi}\Psi$ is then small and $\langle\bar{\Psi}\Psi\rangle$ suffers only a logarithmic renormalization in scaling from Λ_{etc} down to Λ_{tc}. It can then be shown that $\langle\bar{\Psi}\Psi\rangle_{etc} \simeq \Lambda_{tc}^3$ and the ETC scale required to generate a u or d quark mass of 10 MeV is $\Lambda_{etc}/g \simeq 40$ TeV. Unfortunately, the ETC interactions also generally induce effective four-fermion flavor-changing neutral current Lagrangians for quarks and leptons with effective couplings $\sim g^2/\Lambda_{etc}^2$. The most stringent constraints on these couplings come from $|\delta S| = 2$ effects in the neutral kaon system, which require $\Lambda_{etc}/g \gtrsim 500$-$1000$ TeV.

WALKING TECHNICOLOR

A promising mechanism for eliminating large flavor-changing neutral current interactions in ETC models may be traced back to the suggestion[3] that a TC interaction with a nontrivial ultraviolet fixed point could generate a large anomalous-dimension enhancement for $\langle\bar{\Psi}\Psi\rangle_{etc}/\langle\bar{\Psi}\Psi\rangle_{tc}$, so that a larger Λ_{etc} would be required to produce a fixed mass m_f. Because no realistic theory with a nontrivial ultraviolet fixed point has ever been exhibited, this idea must be regarded as speculative. Recently, however, my collaborators and I[4,5] have shown how to obtain naturally the desired result of raising Λ_{etc} without the troublesome assumption of a nontrivial UV fixed point. We assumed that technicolor is described by an asymptotically free theory in which the β function is small so that the TC coupling $\alpha(p)$ runs very slowly for a large range of momenta, $\Lambda_{tc} \lesssim p \lesssim \Lambda$, where $\Lambda_{tc} \ll \Lambda < \Lambda_{etc}$. Near Λ_{tc}, $(\alpha)p$ is expected to be close to a "critical value" α_c which seems to be required for the occurence of technifermion chiral symmetry breakdown and, so, $\alpha(p) \simeq \alpha_c$ over this large momentum range. In the lowest-order computations made so far, $\alpha_c = \frac{\pi}{3\,C_2(R)}$, where $C_2(R)$ is the quadratic Casimir of the technifermion representation R. In this approximation,

$$\gamma_m(\alpha(p)) \simeq 1 - \sqrt{1-\alpha(p)/\alpha_c} \simeq 1, \qquad (1)$$

and $\langle\bar{\Psi}\Psi\rangle_{etc}/\langle\bar{\Psi}\Psi\rangle_{tc} \simeq \langle\bar{\Psi}\Psi\rangle_\Lambda/\langle\bar{\Psi}\Psi\rangle_{tc} \simeq \Lambda/\Lambda_{tc}$. A 10 MeV quark mass may then be generated by $\Lambda_{etc}/g = 1000$ TeV for Λ as low as 160 TeV.[5] What makes this "walking technicolor" scenario so attractive is that ETC models typically contain a large number of technifermions and so a small TC β function above Λ_{etc} is not unlikely.

The idea of "walking technicolor" has stimulated a good deal of recent work by other researchers. There have been several attempts at model building[6] and it was recently pointed out[7] that large quark-lepton mass splittings can naturally arise in this framework.

CONVERGENCE OF THE LADDER EXPANSION

The analysis I have just decribed was carried out using the (renormalization-group improved) ladder approximation to the Dyson-Schwinger gap equation. While this approximation is adequate to determine the asymptotic behavior of the dynamical fermion mass, it is not at all obvious that it can be used at lower momenta where the coupling equals or exceeds α_c. To investigate the validity of the ladder approximation, Ken Lane, Uma Mahanta and I have begun a program to compute the next higher order contributions to the gap equation.[8] The analytic computations completed so far have shown that the higher order corrections produce only a 1%-20% correction depending on the fermion representation. The analysis has so far been restricted to slowly-running theories, in which the running of the coupling can reasonably be neglected for a range of momenta in the neighborhood of the chiral symmetry breaking scale. For these theories at least, it has provided some evidence that the ladder expansion may be a much better description of chiral symmetry breaking than previously thought. The analytic results obtained so far are currently being checked and extended numerically.[9]

A UNITARITY BOUND

While the technicolor mechanism is a possible route to an understanding of quark and lepton masses, it is important to keep an open mind and to approach the problem in general terms as well. Michael Chanowitz and I[10] have made use of unitarity arguments to establish a model independent upper bound on the energy scale of fermion mass generation. We showed that $SU(2)_L \times U(1)$ gauge interactions and partial wave unitarity imply that the energy scale E associated with the generation of fermion mass m_f is bounded by

$$E \leq E_f = \frac{16\pi}{\sqrt{2} G_F m_f} \simeq \frac{\pi}{m_f} \quad , \tag{2}$$

where G_F is the Fermi constant and the final expression yields E_f in TeV for m_f supplied TeV. For example, for $m_t \gtrsim 50$ GeV we have $E_t \lesssim 60$ TeV. If there is a fourth generation of quarks with masses of order 1 TeV, the physics of their mass generation lies at or below a few TeV.

The physical mechanism that comes closest to saturating this bound is walking technicolor. There, the expression for the fermion mass is

$$m_f \simeq \frac{g^2}{\Lambda_{etc}^2} \langle \bar{\psi}\psi \rangle_{etc} \simeq \frac{g^2}{\Lambda_{etc}^2} \Lambda_{tc}^2 \Lambda \quad . \tag{3}$$

Here Λ_{etc} is the scale associated with fermion mass generation, and while it can be made quite large as noted above, it will not completely saturate the unitarity bound. If however, the running is so slow that is can be neglected all the way to Λ_{etc} ($\Lambda = \Lambda_{etc}$), saturation is nearly realized. In that case

$$\Lambda_{etc} \simeq \frac{g^2 \Lambda_{TC}^2}{m_f} \simeq \frac{g^2}{\sqrt{2}\, G_F m_f} \quad . \tag{4}$$

Whether the bound is saturated now depends on the strength of the ETC coupling. Only if the theory is genuinely strongly coupled ($g^2/4\pi^2 = O(1)$) is the bound essentially saturated.

The first problem with this limit is that the non-running of the coupling all the way to Λ_{etc} is very difficult to achieve in a realistic theory. If it is achieved, there is no natural mechanism to keep the weak scale small compared to Λ_{etc}. A fine tuning of the (essentially flat) coupling is required. Finally, if the ETC theory is strongly coupled, it too can destabilize the technicolor hierarchy by pulling the weak scale toward Λ_{etc}. I will return to this point in my concluding comments. At the present time, it would seem that no natural mechanism has been proposed that fully saturates the unitarity bound.

A RELATED IDEA

Beginning with the work of Maskawa and Nakajima and Kugo and Fukuda,[11] some very useful insight into spontaneous chiral symmetry breaking has been gained by the analysis of a simplified field-theoretic model - a quenched U(1) gauge field theory in four spacetime dimensions. The analysis is carried out in the presence of an ultraviolet cutoff Λ to provide a scale and to allow a careful treatment of the conservation of the underlying chiral symmetry. For $\alpha > \alpha_c \equiv \pi/3$, spontaneous chiral symmetry can take place, leading to a dynamical fermion mass and a Goldstone-boson decay constant set by the scale Λ. Unless α is fine tuned very close to α_c, all the physical scales will in fact be of order Λ.

Since the cutoff cannot therefore naturally be made large compared to the physical scales, the model may be regarded as perhaps interesting but unphysical. In an effort to make the model more physical, Miransky[12] suggested that the coupling constant, while taken not to run in the gap equation, might be imagined to depend on the ultraviolet cutoff. If a particular dependence of $\alpha(\Lambda)$ is assumed, such that $\alpha(\Lambda) \to \alpha_c$ as $\Lambda \to \infty$, then an arbitrarily large hierarchy between the physical scales and Λ can be arranged. The problem with this notion is that it has never been derived from any underlying quantum field theory. Instead, one must invoke some unknown high energy physics and imagine that it wll make $\alpha(\Lambda)$ become extremely close to α_c. Recently, Bardeen, Leung and Love[13] reconsidered the quenched cutoff U(1) gauge model in a study of the spontaneous breaking of scale invariance. Their studies, including a four-fermion, Nambu-Jona-Lasinio[14] interaction, have so far concluded that there is no evidence for the spontaneous breaking of scale invariance in this model. This appears to be the case even if the Miransky assumption is adopted, allowing the limit $\Lambda \to \infty$ to be taken.

Even if the model is unphysical, it remains interesting if for no other reason than it is an approximate description, over a limited momentum range of a walking asymptotically free theory. For that reason, Soldate, Takeuchi, Wijewardhana and I[15] have recently reanalyzed the model, including the four-fermion term.

In a recent series of papers, Bando, Yamawaki and collaborators[16] have attempted to build a technicolor phenomenology along the above lines.[17] The attraction of this idea is evident. With the coupling assumed not to run and with $\alpha(\Lambda)$ very close to α_c (either by fine tuning or be a Miransky fixed-point assumption), a large hierarchy between the technicolor scale Λ_{TC} and the ETC scale Λ can be arranged. Flavor-changing neutral currents are well suppressed while quark and lepton masses of more than 100 MeV are easily attained. The problem with the idea is the one I have already mentioned. No quantum field theory I know of will lead to a coupling that doesn't run over momentum scales all the way to some cutoff Λ, and yet become arbitraily close to a finite critical value as the cutoff increases.

CRITICAL BEHAVIOR IN THREE DIMENSIONAL QED

Quantum electrodynamics in 2+1 dimensions (QED3) is a super-renormalizable gauge theory which, when analyzed in a 1/N expansion, strongly resembles walking four dimensional gauge theories. The analysis of QED3 can provide some valuable insight into chiral symmetry breaking in these kinds of theories. In the massless theory, the dimensionful coupling $\alpha = Ne^2/8$, which is kept fixed when the number of fermions N is taken to infinity, provides the only fixed scale in the theory. At momentum scales p beyond α, the theory is rapidly damped. For $p \ll \alpha$, the singular behavior of the loop expansion is softened in the 1/N expansion. To each order in this expansion, the Green's functions of the massless theory can be shown to be infrared finite.[18]

The infrared finiteness is a consequence of an effective low energy scale invariance that the theory exhibits to each order in $1/N$.[19] The effective interaction strength in this limit is proportional to $1/N$.

Despite this exemplary behavior of the massless theory, dynamical symmetry breaking, leading to a nonzero fermion mass, could take place. This question had been examined previously[20] to first and second order in the $1/N$ expansion. Incomplete analytic studies indicated that dynamical fermion mass generation takes place for arbitrarily large N.[15] Numerical studies also demonstrated mass generation, but were restricted to $N < 3$ in order to achieve convergence of the iterative procedure. It was important to settle this question. Because of its low energy scale invariance, QED3 is a useful theoretical laboratory to gain insight into dynamical chiral symmetry breaking in four dimensional gauge theories, especially slowly running ones.

L.C.R. Wijewardhana, Daniel Nash and I[21] have recently shown that to lowest order in the $1/N$ expansion of the Dyson-Schwinger kernel, dynamical mass generation takes place only for $N < N_c = 32/\pi^2$. We also showed that the approach to criticality is universal in the sense of being independent of the details of the ultraviolet physics. To what extent these results are modified by higher order corrections remains an open question. This problem is now being examined by Daniel Nash at Yale.

CONCLUSIONS

At the very least, the idea of walking technicolor has helped to reawaken interest in dynamical electroweak symmetry breaking. It has shown that some of the predictions and apparent problems of these theories are quite sensitive to modest changes in the high energy dynamics of the theory. Whether a realistic theory of electroweak symmetry breaking incorporating this mechanism can be constructed remains to be seen.

Based on the analysis of Ref. 15, Wijewardhana, Takeuchi and I[22] have recently been considering an additional effect that may play a role in developing a realistic theory. Following the analysis of Ref. 15, we have been examining the gap equation for $\Sigma(p)$ in the presence of a four-technifermion interaction. If this term is strong enough, it can dramatically affect the character of the solution and the hierarchies between the ETC scale Λ, the weak scale $\Sigma(0) \approx F \sim 250$ GeV, the technicolor confinement scale Λ_{TC}. It might also play a role in generating large mass hierarchies among the various fermion generations.

What seems clear to me is that the general technicolor framework allows for a very rich structure, far richer than originally imagined. I remain optimistic that it will eventually lead to a realistic description of quark and lepton masses and mixing angles.

ACKNOWLEDGEMENTS

The author acknowledges the support of the Yamada Science Foundation for his visit to Japan on August 23, 1988 through September 3, 1988, during which he attended the Nagoya Workshop.

REFERENCES

1] S. Weinberg, Phys. Rev. D $\underline{13}$, 974 (1976); $\underline{19}$, 1277 (1979); L. Susskind, ibid. $\underline{20}$,, 2619 (1979).

2] S. Dimopoulos and L. Susskind, Nucl. Phys. $\underline{B155}$, 237 (1979); E. Eichten and K. Lane, Phys. Lett. $\underline{90B}$, 125 (1980).

3] B. Holdom, Phys. Rev. D $\underline{24}$, 1441 (1981).

4] T. Appelquist, D. Karabali, and L.C.R. Wijewardhana, Phys. Rev. Lett. $\underline{57}$, 957 (1986); T. Appelquist and L.C.R. Wijewardhana, Phys. Rev. D $\underline{35}$, 774 (1987); T. Appelquist and L.C.R. Wijewardhana, Phys. Rev. D $\underline{36}$, 568 (1987).

5] T. Appelquist, D. Carrier, L.C.R. Wijewardhana, and W. Zheng, Phys. Rev. Lett. $\underline{60}$, 1114 (1988).

6] S. King, University of Southampton preprint, August, 1988.

7] B. Holdom, Phys. Rev. Lett. $\underline{60}$, 1233 (1988).

8] T. Appelquist, K. Lane, and U. Mahanta, Phys. Rev. Lett. $\underline{61}$, 1553 (1988).

9] K. Lane and U. Mahanta, work in progress.

10] T. Appelquist and M. Chanowitz, Phys. Rev. Lett. $\underline{59}$, 2405 (1987).

11] T. Maskawa and H. Nakajima, Prog. Theor. Phys. $\underline{52}$, 1326 (1974); $\underline{54}$, 860 (1976); R. Fukuda and T. Kugo, Nucl. Phys. $\underline{B117}$, 250 (1976).

12] V.A. Miransky, Nuovo Cimento $\underline{90A}$, 149 (1985) and references cited therein.

13] W.A. Bardeen, C.N. Leung, and S.T. Love, Phys. Rev. Lett. $\underline{56}$, 1230 (1986); Nucl. Phys. $\underline{B273}$, 649 (1986).

14] Y. Nambu and G. Jona-Lasinio, Phys. Rev. $\underline{122}$, 345 (1961).

15] T. Appelquist, M. Soldate, T. Takeuchi, and L.C.R. Wijewardhana, Yale preprint YCTP-P19-88, August, 1988; to be published in the Proceedings of the 12th Johns Hopkins Workshop on Current Problems in Particle Theory, Baltimore, Maryland, June 8-10, 1988.

16] K. Yamawaki, M. Bando, and K. Matumoto, Phys. Rev. Lett. $\underline{56}$, 1335 (1986).

17] T. Akiba and T. Yanagida, Phys. Lett. $\underline{169B}$, 432 (1986).

18] T. Appelquist and R. Pisarski, Phys. Rev. $\underline{D23}$, 2305 (1981); R. Jackiw and S. Templeton, Phys. Rev. $\underline{D23}$, 2291 (1981).

19] T. Appelquist and U. Heinz, Phys. Rev. $\underline{D24}$, 2169 (1981).

[20] R. Pisarski, Phys. Rev. $\underline{D29}$, 2423 (1984); T. Appelquist, M. Bowick, D. Karabali and L.C.R. Wijewardhana, Phys. Rev. Lett. $\underline{55}$, 1715 (1985); Phys. Rev. $\underline{D33}$, 3704 (1986).

[21] T. Appelquist, D. Nash and L.C.R. Wijewardhana, Phys. Rev. Lett. $\underline{60}$, 2575 (1988).

[22] T. Appelquist, T. Takeuchi and L.C.R. Wijewardhana, work in progress.

CHIRAL CONDENSATE IN TECHNICOLOR THEORIES

T. NONOYAMA

Department of Physics, Nagoya University,
Nagoya 464-01, Japan

ABSTRACT

I show numerically how the flavor changing neutral current sydrome is solved in the scale invariant version of asymptotically non-free technicolor model. Controversy on the slowly running coupling model is settled and is shown that such model does not solve the flavor changing neutral current problem.

Yamawaki, Bando and Matumoto[1] presented the scale invariant technicolor (TC) theory which is an explicit dynamical model realizing the idea of fixed point scenario of Holdom[2]. In this theory the cutoff in the Schwinger-Dyson (SD) equation is taken to be infinity by following the fixed point theory of Miransky[3]. In this talk(*) I report mainly on the numerical studies of both the scale invariant TC model[1], which is asymptotically non-free, and the asymptotically free TC model with slowly running coupling[5] studied by Holdom[6] and Appelquist et al.[7].

In the previous talks of Yamawaki and Appelquist, the TC models are discussed. TC model provides a beautiful mechanism to break dynamically the weak gauge symmetry. The technifermion condensation plays the role of Higgs field condensation which results as the W/Z gauge boson masses. On the other hand the sideway gauge boson gives the quark/lepton masses and flavor changing neutral current (FCNC). It is well known that TC model which is merely the scale up of QCD failed as a realistic model, because of the severe FCNC problem.

The tale of the disasters of the standard TC models and the recipe by a specific medicine, the scale invariant technicolor model, is explained in detail by Yamawaki in the previous talk. My talk is complementary to that. So let me only shortly sketch the points of TC models.

(*) This talk is based on [4].

A typical TC scale is chosen to be $\Lambda_{TC} = 350 GeV$ in order to obtain the realistic weak gauge boson masses: $m_{W/Z} = \frac{g}{2} F_\pi$, where $F_\pi = f_\pi \frac{\Lambda_{TC}}{\Lambda_{QCD}}$. In order to suppress

$$FCNC \sim \theta_c^2 \quad \sim O(\frac{\theta_c^2}{\Lambda_S^2}) < \frac{1}{2} \times 10^{-6} TeV^{-2} \qquad (1)$$

below the experimental limit, the scale Λ_S of the extended technicolor (ETC) must be $\Lambda_S \geq 350 TeV$. Then the ordinary fermion (quark/lepton) masses are given by

$$m_f = \quad \sim \frac{g^2}{m_S^2} \langle \bar\psi\psi \rangle \sim \frac{\Lambda_{TC}^3}{\Lambda_S^2} \leq 10^{-1} MeV. \qquad (2)$$

This value is insufficient compared to the strange quark mass ($m_s \sim 200 MeV$), which we need because the experimental limit of FCNC is most severe in $K^0 \bar{K}^0$ decay.

An idea to get through this dilemma is proposed by Holdom[2] without showing any explicit dynamics: A large anomalous dimension γ^* of $\langle \bar\psi\psi \rangle$ provides a new mechanism to avoid FCNC syndrome. In this case the quark/lepton mass is given, if $\gamma^* \sim 1$, by

$$m_f \sim \frac{\Lambda_{TC}^3}{\Lambda_S^2} \left(\frac{\Lambda_S}{\Lambda_{TC}} \right)^{\gamma^*} \sim 10^2 MeV. \qquad (3)$$

Recently, Yamawaki, Bando and Matumoto[1] have shown that there exists such a model: The scale invariant version of an asymptotically non-free TC model, based on ladder SD equation, gives a large anomalous dimension $\gamma^* \sim 1$,[(*)] which naturally resolves the FCNC problem.[(†)]

(*) QED is studied as a model of such dynamical system. Historically, such nonperturbative behavior of QED was studied more than a decade ago. Johnson, Baker and Willey[8] studied the Chiral symmetry breaking of QED in the ladder approximation. Maskawa and Nakajima[9] subsequently studied this problem carefully by introducing the ultraviolet cutoff and concluded that the spontaneous chiral symmetry breaking occurs only for the coupling which is larger than some critical value. Fukuda and Kugo[10] studied the model and claimed the confinement of the constituent fermion. Importance is the large anomalous dimension and the presence of the ultraviolet fixed point is stressed by Miransky[3].

(†) Akiba and Yanagida[11] also considered a similar mechanism within the framework of slowly running coupling theory of Holdom[5] with a finite cutoff, where a hierarchy between $\Sigma(0)$ and the cutoff is achieved by fine tuning the gauge coupling.

Then came some controversy.

Appelquist et al.[13] considered that the slowly running coupling theory gives enhanced quark/lepton masses. Their original claim was that the enhancement of m_f in the slowly running coupling models is *more than* 200. Explicit values for m_f was also claimed to be 800MeV *or more*.[*]

This was criticized by the analytical method by Bando, Morozumi, So and Yamawaki.[12] Their estimate of relative enhancement of slowly running coupling is *at most* 10 or the realistic value is *less than* 1MeV.

Recently Appelquist et al.[13] made a numerical analysis and claimed that the enhancement is about 50 ~ 60 and the above analytical result is incorrect.[†]

Anyway, dynamical mechanism beyond the simple scale up of QCD is important. Decisive test is very important for model building.

Then comes our numerical analysis which shows;
1) The original work on slowly running by Appelquist et al. was overestimated.
2) The analytical study by Bando et al. is confirmed numerically.
3) Our conclusion is different from that of Appelquist et al.. However our numerical values are actually consistent with theirs. They therefore should have concluded that their results are consistent with those of Bando et al..
4) Rough order estimate of scale invariant model of Yamawaki et al. has been explicitly confirmed numerically to solve the FCNC syndrome.

Now let us go into the detail of our numerical results.

Physical quantities, F_π and m_f, are given in terms of the dynamical mass $\Sigma(x)$, which is the solution of the Schwinger-Dyson (SD) equation,

$$F_\pi^2 = \frac{N}{4\pi^2} \int \frac{x\Sigma(x)(\Sigma(x) - \frac{1}{2}\frac{d}{dx}\Sigma(x))}{(x + \Sigma^2(x))^2} dx \quad , \tag{4}$$

$$m_f^{(ETC)} = \quad = \frac{N}{8\pi^2} \int \frac{g^2}{x + m_S^2} \frac{x\Sigma(x)}{x + \Sigma^2(x)} dx \quad . \tag{5}$$

Here N is the number of technifermions communicating to the mass of the ordinary

---

[*] 800MeV is modified to be 20MeV afterwards.
[†] However a careful reading of their paper makes it evident that the relative enhancement is *at most* 10 (= 50/5) thus it coincides with the claim of [12].

fermions. Our notation here is $x = p^2$. The latter can be approximated as

$$m_f = \text{[diagram: fermion line with } \langle \bar{F}F \rangle \text{ loop]} = \frac{N}{8\pi^2 \Lambda_S^2} \int^{\Lambda_S^2} \frac{x\Sigma(x)}{x + \Sigma^2(x)} dx \quad . \tag{6}$$

We stress here that in order to obtain physical quantities of (5) and (6), we need no ultraviolet cutoff at all: They are finite unless $\Sigma(x) = $ constant, i.e., if it is a decreasing function of x. Indeed $\Sigma(x) \sim 1/p^2$ for the ordinary running coupling and $\Sigma(x) \sim 1/p$ for $\gamma^* = 1$ case. In the calculations below, we refer to some ultraviolet cutoff, which is introduced only from the technical reason.

In order to give the dynamical mass $\Sigma(x)$, we solve the Schwinger-Dyson equation for the technifermion propagator $S(p)$ in the ladder approximation. In the Landau gauge,[*] $S(p) = \gamma \cdot p + \Sigma(p^2)$ and the equation for $\Sigma(p^2)$ becomes especially simple.

For the fixed coupling constant $\lambda = 3e^2/16\pi^2$,

$$\Sigma(x) = \frac{3\lambda}{x} \int_0^x \frac{y\Sigma(y)}{y + \Sigma^2(y)} dy + \int_x^{\Lambda^2} \frac{3\lambda\Sigma(y)}{y + \Sigma^2(y)} dy \quad . \tag{7}$$

Here the angular variables are integrated out and Λ is the computer cutoff, introduced by the technical reason.

On the other hand for the running coupling case we may replace the coupling by the one which is the solution of one-loop renormalization group equation as did by Higashijima[15]

$$\beta = k\frac{d}{dk}\alpha(k) = -\frac{b_0}{4\pi}\alpha(k) + \cdots \quad . \tag{8}$$

Thus we have the running coupling

$$\lambda(k^2) = \frac{4\pi/b_0}{\ln k^2/\mu^2 + 4\pi/b_0\alpha_\mu} \equiv \frac{A}{\ln k^2/\mu^2 + A/\alpha_\mu} \quad . \tag{9}$$

(*) The study for arbitrary covariant gauges is done recently by myself and M. Tanabashi[14] and quite recently also by K-I. Aoki, et al.

So the equation to solve is

$$\Sigma(x) = \frac{3\lambda(x)}{x}\int_0^x \frac{y\Sigma(y)}{y+\Sigma^2(y)}dy + \int_x^{\Lambda^2} \frac{3\lambda(y)\Sigma(y)}{y+\Sigma^2(y)}dy \quad . \tag{10}$$

This is an approximate expression contrary to the fixed coupling case since the angular integration is performed only approximately and the use of the solution of the renormalization group equation for the coupling is an ansatz.

There is a comment on the running coupling case. $A \leq 1$ is for the ordinary running coupling. For TC and ETC models, $\Lambda_S/\mu = 10^3$ where $\mu \sim \Lambda_{TC}$. This indicates that $A = 2 \sim 6$ corresponds to the slowly running coupling while $A \geq 7$ is almost fixed coupling.

The solution of the linearized SD equation is known. For the fixed coupling case, it is given by the hypergeometric function[3]

$$\Sigma(p^2) = \mu\xi F(\frac{1}{2} + \frac{i}{2}\sqrt{4\lambda-1}, \frac{1}{2} - \frac{i}{2}\sqrt{4\lambda-1}; 2, -\frac{p^2}{\mu^2}) \tag{11}$$

where $m = \Sigma(0)$ and ξ is some numerical constant of order one. The asymptotic behavior of this is given as

$$\Sigma(p^2) \sim \Sigma^{FP} \cdot (\frac{\mu}{p})[\ln\frac{p^2}{\mu^2} - \delta] \quad . \tag{12}$$

This implies $\gamma^* \sim 1$. For the running coupling case, it is given by the Whittaker function[12]

$$\Sigma(p^2) = \Sigma_\mu \cdot (\frac{\mu^2}{p^2})(1 + \frac{1}{2A}\ln\frac{p^2}{\mu^2})^{-\frac{1}{2}} \frac{W_{\frac{A}{2}-\frac{1}{2},\frac{1}{2}}(\ln\frac{p^2}{\mu^2} + 2A)}{W_{\frac{A}{2}-\frac{1}{2},\frac{1}{2}}(2A)} \quad , \tag{13}$$

whose asymptotic form is

$$\Sigma(p^2) = \Sigma_\mu^{AFT}(\frac{\mu^2}{p^2})[1 + \frac{1}{2A}\ln(p^2/\mu^2)]^{A/2-1} \quad . \tag{14}$$

There is no anomalous dimension other than the logarithmic one.

Notice that the integral SD equation can be cast into the differential equation form

$$\left(\frac{\Sigma'(x)}{(\lambda(x)/x)'}\right)' = -\frac{3\lambda(x)x\Sigma(x)}{x+\Sigma^2(x)} , \qquad (15)$$

we obtain a simple formula

$$m_f = \frac{N}{8\pi^2}\left(\frac{\Sigma'(x)}{(\lambda(x)/x)'}\right)\bigg|_{x=\Lambda_S^2} . \qquad (16)$$

We therefore can estimate m_f using the asymptotic solutions of SD equation[12] if we fix $\Sigma^{AF} = \Sigma^{FP} \sim \mu = 350 GeV$. Then we have for the asymptotically free case with $\Sigma(p^2) \sim (1/p^2)(\ln p^2 - \cdots)^{\cdots}$, $m_f \sim 10^{-1}(MeV)$. Indeed we have $(m_f(MeV)/N)_A = (4.8 \times 10^{-2})_1$, $(7.5 \times 10^{-2})_2$, $(1.2 \times 10^{-1})_4$, $(1.6 \times 10^{-1})_6$, and $(1.7 \times 10^{-1})_7$. On the other hand for the asymptotically non-free case with $\Sigma(p^2) \sim (1/p)(\ln p^2 - \cdots)$, $m_f \sim 10^2(MeV)$. Indeed $m_f(MeV)/N = 107MeV$.

So far I reported on the analytical study using the solutions of the linearized SD equation. Then there come questions about normalizations, Σ^{AF} and Σ^{FP}, and the non-linear effect. We therefore solve the full non-linear SD equation numerically and evaluate F_π, m_f and $m_f^{(ETC)}$ using thus obtained solutions.

In order to solve differential SD equation, we must put the infrared (IR) and ultraviolet (UV) boundary conditions (BC's) which depend on the underlying physics. Those BC's are summarized as follows.

(IR-a) $\Sigma(x) = \Sigma(0)$ for $0 \geq x \geq \mu^2$ and vary $\Sigma'(\mu^2)$.

(IR-b) Integral SD equation with IR-cutoff $= 0$.

(IR-c) Integral SD equation with IR-cutoff $= \mu^2$.

(UV-α) $\Sigma(x) \sim$ asymptotic solution at large x.

(UV-β) Integral SD equation with UV-cutoff $= \Lambda^2$.

There are three types of studies combining these:

(I) $= (a + \alpha)$: (Bando, Morozumi, So and Yamawaki[12])

(II) $= (b + \beta)$: (Appelquist, Karabali and Wijewardhana[13])

(III) $= (c + \beta)$: (Holdom[6])

The summary of our numerical calculation is given in Fig.[1]. We studied the running coupling case for $A = 1 \sim 20$.

For the asymptotically free case we conclude from this figure that (1) the dependence on IR and UV BC is not so strong, (2) (I) is in an excellent agreement with the analytical study which means small non-linear effect, (3) (II) is in a good agreement with the analysis of Appelquist et al. who displayed only the ratio $m_f(A)/m_f(1)$, though, and (4) (III) is also in a good agreement with the analysis of Holdom. In addition to these comments, we stress that those m_f's are smaller than $m_s \sim 200 MeV$ by order 4 for $A = 1$ and by order $3 \sim 2$ for $A = 2 \sim 4$. For $A \geq 7$, it should be considered to be asymptotically non-free theory with fixed coupling. Thus it requires a special treatment on the renormalization. Anyway as long as we do not treat $\Lambda \to \infty$, an additional 1 order enhancement is required. On the other hand $F_\pi(GeV)/\sqrt{N}$ is given as

A	1	2	4	6	7	10	16	20
(I)	42	44	46	47	47	48	49	49
(II)	33	38	47	53	55	59	63	65
(III)	34	49	62	68	70	74	77	78

These values are quite sufficient for W/Z boson masses.

We thus conclude that even the slowly running coupling theories do not solve FCNC syndrome as long as we study the asymptotic free theories.

On the other hand for the case of scale invariant version of asymptotic non-free technicolor model, we shall let the UV cutoff, introduced from the technical reason, infinity, $\Lambda \to \infty$. Then we obtain the results:

		$m_f^{(ETC)}(MeV)/N$	$m_f(MeV)/N$	$F_\pi(GeV)/\sqrt{N}$
(I)	($\Lambda = \infty$ theory)	120	66.7	66
(II)	($\Lambda \to \infty$ limit)	125	69.3	77
(III)	($\Lambda \to \infty$ limit)	165	89.0	88

These values are also plotted in Fig.[1] where $1/\sqrt{A} = 0$.[*] We see these values are in a good agreement with each other and indeed m_f is sufficient to give s-quark mass.

We therefore conclude that this model solves the FCNC syndrome.

(*) Some points for finite UV cutoff are also plotted. They are insufficient as s-quark mass.

Here are some comments. For the asymptotically free theories, UV cutoff dependence is small or none for $A \leq 7$. Appelquist et al. considered the cutoff as physical, $\Lambda = \Lambda_S$. This can be considered as one of the models of TC that above the ETC scale the running coupling damps very rapidly. On the other hand for the asymptotically non-free theories, UV cutoff, which is introduced from the technical reason, must tends to infinity. Especially for UV BC (UV-β) ((II) and (III)), such limit, *a la* Miransky,[3] must be taken.

The last point of view is the (attitude of the) definition of the model studied by Yamawaki, Bando and Matumoto. Namely they require that the TC model must be a renormalizable effective theory. Thus the UV cutoff must be removed at the final stage of calculation by letting $\Lambda \to \infty$. This is the same (attitude of the) definition of the theory by Miransky and also by lattice construction.

As a summary, we illustrated both (1) the analytical study of m_f and (2) the numerical study of m_f, $m_f^{(ETC)}$ and F_π.

I conclude from those values that (a) the analytical study of [12] is confirmed numerically, (b-1) the asymptotically free TC model has the FCNC syndrome even for the slowly running coupling theories, (b-2) the slowly running coupling model has a relative enhancement $m_f(A = 2 \sim 4)/m_f(A = 1) = 5 \sim 20$ but the absolute values of m_f are still too small compared with m_s by order $2 \sim 3$ and (c) the scale invariant version of asymptotically non-free TC model solves the FCNC syndrome and $m_f \sim 100 MeV$ $(\sim m_s)$.

REFERENCES

1. Yamawaki, K., Bando, M., and Matumoto, K., Phys. Rev. Lett. <u>56</u>, 1335 (1986); Bando,M., Matumoto, K., and Yamawaki, K., Phys. Lett. <u>178B</u>, 308, (1986).

2. Holdom, B., Phys. Rev. <u>24</u>, 1441, (1981); for subcolor model, Yamawaki, K., and Yokota, M., Nucl. Phys. <u>B223</u>, 144, (1983).

3. Miransky, V.A., Nuovo Cim. <u>90A</u>, 149, (1985) and references therein.

4. Aoki,K-I., Bando, M., Mino, H., Nonoyama, T., So, H., and Yamawaki, K., DPNU-88-7.

5. Holdom, B., Phys. Lett. <u>150B</u>, 301, (1985).

6. Holdom, B., Phys. Lett. <u>198B</u>, 535, (1987).

7. Appelquist, T., Karabali, D., and Wijewardhana, L.C.R., <u>D36</u>, 568, (1987).

8. Johnson, K., Baker, M., and Willey, R., Phys. Rev. <u>136</u>, B1111, (1964); ibid <u>54</u>, 860, (1967).

9. Maskawa, T. and Nakajima, H. Progr. Theor. Phys. <u>52</u>, 1326, (1975); ibid <u>54</u>, 860, (1975).

10. Fukuda, R. and Kugo, T., Nucl. Phys. <u>B117</u>, 250, (1976).

11. Akiba, T., and Yanagida, T., Phys. Lett. <u>169B</u>, 432, (1986).

12. Bando, M., Morozumi, T., So, H. and Yamawaki, K., Phys. Rev. Lett. <u>59</u>, 389, (1987).

13. Appelquist, T., Karabali, D., and Wijewardhana, L.C.R., Phys. Rev. Lett. <u>57</u>, 957, (1986); Appelquist, T., and Karabali, D., Phys. Rev. <u>D35</u>, 774, (1987).

14. Nonoyama, T. and Tanabashi, M., DPNU-88-22.

15. Higashijima, K.. Phys. Rev. <u>D29</u>, 1228, (1984).

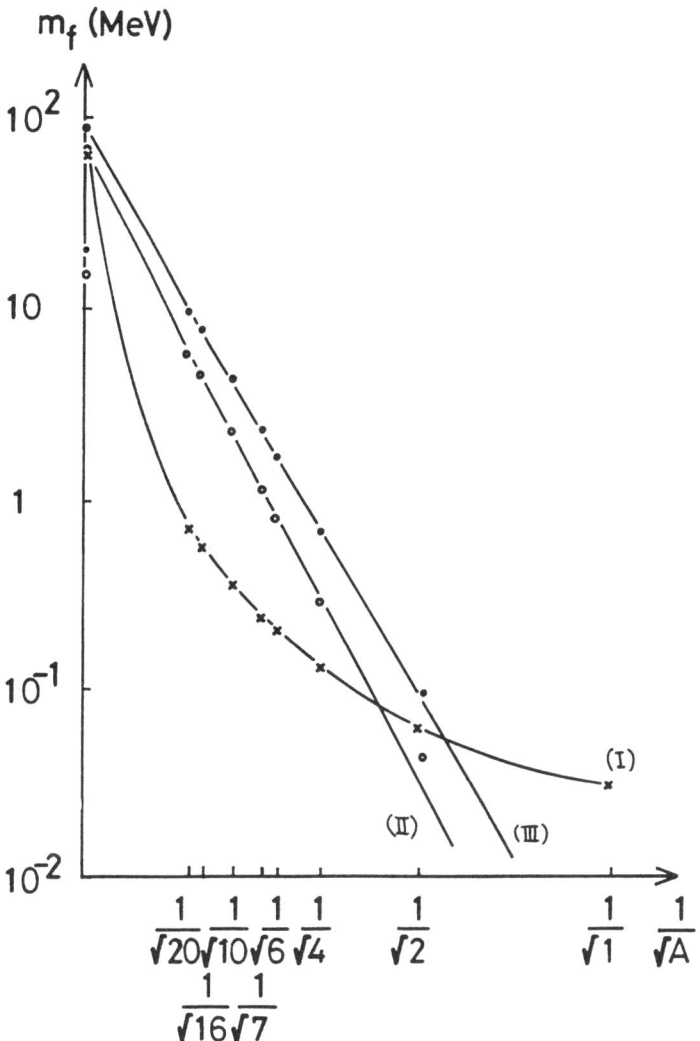

Fig.[1] $m_f(MeV)/N$ vs $1/\sqrt{A}$ is plotted. x, o and $\bullet$ are for (I), (II) and (III), respectively. The curves are drawn as a guide for the eyes. $1/\sqrt{A} = 0$ corresponds to the fixed coupling theory for which points with lower m_f are for the finite cutoff, $\Lambda = \Lambda_s$, while those which are close to 10^2 is for the infinite cutoff.

BEYOND THE LADDER

Bob Holdom
Department of Physics
University of Toronto
Toronto Ontario
CANADA M5S 1A7

ABSTRACT

We study chiral symmetry breaking for vanishing β-function and with ultraviolet cutoff Λ, beyond the ladder approximation. We are able to deduce enough about the Schwinger-Dyson equation to obtain properties of the self-energy $\Sigma(p)$. In particular in the limit $\Lambda/\kappa \to \infty$ ($\kappa \equiv \Sigma(\kappa)$) we obtain the relation between Λ/κ and a function of the coupling α. And in this limit we obtain an explicit expression for $\Sigma(p)$ which is good over most of the range $\kappa < p < \Lambda$.

INTRODUCTION

The subject of this talk is chiral symmetry breaking beyond the ladder approximation. We shall assume that the gauge theory under discussion has a β-function which is small or vanishing. We shall also assume the presence of a physical ultraviolet cutoff Λ. In the context of QED our analysis corresponds to the study of the full quenched approximation. And we note that an SU(N) gauge theory is effectively quenched in a certain limit: $C_2 \gg N$ and $C_2 n \ll N^3/d$ (n flavors, d dim. rep. with Casimir C_2). In this limit chiral symmetry breaking occurs for small coupling $\alpha \approx 1/C_2$. The β-function receives negligible contribution from fermion loops and the coupling is arbitrarily slowly varying to all orders in α since $\beta(\alpha)/\alpha \approx N\alpha$

$\approx N/C_2 \ll 1$. These quenched theories are toy examples of theories with small or vanishing β-function; we hope that a nontrivial fixed point may be a property of more general strongly coupled gauge theories.

By proceeding beyond the ladder approximation our goal is to decide which of the results[1-9] obtained in the ladder approximation are mere artifacts of that approximation.

The fermion self-energy in the linearized ladder approximation is

$$\Sigma(p) = \frac{\kappa^2 \cos(\sqrt{\alpha/\alpha_c - 1}[\ln(p/\kappa) - \frac{1}{2}\ln(\Lambda/\kappa)])}{p \; \cos(\sqrt{\alpha/\alpha_c - 1}[\frac{1}{2}\ln(\Lambda/\kappa)])} \quad \kappa < p < \Lambda \quad (1)$$

Note that $\kappa = \Sigma(\kappa)$. The relation between Λ/κ and α is

$$\cot(\sqrt{\alpha/\alpha_c - 1}[\tfrac{1}{2}\ln(\Lambda/\kappa)]) = \sqrt{\alpha/\alpha_c - 1} \quad (2)$$

In the first half of this talk we will show that the self-energy beyond the ladder takes the form

$$\Sigma(p) = (\kappa^2/p)\varphi(\ln(p/\kappa)) \quad \kappa < p < \Lambda \quad \varphi(0) \equiv 1 \quad (3)$$

$\varphi(x)$ will be an even or odd function about the point $x = \tfrac{1}{2}\ln(\Lambda/\kappa)$. This property of $\varphi(x)$ will be useful to demonstrate the enhancement of fermion condensates in theories with slowly varying gauge couplings.

In the last half of this talk we shall discuss the relation between Λ/κ and α, as $\Lambda/\kappa \Rightarrow \infty$. We will refer to this relation in the ladder approximation as the ladder continuum limit. From (2) as $\Lambda/\kappa \Rightarrow \infty$:

$$\frac{\Lambda}{\kappa} = \exp\left(\frac{\pi}{\sqrt{\alpha/\alpha_c - 1}} - 2\right) \quad (4)$$

We may parameterize the continuum limit beyond the ladder in the same form

$$\frac{\Lambda}{\kappa} = \exp\left(\frac{c_2}{(c_1 - 1)^{c_4}} - c_3\right) \quad (5)$$

We shall be able to find sufficient conditions for a theory to have such a continuum limit. We will express the constants c_i in terms of the kernel in the Schwinger-Dyson (SD) equation. The constant c_1 will be a function of the coupling α and the critical coupling may be defined by $c_1(\alpha_c) = 1$. α_c and the values for c_2 and c_3 turn out to differ from the ladder values; this is perhaps not surprising. The most interesting question is whether $c_4 = 1/2$ remains true. We find that the answer is yes. And finally we shall obtain an expression for $\Sigma(p)$ in the continuum limit, in terms of the constants c_i.

SD EQUATION AND PROPERTIES OF SOLUTIONS BEYOND THE LADDER

We remind the reader of the linearized ladder SD equation in Landau gauge:

$$p\Sigma(p) = \frac{\alpha}{2\alpha_c} \int_\kappa^\Lambda \frac{dk}{k} \min\left(\frac{p}{k}, \frac{k}{p}\right) k\Sigma(k) \tag{6}$$

$$\Sigma(\kappa) \equiv \kappa$$

This linearization is well known to be a good approximation when Λ/κ is large. Note that the information on the overall scale of $\Sigma(p)$ is retained via the self-consistent determination of κ.

We shall show that the SD equation has a similar form beyond the ladder:

$$p\Sigma(p) = \frac{1}{2} \int_\kappa^\Lambda \frac{dk}{k} F(p,k) k\Sigma(k) \tag{7}$$

$$\Sigma(\kappa) \equiv \kappa$$

We shall demonstrate that the kernel $F(p,k)$ is

A) in good approximation independent of Σ
B) symmetric under $p \leftrightarrow k$
C) a function of p/k only

(These properties are consistent with the order α^2 calculation of the kernel in Ref. (10).)

We will obtain these properties by deriving the SD equation from the effective potential $\Gamma(S)$,[11] a functional of the full fermion propagator $S(p)$:

$$S(p) = \delta_{ab}\{Z(p)(\not{p} + \Sigma(p))\}^{-1} \qquad (8)$$

The indices a and b each run over the n flavors and the dimension d of the fermion representation. We choose to renormalize $Z(\Lambda) = 1$.

$$\Gamma(S) = -\text{Tr}\{\ln S^{-1}\} + \text{Tr}\{(S^{-1}-\not{p})S\} - (2\text{PI diagrams}) \qquad (9)$$

$\delta\Gamma(S)/\delta S = 0$ is equivalent to the full SD equation.

We first sketch how the three properties of $F(p,k)$ follow before proceeding with the details. An infrared cutoff will be introduced into all momentum integrations in $\Gamma(S)$, at roughly the momentum where a nonzero $\Sigma(p)$ naturally damps the integrals: $\kappa \equiv \Sigma(\kappa)$. This permits an expansion of $\Gamma(S)$ in powers of $\Sigma(p)/p$. We will find that the Σ^2 terms grow like $\ln(\Lambda/\kappa)$ and come to dominate the Σ dependent piece of $\Gamma(S)$. $\Gamma(\Sigma^2)$ receives equal contributions from the two momentum ranges above and below $\sqrt{\Lambda\kappa}$. This is unlike the terms higher order in $\Sigma(p)/p$ which are suppressed by additional powers of κ/p and only distort the results in the infrared.

We may also mention the possible introduction of local four-fermion operators into the theory. These terms will involve extra factors of p/Λ and thereby only distort the theory in the ultraviolet. Of course if extraneous ultraviolet or infrared effects are made large enough they may destroy our results; we will assume that this is not the case.

The Σ^2 term in (2PI diagrams) is of the form shown below. This term is obviously symmetric under the interchange $p \leftrightarrow k$, and this will be reflected in the kernel $F(p,k)$. This property of the kernel is not so obvious when starting from the full SD equation; and this property is often lost when a specific anzatz is made for the vertex function appearing in the SD equation.

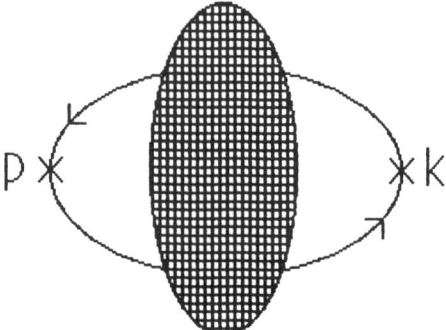

The third property of F(p,k) is nontrivial since in principle anomalous scaling could produce a factor of the form $([p \text{ or } k]/x)^{\chi}$. But we will find that F(p,k) is expressed in terms of certain 4-point and 2-point functions. When renormalization group arguments are applied to these functions we find that the anomalous scaling cancels out. The result is F(σp,σk) = F(p,k), and since F(p,k) is a dimensionless function it must therefore be a function of p/k.

We now derive these results by expanding $\Gamma(S)$ to second order in $\Sigma(p)$ and to all orders in the gauge coupling.

$$\Gamma(\Sigma^2) = \int_\kappa^\Lambda \frac{d^4p}{(2\pi)^4} \delta_{\alpha\alpha} \left(\ln Z(p) + \frac{Z(p)-1}{Z(p)} \right) - (\text{2PI diagrams})_{S^0}$$
$$+ \int_\kappa^\Lambda \frac{d^4p}{(2\pi)^4} \delta_{\alpha\alpha} \frac{\Sigma(p)^2}{p^2} \left(-\frac{1}{2} + \frac{1+A(p)}{Z(p)} \right) \qquad (10)$$
$$- \iint_\kappa^\Lambda \frac{d^4p \, d^4k}{(2\pi)^8} \frac{\Sigma(p)}{Z(p)p^2} \frac{\Sigma(k)}{Z(k)k^2} \delta_{\alpha\beta} \delta_{\gamma\delta} K_{\alpha\beta\gamma\delta}(p,k)$$

(2PI diagrams)$_{S^0}$ are evaluated using the fermion propagator expanded to zeroth order in $\Sigma(p)$:

$$S^0{}_{\alpha\beta}(p) \equiv \frac{\delta_{ab}}{Z(p)\not{p}} \qquad (11)$$

Each Greek subscript denotes flavor, color, and Lorentz index. The expansion of the fermion propagator contains a term $\delta_{ab}/(Z(p)\not{p})(-\Sigma(p)^2/p^2)$; the set of diagrams with one $S^0(p)$ replaced with this term gives rise to the $A(p)$ term in $\Gamma(\Sigma^2)$, where

$$\delta_{ab} A(p)\not{p} \equiv \frac{\delta(\text{2PI diagrams})}{\delta S^0{}_{\alpha\beta}(p)}\bigg|_{S^0} \qquad (12)$$

The expansion of the fermion propagator also has the term $\delta_{ab}\Sigma(p)/(Z(p)p^2)$. Thus there is a set of diagrams with one propagator $S^0(p)$ replaced with $\Sigma(p)/(Z(p)p^2)$ and another propagator $S^0(k)$ replaced with $\Sigma(k)/(Z(k)k^2)$. This gives rise to the last term in $\Gamma(\Sigma^2)$, where

$$K_{\alpha\beta\gamma\delta}(p,k) \equiv \frac{\delta(\text{2PI diagrams})}{\delta S^0{}_{\alpha\beta}(p)\delta S^0{}_{\gamma\delta}(k)}\bigg|_{S^0} \qquad (13)$$

K is the standard amputated 4-point 2PI kernel which appears in the Bethe-Salpeter equation, but evaluated using S^0 and with external incoming momenta p, -p, k, -k.

$Z(p)$ and $\Sigma(p)$ are to be determined by finding the stationary point of $\Gamma(\Sigma^2)$.

$$\delta\Gamma(\Sigma^2)/\delta Z(p) = 0 \qquad \delta\Gamma(\Sigma^2)/\delta\Sigma(p) = 0 \qquad (14)$$

In the first equation we may safely ignore the Σ dependent terms and keep only the Σ independent terms. This gives

$$Z(p) - 1 = A(p) \qquad (15)$$

This in turn may be used to eliminate $A(p)$ from the Σ^2 terms in $\Gamma(\Sigma^2)$. And then after dropping the terms independent of $\Sigma(p)$ the result is

$$\Gamma(\Sigma^2) = (nd/4\pi^2)\{ \int_\chi^\Lambda dp\, p\Sigma(p)^2 - \tfrac{1}{2}\iint_\chi^\Lambda dp\, dk\, \Sigma(k)\Sigma(p) F(p,k) \} \qquad (16)$$

$$F(p,k) = pk(2\pi^2 n d Z(p) Z(k))^{-1} \int_0^\pi d\theta \sin^2\theta \delta_{\alpha\beta}\delta_{\gamma\delta} K_{\alpha\beta\gamma\delta}(p,k) \tag{17}$$

θ is the angle between the four-vectors p and k, appearing only in $K_{\alpha\beta\gamma\delta}(p,k)$.

The renormalization group equations greatly simplify in Landau gauge since a vanishing gauge parameter does not "evolve". And when the β-function vanishes, the RG equation for an amputated fermion n-point function is simply

$$[\sigma\partial/\partial\sigma + (^3/_2 n - 4 + n\gamma)]\, \Gamma^{(n)} = 0 \tag{18}$$

This yields

$$Z(\sigma p) = \sigma^{-2\gamma} Z(p)$$
$$K_{\alpha\beta\gamma\delta}(\sigma p, \sigma k) = \sigma^{-(2+4\gamma)} K_{\alpha\beta\gamma\delta}(p,k) \tag{19}$$

which implies the stated result, $F(\sigma p, \sigma k) = F(p,k)$.

The three properties of $F(p,k)$ are now evident. The SD equation in (7) is finally obtained from $\delta\Gamma(\Sigma^2)/\delta\Sigma(p) = 0$. It is convenient to change variables and define

$$x \equiv \ln(p/\kappa), \quad y \equiv \ln(k/\kappa), \quad L \equiv \ln(\Lambda/\kappa) \tag{20}$$

The properties of $F(p,k)$ allow us to write the SD equation as

$$\psi(y) = \tfrac{1}{2}\int_0^L dx\, V(|x-y|)\psi(x) \qquad \psi(0) \equiv 1 \tag{21}$$

where

$$\psi(x) \equiv p\Sigma(p)/\kappa^2 \tag{22}$$
$$V(|x-y|) \equiv F(p,k) \tag{23}$$

V is determined by the theory, and thus we look for solutions of this equation by adjusting L. Each solution will in general occur for a different value of L, and of these solutions we must seek the solution giving the minimum vacuum energy. We expect by dimensional arguments that the vacuum energy density is proportional to $-\kappa^4$, with the sign determined by the requirement that a symmetry breaking solution be preferred over the trivial solution $\Sigma(p) = 0$. Thus we seek the largest κ, and for a fixed physical scale Λ we therefore seek the solution with smallest value of $L \equiv \ln(\Lambda/\kappa)$.

An important property of $\psi(x)$ follows from the form of the kernel in (21). If a solution exists for some value of L then that solution is either even or odd about L/2. (We ignore the unnatural accident where both even and odd solutions occur at the same value of L.) This implies for example that $\psi(L) = \pm\psi(0) = \pm 1$, which in turn gives $\Sigma(\Lambda) = \pm \kappa\Sigma(\kappa)/\Lambda$.

With this result we may demonstrate the enhancement of condensates. In technicolor, of interest are vacuum expectation values of operators generated by physics at scale Λ. We thus consider the following quantity.

$$\frac{\langle\bar{\Psi}\psi\rangle}{\kappa^3} \propto \int^{\Lambda} \frac{dk k \Sigma(k)/Z(k)}{\kappa^3} \qquad (24)$$

$$= \int_0^L dx e^x \psi(x)/Z(x), \text{ where } \psi(L) = \pm 1 \text{ and } Z(L) = 1$$

For any reasonable kernel in (21) the resulting $\psi(x)$ will give this integral an enhancement factor of order $e^L = \Lambda/\kappa$. This conclusion may only be avoided by introducing and fine-tuning additional ultraviolet physics.

CONTINUUM LIMIT BEYOND THE LADDER

We will now find the analog of the ladder continuum limit when $\Lambda/\kappa \to \infty$. The ladder approximation corresponds to the particular choice $V(x) = \alpha/\alpha_c e^{-x}$. Note that $V(|x-y|)$ then is not smooth at $x = y$. Going beyond the ladder approximation will surely modify $V(x)$ for small x, as well as the asymptotic behavior. Does the continuum limit depend crucially on the choice $V(x) \propto e^{-x}$?

We will find that the following three conditions are sufficient for the existence of a continuum limit.

    A) $V(x)$ is exponentially small for large x

    B) $\int_0^\infty dx V(x) \to 1+$

    C) $\int_0^\infty dx V(x) x^2 > 0$

The first condition occurs for example if, when p or k is much larger than the other,

$$F(p,k) \propto (p/k)^\nu \text{ for } k \gg p \quad (\nu > 0) \qquad (25)$$
$$ (k/p)^\nu \text{ for } p \gg k$$

This corresponds to $V(x) \propto e^{-\nu x}$, for large x only. Note that ν is not determined by the previous RG analysis. (The order α^2 calculation of the kernel in Ref. (10) finds $\nu = 1+2\delta$.)

We first consider the case that $\psi(x)$ is even about $x = L/2$. We expand $\psi(x)$ about this point.

$$\psi(L/2+x) = \psi^{(0)} + \frac{\psi^{(2)} x^2}{2!} + \frac{\psi^{(4)} x^4}{4!} + \ldots \qquad (26)$$
$$\psi^{(i)} \equiv d^i \psi(x)/dx^i \big|_{x=L/2}$$

If we take 2n derivatives of $\psi(y) = \tfrac{1}{2}\int_0^L dx V(|x-y|)\psi(x)$ and set $y = L/2$ we obtain

$$\psi^{(2n)} = V^{(1)}(0)\psi^{(2n-2)} + V^{(3)}(0)\psi^{(2n-4)} + \ldots + V^{(2n-1)}(0)\psi^{(0)} + \int_0^{L/2} dx V^{(2n)}(x)\psi(L/2+x) \qquad (27)$$

Inserting (26) into the integral and integrating by parts we find that the first n terms inside the integral cancel the n terms outside the integral, if we make use of condition A) and neglect all exponentially small boundary terms. Replacing the upper limits in the remaining integrals by ∞ gives

$$\psi^{(2n)} = \psi^{(2n)}\int_0^\infty dx V(x) + \psi^{(2n+2)}\int_0^\infty dx V(x) x^2/2 + ... \qquad (28)$$

We will see that the ... terms may be neglected as the continuum limit is approached.

If we define the positive quantities

$$b_1 \equiv \int_0^\infty dx V(x) \qquad (29)$$

$$b_2 \equiv \frac{\int_0^\infty dx V(x) x^2}{2\int_0^\infty dx V(x)} \qquad (30)$$

then we may write (28) in the form

$$1 = \{1 + (b_1 - 1)\}[1 + b_2 \psi^{(2n+2)}/\psi^{(2n)}] \qquad (31)$$

The continuum limit will be obtained by allowing b_1 to approach unity from above. Then $\psi^{(2n+2)}/\psi^{(2n)}$ is a small negative quantity, independent of n.

$$\psi^{(2n+2)}/\psi^{(2n)} = -(b_1 - 1)/b_2 \qquad (32)$$

Thus our original expansion of $\psi(L/2+x)$ takes the form of a cosine.

$$\psi(x) = \psi^{(0)}\cos\{\sqrt{(b_1 - 1)/b_2}(x-L/2)\} \qquad (33)$$

In this derivation we have ignored the boundary terms since they were exponentially small. But this is clearly not possible if we instead consider the SD equation for $\psi(x)$ when x is near 0 or L. Thus (33) is expected to be good everywhere except possibly for x near 0 or L.

This analysis does not determine $\psi^{(0)}$. On the other hand the SD equation along with the boundary condition $\psi(0) = 1$ does determine $\psi^{(0)}$, as can be easily checked numerically. So $\psi^{(0)}$ must be determined by what we have neglected, ie. by the exponentially small boundary terms. That is, $\psi^{(0)}$ must depend on the asymptotic behavior of V(x).

Although we have not proved it we may easily test the hypothesis that $\psi^{(0)}$ depends *only* on the asymptotic behavior of V(x). If this is the case then we may determine $\psi^{(0)}$ for *any* $V(x)$ with the asymptotic behavior of interest. For example for the asymptotic behavior $e^{-\nu x}$ we may choose $V(x) = \alpha e^{-\nu x}$ for all positive x. This is convenient since in this case we may solve the SD equation analytically. The result (see appendix) is

$$\psi^{(0)} = \nu L/\pi \qquad (34)$$

Since $\psi^{(0)}$ is large for large L, the boundary condition $\psi(0) = 1$ applied to (33) takes the simple form

$$\sqrt{(b_1 - 1)/b_2}(L/2) = \pi/2 - 1/\psi^{(0)} \qquad (35)$$

Combining the last two equations and rearranging gives our main result. We obtain the continuum limit relation in (5) with

$$c_1 = \int_0^\infty dx V(x) \qquad (36)$$

$$c_2 = \pi \left[\frac{\int_0^\infty dx V(x) x^2}{2\int_0^\infty dx V(x)} \right]^{1/2} \qquad (37)$$

$$c_3 = 2/\nu \qquad (38)$$

$$c_4 = 1/2 \qquad (39)$$

We now see exactly how the constants c_1, c_2, c_3, c_4 introduced in (5) are related to the kernel $V(|x-y|) \equiv F(p,k)$ of the SD equation. Most interesting is that $c_4 = 1/2$ is not just an artifact of the ladder approximation; this result may be traced to the fact that only every other power of x appears in the expansion of $\psi(L/2+x)$.

We may write $\Sigma(p)$ for this continuum limit in terms of the $L \equiv \ln(\Lambda/\kappa)$ which satisfies (5).

$$\Sigma(p) = \frac{\kappa^2 \nu L \cos([\pi/L][1-2/(L\nu)][\ln(p/\kappa)-L/2])}{p\pi} \quad \kappa < p < \Lambda \qquad (40)$$

The results (34) and (36-39) have been checked by solving the SD equation numerically for a variety of $V(x)$. The error in relations (5) and (34) for $\Lambda/\kappa \gtrsim 10^4$ is typically less than a percent, although the error tends to increase as c_2 becomes small. The accuracy is surprising given that the derivation of $\varphi^{(0)}$ and c_3 is not on as firm a footing as the other constants.

Another check is to compare to the special case $V(x) = \alpha e^{-\nu x}$ described in the appendix. For this $V(x)$, (36) and (37) give $c_1 = \alpha/\nu$ and $c_2 = \pi/\nu$ which agree with the analytical solution.

If we consider the contant coupling α as a parameter of the theory, then $V(x)$ should be considered as a function of this parameter, $V(x,\alpha)$. The critical coupling α_c in the continuum limit $c_1 \Rightarrow 1$ is determined implicitly by

$$\int_0^\infty dx V(x,\alpha_c) = 1 \qquad (41)$$

Of course α_c here will be different from the ladder value. To write the expression (5) for Λ/κ as a function of the gauge coupling in the continuum limit, we may replace $c_1(\alpha)-1$ by $(\alpha-\alpha_c)c_1'(\alpha_c)$. (41) shows that α_c is sensitive to $F(p,k)$ for $p \approx k$, as well as for $p \gg k$. We also see that c_2 depends on more than just the asymptotic behavior of $V(x)$. It is of interest that the larger c_2 happens to be, the less "fine-tuning" of the quantity c_1 is required to obtain a large Λ/κ.

We may consider the more general asymptotic behavior $V(x) \propto x^\xi e^{-\nu x}$, which corresponds to $F(p,k) \propto (\ln p/k)^\xi (k/p)^\nu$ for $p \gg k$. But only the formulas for $\varphi^{(0)}$ and c_3 would be altered by this change. (But it may well be that these powers of logs do not appear; eg. when the β-function has a simple zero an operator product expansion analysis indicates that they do not occur.)

We note also that our results are modified if four-fermion terms of the form $(G/\Lambda^2)\bar{\psi}\psi\bar{\psi}\psi$ are introduced. In the ladder approximation it is found that $\varphi(x)$ is still a cosine for $\alpha > \alpha_c$, and that only the π is changed in (4) due to a different boundary condition. But this is just a peculiarity of the ladder approximation; in general the cosine solution no longer applies in the continuum limit when large four-fermion terms are introduced. On the other hand for small enough G we find numerically that the cosine solution is only distorted in the ultraviolet near $x \approx L$. In this case it seems that the effect on (5) may be absorbed into a new c_3.

We may repeat the whole analysis for a solution which is odd about L/2. Now only odd powers of x appear in (26). The result is

$$\varphi(x) = A\sin\{\sqrt{(b_1 - 1)/b_2}(L/2-x)\} \tag{42}$$

with b_1 and b_2 as before. From the analytical solution with one zero in the special case $V(x) = \alpha e^{-\nu x}$ we obtain $A = \nu L/(2\pi)$. And the condition $\varphi(0) = 1$ yields $\sqrt{(b_1 - 1)/b_2}(L/2) = \pi - 1/A$. The result is that of the constants c_i, only c_2 is changed; its value for the odd solution is twice that for the even solution. This implies that for the same α, L is smaller for the even solution. From the previous discussion of vacuum energy, we therefore find that the even solution is the appropriate one for the continuum limit.

We conclude by stressing the one further nontrivial requirement for a continuum limit, besides the three conditions on $V(x)$. This is the requirement that the function $F(p,k)$ given in (17) is only a function of p/k. In a gauge theory we have shown that this property follows from a vanishing β-function.

But since we are discussing a strongly interacting theory we must be careful not to presume to know the true effective degrees of freedom well below the ultraviolet cutoff. Perhaps the gauge field is not included among these strongly interacting degrees of freedom. This type of question may hopefully be addressed by lattice calculations. But if the lattice results should confirm the existence of a continuum limit relation with $c_4 = 1/2$ in an unquenched theory, the results presented here would have interesting implications for the effective theory throughout the momentum range $\kappa < p < \Lambda$.

APPENDIX

Here we derive the analytical solution to the SD equation (21) in the special case $V(x) = \alpha e^{-\nu x}$ (for positive x and $\nu > 0$). Then the SD equation is equivalent to a trivial differential equation plus boundary conditions.

$$\psi''(x) + 1/4[\alpha\nu - \nu^2]\psi(x) = 0 \qquad (A1)$$
$$\psi'(0) = \nu/2\,\psi(0), \quad \psi'(L) = -\nu/2\,\psi(L), \quad \psi(0) = 1$$

Solutions only exist for $\alpha > \nu$ and the solution even about L/2 takes the form

$$\psi(x) = \frac{\cos(b/2(x-L/2))}{\cos(bL/4)} \quad \text{with } b = \sqrt{\alpha\nu - \nu^2} \qquad (A2)$$

The relation between L and α (for the solution with no zeros) follows from the boundary conditions.

$$\cot(bL/4) = b/\nu \qquad (A3)$$

In the limit $\alpha \to \nu+$ the $\psi^{(0)}$ in (33) is given by

$$\psi^{(0)} = 1/\cos(bL/4) = \nu L/\pi \qquad (A4)$$

and (A3) corresponds to

$$\Lambda/\kappa = \exp[(\pi/\nu)/\sqrt{\alpha/\nu - 1} - 2/\nu] \qquad (A5)$$

ACKNOWLEDGEMENT

I thank the Research Institute for Fundamental Physics, Kyoto University for their hospitality and support during my three month visit.

REFERENCES

1) T. Maskawa and H. Nakajima, Prog. Theor. Phys. $\underline{52}$ (1974) 1326; $\underline{54}$ (1976) 860.

2) R. Fukuda and T. Kugo, Nucl. Phys. $\underline{B117}$ (1976) 250.

3) V.A. Miransky, Sov. Phys. JETP $\underline{61}$ (1985) 905; Nuovo Cim. $\underline{90A}$ (1985) 149 (and earlier references therein).

4) B. Holdom, Phys. Lett. $\underline{150B}$ (1985) 301; Phys. Lett. $\underline{198B}$ (1987) 535.

5) W. Bardeen, C. Leung and S. Love, Phys. Rev. Lett. $\underline{56}$ (1986) 1230; Nucl. Phys. $\underline{B273}$ (1986) 649.

6) T. Akiba and T. Yanagika, Phys. Lett. $\underline{169B}$ (1986) 432.

7) K. Yamawaki, M. Bando and K. Matumoto, Phys. Rev. Lett. $\underline{56}$ (1986) 1335; M. Bando, T. Morozumi, H. So, K. Yamawaki, Phys. Rev. Lett. $\underline{59}$ (1987) 389.

8) T. Appelquist, D. Karabali and L.C.R. Wijewardhana, Phys. Rev. Lett. $\underline{57}$ (1986) 957; T. Appelquist and L.C.R. Wijewardhana, Phys. Rev. $\underline{D35}$ (1987) 774; Phys. Rev. $\underline{D36}$ (1987) 568; T. Appelquist, D. Carrier, L.C.R. Wijewardhana, W. Zheng, Phys. Rev. Lett. $\underline{60}$ (1988) 1114.

9) A. Cohen and H. Georgi, Harvard preprint, HUTP-88/A007, March 1988.

10) T. Appelquist, K. Lane, U. Mahanta, Yale preprint, YCTP-P10-88, June 1988.

11) J. Cornwall, R. Jackiw and E. Tomboulis, Phys. Rev. $\underline{D10}$ (1974) 2428; M. Peskin, in "Recent Advances in Field Theory and Statistical Mechanics", edited by J. Zuber and R. Stora, Les Houches Summer School Proceedings, Vol. 39 (North-Holland, Amsterdam, 1984).

DYNAMICS OF THE NONTRIVIAL FIXED POINT IN GAUGE FIELD THEORIES

V.A. Miransky[†]
Department of Physics, Nagoya University,
Nagoya 464-01, Japan

ABSTRACT

The dynamics of the breakdown of chiral and scale symmetries in the critical (and nearcritical) regime with a nontrivial fixed point in gauge theories is considered. The emphasis is on the discussion of the possibility of the existence of nontrivial non-asymptotically free continuum theories and of the character of scale symmetry breakdown in gauge theories.

1. INTRODUCTION

We are in the very begining of the understanding of the nonperturbative dynamics in quantum field theories. So we should be ready to find new surprising result there.

In the present paper we discuss the dynamics of chiral and scale symmetry breakdown in the critical (and nearcritical) regime with a nontrivial ($\alpha_c > 0$) ultraviolet fixed point. We consider both asymptotically free (AF) and non-asymptotically free (NAF) vector-like gauge theories (the latter can be either Abelian QED-like theories or non-Abelian theories with a sufficiently large number of fermions). In particular, we discuss the possibility that NAF gauge theories with a nontrivial ultraviolet fixed point $\alpha_c > 0$ separating two phases <u>with</u> <u>different</u> <u>structures</u> <u>of</u> <u>renormalizations</u> can give an example of four dimensional continuum field theories with nontrivial S-matrix[1,2,3].

---

† Permanent address: Institute for Theoretical Physics, 252130 Kiev 130, USSR

The background of the present approach is the collapse phenomenon[1,2,3,4] which in quantum mechanics is named "fall into the centre" one.[5] I remind that quantum mechanics began with the question: why doesn't electron fall into a nuclei? Here we will attempt to show that sometime it does "fall" and as the result interesting consequenses arise.

2.   RENORMALIZATIONS: GENERAL FEATURES.

Continuum four dimensional field theories are not completely defined systems. We need in renormalizations to define completely them. Let us remind some basic features of renormalizations in perturbation theory.

Let us consider some gauge theory (AF or NAF one). We start from the theory with an ultraviolet cutoff Λ and the bare coupling constant $\alpha^{(0)}$. Our aim is to find a continuum ($\Lambda\to\infty$) solution with the "physical" coupling $\alpha_\mu \neq 0$ <u>independent</u> of Λ. The coupling α_μ determines the intensity of the interaction of the initial fields of the Lagrangian at distances $r\sim 1/\mu$ (for the sake of simplicity we consider the chiral limit in which the bare fermion mass equals zero).

After this and wavefunction renormalizations the theory becomes completely defined. All Green's functions are expressed through the free parameter α_μ. Also, the procedure determines the form of the bare coupling $\alpha^{(0)}$ as the function of Λ. For example, in the one-loop approximation we have:

$$\alpha^{(0)}(\Lambda) = \frac{\alpha_\mu}{1+C\alpha_\mu \ln \frac{\Lambda}{\mu}}, \tag{1}$$

where C is numeral constant.

Why do we determine renormalizations by the choice of α_μ as <u>the basic</u> (i.e. independent of Λ) parameter? Historically, this choice was apparently only possible. In perturbative QED the physical charge $\alpha_{ph} \simeq 1/137$ and we need in such a renormalization to explain experimental results. In AF gauge theories this choice leads to a nontrivial spectrum of colourless states ("hadrons") and to nontrivial S-matrix (although this fact is not yet proved it is commonly believed).

From the general point of view the choice of α_μ as the basic parameter is justified due to that it provides the nontrivial character of S-matrix (at least in perturbation theory). In AF theories this choice is apparently the only possible due to the point (in fact, yet unproved) that there is only one phase in coupling constant there. Can other choices of basic parameters (providing nontrivial S-matrix) exist in NAF theories?

Let us consider the relation

$$\alpha_\mu = Z_3(\mu) \alpha^{(0)} \qquad (2)$$

in QED. Is this relations valid always? The answer is "yes". It follows from the equations of motion and it determines the connection between the intensity of the interaction at the distances $r \sim \frac{1}{\Lambda}$ and $r \sim \frac{1}{\mu}$.

Does this relation always determine the renormalization of the bare coupling constant $\alpha^{(0)}$? The answer is the following: it determines the renormalization of $\alpha^{(0)}$ only if we assume that α_μ is independent of Λ, i.e. if α_μ is chosen as the basic parameter. In this case

$$\alpha^{(0)}(\Lambda) = \alpha_\mu Z_3(\mu)^{-1}(\Lambda), \qquad (3)$$

$$\beta = \frac{\partial \alpha^{(0)}}{\partial \ln \Lambda} = \alpha_\mu \frac{\partial Z_3(\mu)^{-1}}{\partial \ln \Lambda}. \qquad (4)$$

But if we look for a continuum solution determined by the choice of another basic parameter, the coupling α_μ can in principle depend on Λ itself. In this case the form of β-function will be different:

$$\beta = \frac{\partial \alpha^{(0)}}{\partial \ln \Lambda} = \frac{\partial \alpha_\mu}{\partial \ln \Lambda} Z_3(\mu)^{-1} + \alpha_\mu \frac{\partial Z_3(\mu)^{-1}}{\partial \ln \Lambda}. \qquad (5)$$

Below, in the framework of a simple model, we shall demonstrate that such a situation is in principle possible.

3. COLLAPSE PHENOMENON AND RENORMALIZATIONS IN FIELD THEORIES

Let us consider the ladder Schwinger-Dyson equation for the fermion propagator

$$G_{ij}(p) = \frac{\delta_{ij}}{\hat{p}A(p^2) - B(p^2)} \qquad (6)$$

in QED ($i=1,2,\cdots,N_f$ is the flavour index). In the Landau gauge the ladder Schwinger-Dyson equation takes the following form in the chiral limit (after Wick rotation and angular integration):

$$B(p^2) = \lambda \int_0^{\Lambda^2} dk^2 K(p^2,k^2) \frac{k^2 B(k^2)}{k^2 + B^2(k^2)}, \qquad (7)$$

where $\lambda = \frac{3\alpha(0)}{4\pi}$, $K(p^2,k^2) = \frac{\theta(p^2-k^2)}{p^2} + \frac{\theta(k^2-p^2)}{k^2}$ (in this approximation $A(p^2)=1$). Note that the choice of the Landau gauge in this approximation is not accidental: just in the Landau gauge the Ward identities are satisfied in this approximation[3]. The transition to other gauges implies essential changing of the fermion-antifermion-photon vertex (the recent discussion of this problem see in Ref.[6])

Although the analytical form of the solutions of the equation (7) is not yet found, some thier important properties are known.[3,7,8] It is known that the value $\lambda=\frac{1}{4}$ ($\alpha(0)=\frac{\pi}{3}$) is the critical value separating the perturbative ($\lambda<\frac{1}{4}$) phase and the phase with spontaneous chiral symmetry breaking. In the perturbative phase there is only the trivial solution $B(p^2)=0$. In the supercritical, $\lambda>\frac{1}{4}$, phase the asymptotics of the nontrivial solution takes the following form:

$$B(p^2) \underset{p^2 \to \infty}{\simeq} \eta^{1/2} \frac{2}{\pi\nu} \frac{m_d^2}{p} \sin(2\nu \ln \frac{p}{m_d} + C\nu), \qquad (8)$$

where $\nu=(\lambda-\frac{1}{4})^{1/2}$ and the "dynamical mass" $m_d \equiv B(0)$. The numeral constants η and C are $\eta \simeq 1$ and $C \simeq 2$.

The dynamical mass m_d is expressed through the cutoff parameter Λ in the following way:

$$m_d \underset{\nu \ll 1}{\simeq} 4\Lambda \exp(-\frac{\pi}{2\nu}) \ . \qquad (9)$$

Thus, in the supercritical phase a new nonperturbative divergence appears (in perturbation theory the mass divergence is of course absent in the chiral limit). We cannot remove this divergence if the perturbative renormalization relations are not changed: in the ladder approximation $Z_1 = Z_2 = Z_3 = 1$ and hence if the coupling α_μ is the basic parameter, the bare coupling $\alpha^{(0)} = Z_3^{-1} \alpha_\mu = \alpha_\mu$ must not depend on the cutoff Λ too.

So in the present case we must use the notion of renormalizations in a more general sense.

What is the general meaning of renormalizations? Performing a renormalization means making the bare parameters (in our case $\alpha^{(0)}$) to depend on Λ in such a way that physical parameters (in our case m_d) remain finite in the limit $\Lambda \to \infty$. Eq.(9) implies that

$$\alpha^{(0)}(\Lambda) = \frac{\pi}{3}(1 + \frac{\pi^2}{\ln^2 \frac{4\Lambda}{m_d}}) \xrightarrow[\Lambda \to \infty]{} \alpha_c = \frac{\pi}{3} \ . \qquad (10)$$

The β-function is

$$\beta = \frac{\partial \alpha^{(0)}}{\partial \ln \Lambda} = -\frac{2}{3}(\frac{3\alpha^{(0)}}{\pi} - 1)^{3/2} \ . \qquad (11)$$

This β-function has a nontrivial ultraviolet stable zero at $\alpha^{(0)} = \frac{\pi}{3}$.

In this approximation the phase diagram in the coupling constant is the following: at $\alpha^{(0)} < \alpha_c = \frac{\pi}{3}$ β-function equals zero, and all these values of $\alpha^{(0)}$ form the line of the fixed points; in the supercritical phase the additional renormalization of the charge takes place which results in the appearance of the nontrivial ultraviolet stable fixed point $\alpha^{(0)} = \alpha_c = \frac{\pi}{3}$ separating the massless and the massive phases (see Fig.1).

The crucial point providing the possibility of such a renormalization in the supercritical is that at $\alpha^{(0)} = \alpha_c$ the second order phase transition takes place: the correlation length

$$\xi = \frac{\Lambda}{m_d} \simeq \frac{1}{4} \exp(\pi/(\frac{\alpha^{(0)}}{\alpha_c} -1)^{1/2}) \qquad (12)$$

is infinite at the critical value of $\alpha^{(0)}$.

The main lesson of this exercise are the following:

1) In the supercritical phase the relation

$$\alpha_\mu = Z_3^{(\mu)} \alpha^{(0)}$$

is valid. In the present approximation it states that $\alpha_\mu = \alpha^{(0)}$. But due to the choice of the dynamical mass m_d as the basic parameter the coupling α_μ depends on Λ in the same manner as $\alpha^{(0)}$.

2) The choice of m_d as the basic parameter is not accidental. When spontaneous chiral symmetry breaking takes place, the massless pseudoscalar fermion-antifermion bound states (Goldstone bosons) must appear. In the low energy region their interaction is described by the Lagrangian

$$L = \frac{(\partial_\mu \vec{\pi}^2)^2}{8F_\pi^2} ; \qquad \vec{\pi} = (\pi_1, \cdots, \pi_{N_f^2-1}) \qquad (13)$$

The pseudoscalar decay coupling $F_\pi \sim m_d$. Thus, the condition $0 < m_d < \infty$ provides the nontrivial character of S-matrix in the theory.

Thus, the general strategy of the search of the continuum solution with nontrivial S-matrix in NAF gauge theories should be as follows: we must look for the second order phase transition at $\alpha^{(0)} = \alpha_c > 0$ connected with spontaneous chiral symmetry breaking.

We also stress that the nonperturbative divergence considered above has nothing to do with the loop divergences of field theories. In fact, it is field analog of the divergence which appears in the quantum-mechanical problems with singular at r=0 potentials (collapse phenomenon).[1,2,3,4] In particular, the present model is very close to the problem of the Dirac equation with the supercritical Coulomb potential $V(r) = -\frac{\alpha}{r}$, $\alpha > 1$. Also a similar phenomenon takes place in some soluble two-dimensional models (in particular in the sine-Gordon model).

The detailed discussion of this analogy can be found in Refs.[2,3]

4. RENORMALIZATION GROUP AND SCALE TRANSFORMATIONS IN NAF THEORIES

In the model discussed above the β-function is different in the subcritical and supercritical phases. Since the β-function determines the renormalization group transformations, this means that the renormalization groups (and therefore scale transformations of Green's functions) are different in these phases.

In the subcritical phase the β-function $\beta \equiv 0$. Therefore scale symmetry is exact and Green's functions have a simple form there.

On the other hand, in the supercritical phase the Callan-Symanzik equation for unrenormalized Green's functions takes the following forms:

$$\left(\frac{\partial}{\partial \ln \Lambda} + \frac{\partial}{\partial \alpha^{(0)}(\Lambda)} - \gamma_\Gamma\right)\Gamma^{(n)}(p_1,\ldots,p_n,\alpha^{(0)}(\Lambda),\Lambda)\Big|_{\Lambda \to \infty} = 0, \quad (14)$$

where β is the β-function (11) and γ_Γ is the anomalous dimension of the Green's function $\Gamma^{(n)}$. The scale transformation of Green's function takes the form:

$$\Gamma^{(n)}(\sigma p_1,\ldots,\sigma p_n; \alpha^{(0)}(\Lambda),\Lambda) =$$
$$= \Gamma^{(n)}(p_1,\ldots,p_n; \bar{\alpha}(t,\alpha^{(0)}(\Lambda)))\exp\int_0^t dx(D_\Gamma - \gamma_\Gamma(x)), \quad (15)$$

where $t = \ln\sigma$, D_Γ is the canonical dimension of the function $\Gamma^{(n)}$ and the running coupling $\bar{\alpha}$ is determined from the standard relation:

$$\int_{\alpha^{(0)}(\Lambda)}^{\bar{\alpha}(t,\alpha^{(0)}(\Lambda))} \frac{dx}{\beta(x)} = t. \quad (16)$$

Of course, this running coupling does not coincide with the perturbative running coupling which is constant in this approximation.

Due to the nonzero β-function (11) the scale transformation (15) is nontrivial. It admits Green's functions of a rather complicated form (see Eq.(8)).

Moreover, as it has been shown in the papers[9], the conventional relation for the devergence of the dilaton current (trace of the energy-momentum θ^μ_μ) takes place in the supercritical phase:

$$\partial_\mu D^\mu = \theta^\mu_\mu = \frac{\beta(\alpha^{(0)})}{4\alpha^{(0)}} F_{\mu\nu} F^{\mu\nu} + (1+\gamma_m) \sum_{i=1}^{N_f} m_i^{(0)} \bar\psi_i \psi_i , \qquad (17)$$

where $\beta(\alpha^{(0)})$ is the β-function (11), $m_i^{(0)}$ is the bare mass of the i-th fermion, γ_m is the anomalous dimension of the operatores $\bar\psi_i \psi_i$. In the chiral limit

$$\theta^\mu_\mu = \frac{\beta(\alpha^{(0)})}{4\alpha^{(0)}} F_{\mu\nu} F^{\mu\nu} . \qquad (18)$$

In such a theory, even though $\beta(\alpha^{(0)})$ vanishes in (11) as $\alpha^{(0)} \to \alpha_c$, $\Lambda \to \infty$, the operator $F_{\mu\nu} F^{\mu\nu}$ diverges there, and the divergence of the dilation current remains nonvanishing and finite in (18). In particular, the gluon condensates is[9]

$$H \equiv -\langle 0|\theta^\mu_\mu|0\rangle = - \lim_{\substack{\Lambda \to \infty \\ \alpha^{(0)} \to \alpha_c}} \frac{\beta(\alpha^{(0)})}{4\alpha^{(0)}} \langle 0|F_{\mu\nu} F^{\mu\nu}|0\rangle \simeq \frac{4N_f}{\pi^4} m_d^4 , \qquad (19)$$

where the symbol $\substack{\Lambda \to \infty \\ \alpha \to \alpha_c}$ means that the transition to the continuum limit is realized together with the renormalization procedure (10).

Thus, there are different renormalization groups (and hence scale transforamtions) in different phases. In particular, the form of the scale anomaly depends on the type of the phase to which it relates.

Of course, the central question is what is the form of the β-function beyond the ladder approximation? We shall discuss this question in Sec.6 but beforehand we will discuss the problem of scale symmetry breaking in the critical regime with $\alpha_c > 0$ in more detail.

5. ON THE CHARACTER OF SCALE SYMMETRY BREAKING IN GAUGE THEORIES

The relation (19) implies that in the supercritical phase the breakdown of scale invariance is explicit. Therefore, the scalar fermion-antifermion bound state (σ) connected with the operator θ^μ_μ has to be massive.

To estimate its mass one can use the Dashen-like mass formula:[10]

$$M^2_\sigma = \frac{4H}{F^2_\sigma}, \qquad (20)$$

where the parameter F_σ is determined from the relation

$$\langle 0|\theta^\mu_\nu|\sigma; q\rangle = \frac{1}{3}(\delta^\mu_\nu M^2_\sigma - q^\mu q_\nu)F_\sigma. \qquad (21)$$

Moreover, the scale Ward identities imply that[11]

$$F_\sigma = d\left(\frac{N_f}{2}\right)^{1/2} F_\pi, \qquad (22)$$

where d is the dynamical dimension of the field $\bar\psi\psi$ which is (with some normalization factor) interpolating field for $|\sigma\rangle$. Taking into account that in the critical regime $d=2$[2,12] we obtain the following mass relation from Eqs.(19), (20) and (22)[11]:

$$M^2_\sigma \approx \frac{8m^4_d}{\pi^4 F^2_\pi}. \qquad (23)$$

Hence the mass of scalar $M_\sigma \sim m_d$ (a somewhat different approach of the papers[13] leads to a similar conclusion).

On the other hand, in the papers[12,14] an appealing possibility when the spontaneous breaking of chiral symmetry may imply the spontaneous breaking of scale symmetry has been considered. This possibility could lead to interesting phenomenological applications, in particular, to the existence of the Goldstone boson of spontaneously broken scale invariance (the dilaton)[12,14].

Recently, the no-go theorem showing that such a possibility could be realized only in rather sophisticated theories has been proved[15]. This theorem is as follows: the phenomenon of spontaneous scale symmetry breaking is not compatible with the standard realization of the PCAC dynamics.

By the standard realization of PCAC we mean the scheme in which pseudoscalar bosons ("pions") dominate matrix elements of the axial currents in the low energy region and the fermion bare mass $m^{(0)}$ appears in the Lagrangian only through the standard mass term

$$m^{(0)} \sum_{i=1}^{N_f} \bar{\psi}_i \psi_i = m_c \sum_{i=1}^{N_f} (\bar{\psi}_i \psi_i)_r , \qquad (24)$$

where $(\bar{\psi}_i \psi_i)_r = Z_m \bar{\psi}_i \psi_i$ is the renormalized mass operator and $m_c = Z_m^{-1} m^{(0)}$ is the current mass of fermions (for the sake of simplicity, we consider the case when the current mass is the same for all N_f flavours).

In this case the spontaneous breaking of scale symmetry in the chiral limit implies that

$$\theta_\mu^\mu = (1+\gamma_m) m_c \sum_{i=1}^{N_f} \bar{\psi}_i \psi_i , \qquad (25)$$

i.e. the scale anomaly term is absent (compare with Eq.(17)). The strategy of the proof of the theorem is to show that the equality (25) can not be satisfied on the one-particle pion states.

Here we omit the proof which can be found in the paper[15] and discuss the consequences of this theorem.

As any no-go theorem it does not completely close the possibility of the realization of the situation which is considered but shows the way to overcome itself.

The present theorem could be overcome if
a) the hypothesis about the pion pole dominance is wrong
or (and)
b) the assumption concerning the dependence of the Lagrangian on the current mass only through the mass term $m_c \sum_{i=1}^{N_f} (\bar{\psi}_i \psi_i)_r$ is wrong.

The last point means that a theory with the mass term

$$L_m = -m_c \sum_{i=1}^{N_f} (\bar{\psi}_i \psi_i)_r$$

is badly defined and one must introduce a "counter-term" proportionate to m_c to make the theory to be selfconsistent. In fact such a situation takes place for the interesting effective Lagrangian suggested recently by Bardeen[16]. In the tree approximation the Bardeen Lagrangian (including spinless $\vec{\pi}$ and σ fields) does realize spontaneous scale symmetry breaking in the chiral limit. But the addition of the standard mass term leads to the absence of the vacuum in the theory. To have a stable vacuum one must introduce an additional (scale and chiral invariant) term proportionate to m_c.[*]

In the microsopic gauge theory such an additional term should apparently have the following form at $d=3-\gamma_m=2$:

$$\Delta L_m = \kappa m_c [(\bar{\psi}\psi)_r^2 - (\bar{\psi}\gamma_5\psi)_r^2] , \qquad (26)$$

where $\kappa \sim \dfrac{1}{m_d^3}$ is some dimensional constant.

Thus, in a theory realizing spontaneous scale symmetry breaking in the chiral limit the ordinary mass term can be a "hard" operator. The origin of such an unusual behaviour is in that in the chiral limit the divergence $\partial_\nu D^\nu = 0$ and hence the vacuum energy density

$$\varepsilon_v = \frac{<0|\theta_\mu^\mu|0>}{4} = \frac{<0|\partial_\mu D^\mu|0>}{4}$$

equals zero in such a theory, i.e. there is degeneration of ε_v with respect to the value of the order parameter $<0|(\bar{\psi}\psi)_r|0> \sim m_d^3$ (the flat direction in the effective potential).

It would be very interesting to construct a "microsopic" theory with such unusual properties. Needless to say that at present the question about the existence of such theories is completely open.

6. THE PHYSICAL MEANING OF NAF GAUGE THEORIES WITH THE NONTRIVIAL FIXED POINT

The standard β-function of QED

---

[*] I thank W.A. Bardeen for discussions of this point.

$$\beta = \alpha_\mu \frac{\partial Z_3(\mu)^{-1}}{\partial \ln \Lambda} \tag{27}$$

is nonnegative (this is the consequence of the Källen-Lehmann representation for the photon propagator). Therefore, when the β-function (27) has a nontrivial ultraviolet stable zero, its form should be as that shown in Fig.2.

Originally such a situation was considered by Gell-Mann and Low in 1954.[17] Later the possibility of the existence of a nontrivial zero of the β-function (27) was studied in the framework of the "finite QED" program[18]. However, no definite answer has been obtained there.

In fact, in 1955 Landau and Pomeranchuk[19] and Fradkin[20] argued that the β-function (27) has no a nontrivial ultraviolet stable zero and, as the consequence, QED is transformed into a free-field theory in the continuum limit. More precisely, they argued that the one-loop relation for the running coupling,

$$\bar{\alpha}(r) = \frac{\alpha^{(0)}}{1 + \frac{2N_f \alpha^{(0)}}{3\pi} \ln \Lambda r} , \tag{28}$$

remains qualitativley correct at large values of $\alpha^{(0)}$ too. Then, when $\Lambda \to \infty$, the running coupling is equal to zero at all nonzero distances: at $\Lambda \to \infty$ $\bar{\alpha}(r) = 0$ at $r > 0$ and $\alpha(0) \equiv \lim_{\Lambda \to \infty} \bar{\alpha}(1/\Lambda) = \alpha^{(0)}$. Thus, in such a system a complete screening of the charge takes place ("zero-charge" situation).

However, as it was already discussed above in Secs.2 and 3, the situation may be essentially different in the strong-coupling regime. In this case the collapse-like effects lead to additional ultraviolet divergences which change qualitatively the physical picture.[1,2,3] In particular, the β-function is determined by the relation (5) here and hence it can in principle by negative in this phase.

Two possible forms of the β-function with the nontrivial ultraviolet stable zero in QED are shown in Figs.3 and 4.

Fig.3 corresponds to the situation when the complete screening of the charge takes place at all nonzero distances in the continuum limit ($\alpha(r)=0$ at $r>0$ and $\alpha(0)=\alpha_c$; "zero-charge" situation), but despite of this screening induced interactions provide the existence of a continuum theory with nontrivial S-matrix for bound states.[2]

This rather paradoxical possibility can be imagined as follows.

Let us assume (following Landau and Pomeranchuk[19]) that the relation (28) is qualitatively correct at large values of $\alpha^{(0)}$ too. Due to this relation, at any large value of the bare coupling constant $\alpha^{(0)}$ the vacuum polarization effects reduce the value of the running coupling $\bar{\alpha}(r)$ to a value of the order O(1) already at the distance $r=\rho/\Lambda$ where the parameter $\rho>1$ (due to Eq.(28) it is not too large, $\rho<10$ at $N_f=3$). Thus, the supercritical dynamics is formed in the region $\frac{1}{\Lambda} \leq r \leq \rho/\Lambda$. So this screening effect can crudely be imitated by introducing the <u>infrared</u> cutoff $\delta=\Lambda/\rho$ in the equation (7).

It is possible to show[2] that the equation (7) with such two cutoffs admits the nontrivial solution when the coupling constant $\alpha>\alpha_c(\rho)$ (the critical coupling $\alpha_c(\rho)$ goes to $\alpha_c=\frac{\pi}{3}$ when $\rho\to\infty$ ($\delta\to 0$) and $\alpha_c(\rho)\to\infty$ when $\rho\to 1$). Moreover, at the critical value $\alpha^{(0)}=\alpha_c(\rho)$ the second order phase transition takes place:

$$m_d = \Lambda f(\alpha^{(0)}, \rho) , \qquad (2\cdot)$$

where the function $f(\alpha^{(0)},\rho)$ has a zero at $\alpha^{(0)}=\alpha_c(\rho)$.

Due to arguements of Sec.3, in this case the continuum theory with $\alpha^{(0)}=\alpha_c$ has nontrivial S-matrix.

Needless to say, at present this picture should be considered only as a heuristic one (the recent study of this scenario see in the paper[21]).

Fig.4 corresponds to the situation when both subcritical and supercritical phases have the nontrivial fixed point at $\alpha^{(0)}=\alpha_c$ (there is no complete screening of the charge in such a situation). Due to the appearance of the additional ultraviolet divergences in the supercritical phase, one can expect that the β-function has a singularity at $\alpha^{(0)}=\alpha_c$ (in fact such a situation takes plase for the β-function (11)).

At present the only reliable way to study the phase transition in QED beyond the ladder approximation is computer simulations in lattice theories. The computer simulations shows that in the quenched approximation the second order phase transition connected with spontaneous chiral symmetry breakdown really takes place in both compact and noncompact versions of lattice QED.[22] Moreover, the recent results by Kogut, Dagotto and Kocić in quenched QED indicate that the scaling behaviour of the correlation length near the critical point agrees with the relation (9).[23]

The problem becomes much more complicated if the fermion determinant is taken into account. The computer simulations realized by Kogut, Dagotto and Kocić suggests the existence of the second order phase transition in noncompact QED with a sufficiently small number of fermion flavours[24]. These results also exhibit the strong screening of the charge near the critical point.

The results of computer simulations in compact QED depend on the fermion contents of the theory[25,26]. The recent results of Okawa indicate that in compact QED with a certain fermion contents the second order phase transition takes place[26]. These results can be important for the future studies of non-Abelian NAF gauge theories.

In fact, the renormalization of the bare coupling $\alpha^{(0)}$ is not the whole story for NAF gauge theories. If one starts from the lattice version of a NAF gauge theory, the continuum theory with $\alpha^{(0)}=\alpha_c$ can have additional (with respect to the naive continuum limit) interaction vertices. In the renormalization group language this point means that there are different types of relevant induced operators in the theory. As it has been pointed out by Bardeen, Leung and Love[12], due to the large anomalous dimension $\gamma_m=1$ at the fixed point, a natural candidate for the role of the relevant induced operator in QED with $\alpha^{(0)}=\alpha_c$ is the chiral invariant combination

$$G[(\bar{\psi}\psi)^2 - (\bar{\psi}\gamma_5\psi)^2] , \qquad (30)$$

where $G=g/\Lambda^2$ and in the continuum limit the dimensionless coupling constant g has to be fixed, $g=g_c$. At $\alpha^{(0)}=\alpha_c$ the dynamical dimension of the operator (30) is equal to four and hence the appearance of this vertex should not destroy the renormalization properties of the theory.

The operator (30) is not apparently the only relevant operator here (in principle, there could be infinite number of such operators in the theory). The problem of primary importance is to describe all them.

Where will NAF gauge theories can find their applications? One such a possibility is the technicolour dynamics. Yamawaki, Bando and Matumoto[14] presented the NAF technicolour model in which the notorious flavour-changing neutral-current problem can be solved (this model explicitly realizes the idea of the fixed point scenario of Holdom[27]). Also, it is clear that the existence of a new class of continuum nontrivial theories opens new possibilities in model building in elementary particle physics. It would be rather ironic if these theories returned us to the Heisenberg program for GUT.

7. THE NEARCRITICAL REGIME WITH THE NONTRIVIAL FIXED POINT IN AF GAUGE THEORIES

Up to the present moment we considered the continuum limit in NAF gauge theories. However, it is clear that when the condition $\frac{m_d}{\Lambda} \ll 1$ holds (i.e. the nearcritical regime, $\alpha^{(0)} - \alpha_c \ll 1$, takes place), this dynamical approach can with good accuracy be applied in theories with finite cutoff too. If we assume that the regime with a near-critical slowly running coupling ($\bar{\alpha}_{AF}(q^2) - \alpha_c \ll 1$) is realized in a region $M^2 < q^2 < \Lambda_{ph}^2$, $M^2 \ll \Lambda_{ph}^2$, of an asymptotically free theory, then the dynamics in this region can be considered as an imitation of the fixed point dynamics. Here Λ_{ph} is the physical scale at which the change of the dynamics regime takes place (the dynamics of spontaneous chiral symmetry breaking is switched on).

In this case the meaning of the relations (18) and (19) for the scale anomaly is somewhat different: here they reproduce the part of the scale anomaly connected with the presence of the scale Λ_{ph} at which the nonperturbative dynamics imitating the fixed point dynamics is switched on. In accordance with the observation in Sec.2 the β-function (11) determines the dependence of physical parameters on the scale Λ_{ph} but does not determine the behaviour of the running coupling $\bar{\alpha}_{AF}(q^2)$ which is nearly constant ("frozen") in this regime. It should not be mixed with the β-function of AF theories β_{AF} determining the behaviour of $\bar{\alpha}_{AF}(q^2)$. In AF theories the β-function (11) has to be considered as an auxiliary quantity allowing to separate the part of the scale anomaly caused by the nonperturbative dynamics with $\bar{\alpha}_{AF}(q^2) \simeq \alpha_c$ from the whole scale anomaly

$$\frac{\beta_{AF}}{4\alpha} <0|F_{\mu\nu}F^{\mu\nu}|0> \ .$$

The dynamical regime with the supercritical coupling was being applied to AF gauge theories during a long period[3,28-37].

In the papers[3,28,30-34] it was used to describe spontaneous chiral symmetry breaking in QCD. Here the application of this dynamics is based on the hypothesis about the existence of the region intermediate between that with the confinement forces and that with the perturbative dynamics. In this region the spontaneous breakdown of chiral invariance is formed. Although at present this scenario in QCD is not yet justifed, it works rather well.

In the papers[3,29] the supercritical dynamics was applied to tumbling gauge theories.

Recently, the dynamics with slowly running nearcritical coupling in AF theories has been applied to the technicolour dynamics[35,36]. It in principle allows to solve the flavour-changing neutral current problem there (some subtle points in this scenario are discussed in the recent papers[38]).

8. CONCLUSIONS

The fixed point dynamics, and the dynamics imitating it, can lead to a new field-theoretical insight, on the one hand, and to interesting phenomenological applications, on the other.

At present there exist theories for which the mechanism of scale symmetry breaking is primarily important (finite supersymmetric theories, conformal gravity). It would be interesting to examine the possibility of the realization of the collapse-like phenomena there.

Of course, to get the complete understanding of the dynamics with the nontrivial fixed point, many difficult questions have to be still answered.

The author thanks Japan Society for the Promotion of Science for the financial support.

REFERENCES
1. Miransky, V.A., Phys. Lett B91, 421 (1980).
2. Miransky, V.A., Nuovo Cim. 90A, 149 (1985); Sov. Phys. JETP 61, 905 (1985).
3. Fomin, P.I., Gusynin, V.P., Miransky, V.A. and Sitenko, Yu.A., Rivista del Nuovo Cim. 6, 1 (1983); Miransky, V.A. and Fomin, P.I. Sov. J. Part. Nucl. 16, 203 (1985).
4. Fomin, P.I. and Miransky, V.A., Phys. Lett. B64, 166 (1976); Fomin, P.I., Gusynin, V.P. and Miransky, V.A., Phys. Lett. B78, 136 (1978).
5. Landau, L.D. and Lifshits, Quantum Mechanics (Pergamon Press, 1977).
6. Atkinson, D., Johnson, P.W. and Stam, K., Phys. Lett. B201, 105 (1988); Phys. Rev. D37, 2996 (1988); Appelquist, T., Lane, K. and Mahanta, U. YCTP-P10-88.
7. Maskawa, T. and Nakajima, H., Prog. Theor. Phys. 52, 1326 (1974).
8. Fukuda, R. and Kugo, T., Nucl. Phys. B117, 250 (1976).
9. Gusynin, V.P. and Miransky, V.A., Phys. Lett. B193, 79 (1987); ibid B198, 362 (1987).
10. Migdal, A.A. and Shifman, M.A., Phys. Lett. B114, 445 (1982).
11. Miransky, V.A. and Scadron, M.D. Kiev ITP-87-138E; Miransky, V.A. and Gusynin, V.P., Kiev ITP-87-140E.
12. Bardeen, W.A., Leung, C.N. and Love, S.T., Phys. Rev. Lett. 56, 1230 (1986); Nucl. Phys. B273, 649 (1986).
13. Holdom, B. and Terning, J., Phys. Lett. B187, 357 (1987); ibid B200, 338 (1988).
14. Yamawaki, K., Bando, M. and Matumoto, K., Phys. Rev. Lett. 56, 1335 (1986); Bando, M., Matumoto, K. and Yamawaki, K., Phys. Lett. B178, 308 (1986).

15. Gusynin, V.P., Kushnir, V.A. and Miransky, V.A. Kiev ITP-88-84E (to appear in Phys. Lett. B).
16. Bardeen, W.A., The paper in these Proceedings.
17. Gell-Mann, M. and Low, F., Phys. Rev. <u>95</u>, 1300 (1954).
18. Johnson, K., Baker, M. and Willey, R., Phys. Rev. <u>B136</u>, 1111 (1964); Adler, S.L., Phys. Rev. <u>D5</u>, 3021 (1972).
19. Landau, L.D. and Pomeranchuk, I.Ya., Dokl. Acad. Nauk SSSR <u>102</u>, 489 (1955); Landau, L.D: in Niels Bohr and the Development of Physics, ed. W. Pauli (Pergamon Press, London, 1955).
20. Fradkin, E.S., Zh. Eksp. Teor. Fiz. <u>28</u>, 750 (1955).
21. Kondo, K.-I., Kikukawa, Y. and Mino, H., DPNU-88-21.
22. Bartholomew, J., Kogut, J.B., Shenker, S.M. et al., Nucl. Phys. <u>B230[FS10]</u>, 222 (1984); Azcoiti, V., Cruz, A., Dagotto, E. et al., Phys. Lett. <u>B175</u>, 202 (1986).
23. Kogut, J.B., Dagotto, E. and Kocić, A., Ill-(TH)-88-14.
24. Kogut, J.B., Dagotto, E. and Kocić, A., Phys. Rev. Lett. <u>60</u>, 772 (1988).
25. Kogut, J.B. and Dagotto, E., Phys. Rev. Lett. <u>59</u>, 617 (1987).
26. Okawa, M., KEK-TH-204.
27. Holdom, B., Phys. Rev. <u>24</u>, 1441 (1981).
28. Miransky, V.A., Gusynin, V.P. and Sitenko, Yu.A., Phys. Lett. <u>B100</u>, 157 (1981); Miransky, V.A. and Fomin, P.I., Phys. Lett. <u>B105</u>, 387 (1981); Sov. J. Nucl. Phys. <u>35</u>, 913 (1982).
29. Gusynin, V.P., Miransky, V.A. and Sitenko, Yu.A., Phys. Lett. <u>B123</u>, 407 (1983); ibid. <u>B123</u>, 428 (1983); Sov. J. Nucl. Phys. <u>38</u>, 309 (1983).
30. Finger, J. and Mandula, J.E., Nucl. Phys. <u>B199</u>, 168 (1982).
31. Goldman, T.J. and Haymaker, R.W., Phys. Rev. <u>D24</u>, 724 (1981).
32. Higashijima, K., Phys. Rev. <u>D29</u>, 1228 (1984).
33. Casalbuoni, R., De Curtis, S., Dominici, D. and Gatto, R., Phys. Lett. <u>B150</u>, 295 (1985).
34. Atkinson, D. and Johnson, P.W., Phys. Rev. <u>D37</u>, 2290 (1988); Atkinson, D., Johnson P.W. and Stam, K., Phys. Rev. <u>D37</u>, 2996 (1988).

35. Appelquist, T. and Wijewardhana, L.C.R., Phys. Rev. $\underline{D36}$, 568 (1987); Appelquist, T., Carrier, D., Wijewardhana, L.C.R. and Zheng. W, Phys. Rev. Lett. $\underline{60}$, 1114 (1988).
36. Holdom, B., Phys. Lett. $\underline{B198}$, 535 (1987).
37. Cohen, A. and Georgi, H. HUTP-88/A007.
38. Bando, M., Morozumi, T., So, H. and Yamawaki, K., Phys. Rev. Lett. $\underline{59}$, 389 (1987); Aoki, K.I., Bando, M., Mino, M., Nonoyama, T., So, H. and Yamawaki, K., DPNU-88-7; Nonoyama, T., The paper in these Proceedings.

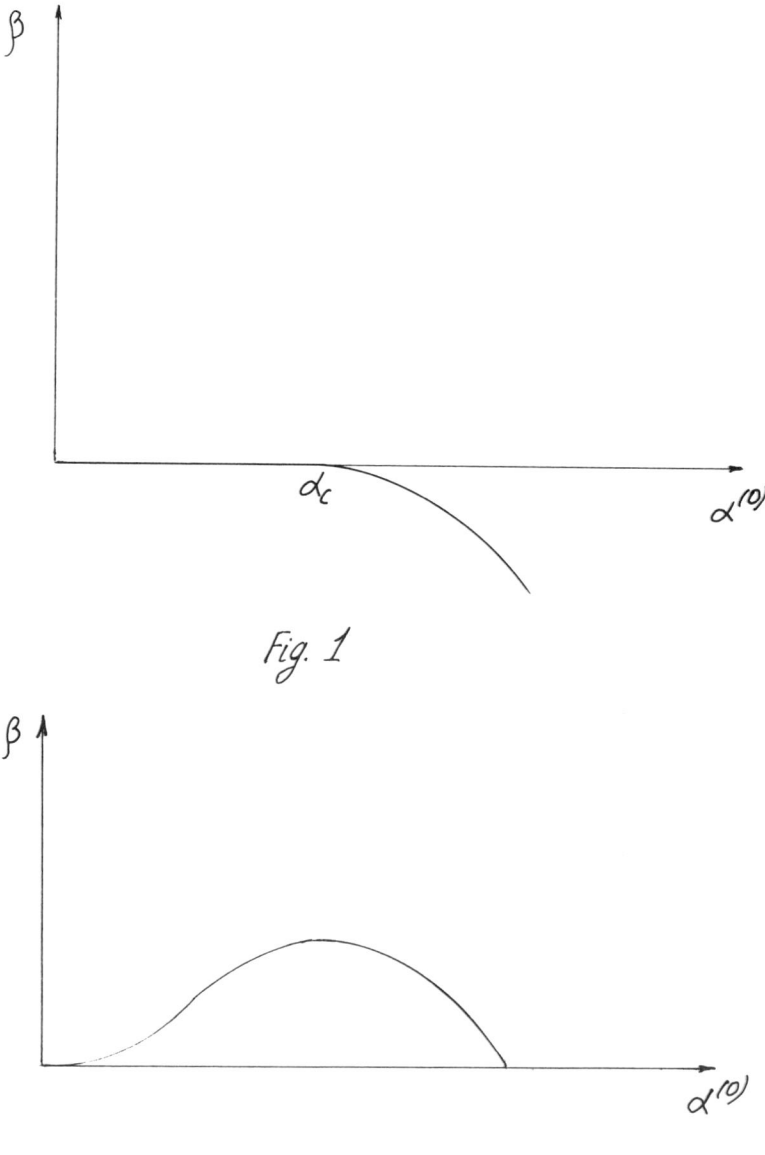

Fig. 1

Fig. 2

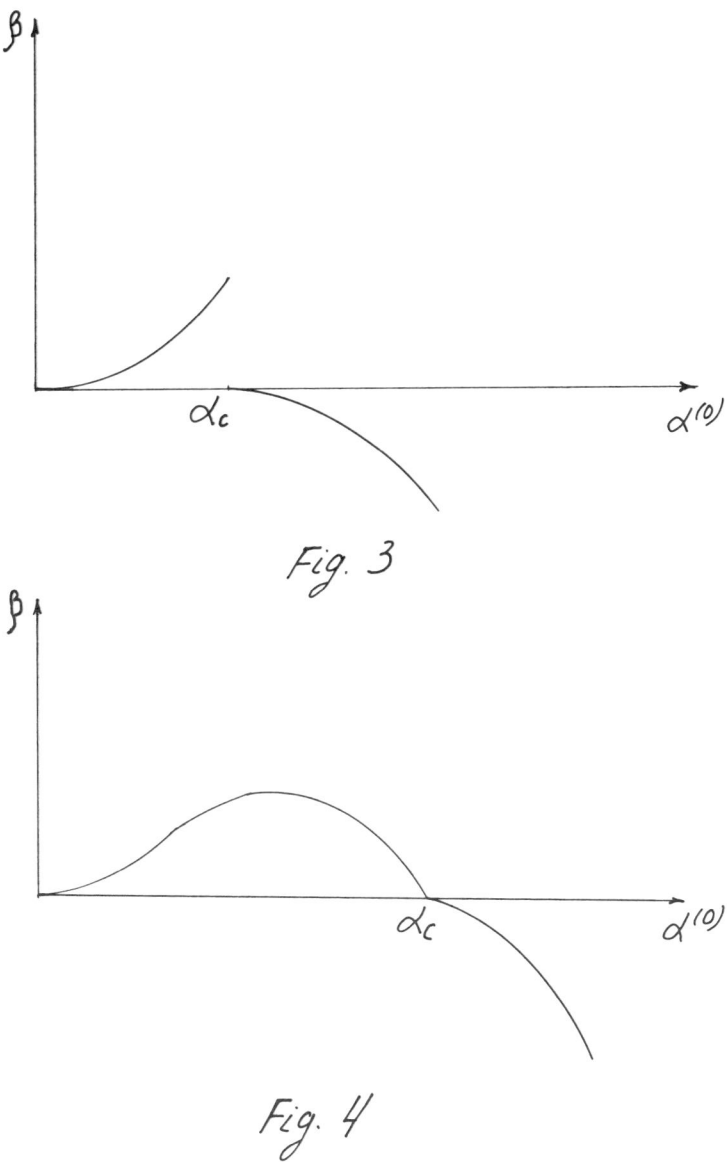

Fig. 3

Fig. 4

THE APPLICATION OF BIFURCATION THEORY
TO DYNAMICAL SYMMETRY BREAKING

D. Atkinson
Institute for Theoretical Physics
P.O. Box 800
9700 AV Groningen
The Netherlands

ABSTRACT

A theorem in bifurcation theory is applied to QED_3 with N flavours as an example. The importance of an infra-red cut-off is emphasized, and the relevance of wave-function renormalization is pointed out.

I wish to explain the application of bifurcation theory to the nonlinear Dyson-Schwinger equations that arise in the discussion of dynamical chiral symmetry breaking. The general method can be applied to asymptotically free theories, with or without confinement, or to non-asymptotically free theories, in any number of dimensions. Under certain conditions[1], one finds the existence of a critical coupling, which is a bifurcation of the trivial solution to a nontrivial branch: these conditions depend crucially on the infra-red and ultra-violet asymptotics. It is important to realize that the linear equations for the Fréchet differentials are *not* approximate linearizations of a nonlinear system, but rather exact characterizations of the bifurcation (i.e. critical) points of the nonlinear equations.

Instead of explaining the theory of bifurcations in general, I will apply it to a specific equation, namely QED$_3$ with N fermion flavours. Appelquist et al.[2], developing an idea of Pisarski[3], considered the lowest order of the $\frac{1}{N}$ expansion, and they modified the photon propagator by insertion of a massless fermion loop. Writing the fermion propagator

$$S(p) = [\alpha(-p^2) + \not{p}\,\beta(-p^2)]^{-1}, \tag{1}$$

one finds for the mass-function,

$$m(-p^2) = \alpha(-p^2)/\beta(-p^2), \tag{2}$$

after Wick rotation to Euclidean space, and the computation of the angular integrals,

$$\beta(p)m(p) = \frac{\lambda}{p}\int_0^\infty dk\, \frac{km(k)}{k^2 + m^2(k)}\log\frac{1+k+p}{1+|k-p|}, \tag{3}$$

in the Landau gauge. Here $\lambda = \frac{4}{N\pi^2}$, and the question is whether a critical value, $\lambda_c > 0$, exists, such that, for $\lambda < \lambda_c$, there is only the trivial solution, $m(p) \equiv 0$, while for $\lambda > \lambda_c$, there is also a nontrivial solution.

Appelquist et al. assumed that $\beta(p) \equiv 1$, which, however, is only true in the Landau gauge if the photon propagator is bare. It is not at all the case when the photon propagator is modified by the fermion-loop insertion. We shall first follow Appelquist et al. in setting $\beta(p) = 1$; but we shall then explain what happens when one determines $\beta(p)$ also from the Dyson-Schwinger equation.

The idea behind bifurcation analysis is to consider a small perturbation of the *function*, $m(p) \to m(p) + \delta m(p)$:

$$\delta m(p) = \frac{\lambda}{p} \int_0^\infty k \, dk \, \log \frac{1 + k + p}{1 + |k - p|} \frac{k^2 - m^2(k)}{[k^2 + m^2(k)]^2} \delta m(k). \tag{4}$$

This is now evaluated at the trivial solution, $m(p) \equiv 0$, giving

$$\delta m(p) = \frac{\lambda}{p} \int_0^\infty \frac{dk}{k} \log \frac{1 + k + p}{1 + |k - p|} \delta m(k). \tag{5}$$

Roughly speaking, if the homogeneous, linear equation (5) has a nontrivial solution, for a given λ, then that λ is a bifurcation point of the nonlinear equation (3), at which a nontrivial branch sprouts from the trivial solution. If the kernel of (5) is compact, we know that there is a smallest positive (eigen-) value, $\lambda = \lambda_c > 0$, for which a nontrivial solution of (5) exists.

The problem is that (5), as it stands, does not have a compact kernel. Let us write, with $0 < \alpha < 1$,

$$M(p) = p^{1-\alpha} \delta m(p), \tag{6}$$

and

$$K(p,k) = (pk)^{-\alpha} \log \frac{1 + k + p}{1 + |k - p|}, \tag{7}$$

so that (5) can be written

$$M(p) = \lambda \int_0^\infty dk \, K(p,k) \, k^{2(\alpha-1)} M(k). \tag{8}$$

It can be shown that K is a square-integrable kernel, if $\frac{1}{2} < \alpha < 1$, but not if $\alpha = 1$. However, in this case, the factor $k^{2(\alpha-1)}$ spoils things, and (8) is not a Fredholm equation. We are forced to introduce an infra-red cut-off as the lower limit of the integral in (8). The equation is then Fredholm, and bifurcation analysis can be used. The precise statement of the theorem can be found in the Appendix of Ref. 4.

It is important to realize that, although an infra-red cut-off is necessary at an intermediate step, it can be taken to zero at the end. For $\lambda > \lambda_c$, and $m(0) \neq 0$, the term $m^2(k)$ in the denominator of (3) acts as an automatic infra-red shielding, and no artificial infra-red cut-off is needed in the solution of the actual nonlinear equation (3).

Since there is a critical value of λ, there is also a critical value of N, the number of flavours,

$$N_c = \frac{4}{\pi^2 \lambda_c}. \qquad (9)$$

It appears then, that mass-generation can only occur if $N < N_c$.

The above results are based on the simplification $\beta(p) = 1$; but in fact β can also be determined from the Dyson-Schwinger equation,

$$\beta(p) = 1 - \frac{\lambda}{4p^2} \int_0^\infty dk \, \frac{k^2}{k^2 + m^2(k)} L(p,k), \qquad (10)$$

where

$$L(p,k) = 2\left[1 - \frac{1 + |k^2 - p^2|}{p_>}\right] + \frac{1}{pk} \log \frac{1 + k + p}{1 + |k - p|}$$

$$- \frac{(k^2 - p^2)^2}{pk} \log \frac{1 + (k + p)^{-1}}{1 + |k - p|^{-1}}, \qquad (11)$$

where $p_> = \max(p,k)$. It has been shown recently[6] that, for large N (small λ), $\beta(p)$, as determined from (10) and (3), is not $1 + 0(1/N)$ uniformly; but rather that $\beta(0) = 0(1/N)$, while $\beta(\infty) = 1$. The consequence for the bifurcation analysis is a delicate one. The crucial domain in (3) is the infra-red, and there the effective coupling parameter is not λ, but $\lambda/\beta(0)$, which remains finite as $N \to \infty$ (it turns out from numerical analysis that it is about 0.47, and this implies the existence of a nontrivial mass-function for *any* value of N). Details of the results can be found in Ref. 6.

REFERENCES

1. Atkinson, D. and Johnson, P.W., Phys. Rev. D37, 2290 (1988).

   Atkinson, D. and Johnson, P.W., Phys. Rev. D37, 2296 (1988).

   Atkinson, D., Johnson, P.W. and Stam, K., Phys. Rev. D37, 2996 (1988).

   Atkinson, D., Johnson, P.W. and Stam, K., Phys. Lett. 201, 105 (1988).

2. Appelquist, T.W., Nash, D. and Wijewardhana, L.C.R., Phys. Rev. Lett. 60, 2575 (1988).

3. Pisarski, R.D., Phys. Rev. D29, 2423 (1984).

4. Atkinson, D., Jour. Math. Phys. 28, 2494 (1987).

5. Fomin, P.I., Gusynin, V.P., Miransky, V.A. and Sitenko, Yu.A., Rev. del Nuovo Cim. 6, 1 (1983).

   Miransky, V.A., Phys. Lett. 165B, 401 (1985).

6. Pennington, M.R. and Webb, S.P., Hierarchy of scales in three-dimensional QED, BNL-40886.

   Atkinson, D., Johnson, P.W. and Pennington, M.R., Dynamical mass-generation in three-dimensional QED, Brookhaven preprint (August 1988).

CHIRAL SYMMETRY BREAKING IN QED BEYOND THE LADDER APPROXIMATION *

Hidetoshi MINO
Dpartment of Physics, Nagoya University
Nagoya, Japan

ABSTRACT

Spontaneous chiral symmetry breaking in QED is studied by solving the Schwinger-Dyson equation beyond the ladder approximation. We make improvement on vertex function and also include vacuum polarization effect. As a result, we found that chiral symmetry breaking occures even in the presense of vacuum polarization.

INTRODUCTION

Spontaneus chiral symmetry breaking in QED, so far, has been studied by solving the Schwinger-Dyson (S-D) equation in ladder approximation in Landau gauge. In this scheme, vacuum polarization effect is completely ignored and vertex function is replaced by bare one γ_μ. In this simple approximation, one can solve the system and a lot of interesting results were obtained. The existence of phase transition was discovered by Maskawa - Nakajima[1]. And recently, construction of sensible (i.e. free from Landau's ghost) continuum QED is discussed by Miransky and co-workers[2].

It is, however, questionable whether these aspects remain in the full (untruncated) QED or not. It might be an artifact of the approximation. In fact, there are some observation which inspire us with distrust about the ladder approximation. Some auther reported the strong gauge dependence of critical coupling[3]. Others discussed the inconsistency of ladder approximation in non-Landau gauge[4,5]. So it is of great interest to study the chiral symmetry breaking beyond the ladder approximation.

Recently, Monte Carlo study of QED, which is certainly one of the promising approaches, has been proceeded, and existence of second order phase transition is reported[6,7]. On the other hand, attempts to improvement S-D equation have also attracted interest[8,9], and this approach is also expected to afford insight on this problem.

---

* This talk is based on the collaboration with K.-I. Kondo of Chiba University and Y. Kikukawa of Nagoya University

In this talk, I would like to report the result of our attempt to go beyond the ladder approximation within the flamework of S-D equation. Our work consists of following two parts.
1. Improvement of vertex function $\Gamma_\mu(p^2)$ (which is taken to γ_μ in ladder approximation)
2. Inclusion of vacuum polarization effect

IMPROVEMENT OF VERTEX FUNCTION

Let us begin with the improvement of vertex function. We write the S-D equation for the fermion propagater as

$$S^{-1}(p) = S_0^{-1}(p) + \Sigma(p)$$
$$\Sigma(p) = e^2 \int \frac{d^4q}{(2\pi)^4} \gamma_\mu S(q) \Gamma_\nu(q,p) D_{\mu\nu}(k), \quad k \equiv q - p \quad (1)$$

where e is gauge coupling constant and Γ is full vertex function and D is full photon propagater. In ladder approximation, Γ is replaced by γ and D is replaced by bare one.

As a first step of improvement, we require the Ward-Takahashi (W-T) identity

$$k_\mu \Gamma_\mu(q,p;k) = S(q)^{-1} - S(p)^{-1}. \quad (2)$$

In ladder approximation, this identity can be satisfied only in Landau gauge. We expect that by requiring this relation gauge invariance will recover. This relation, however, can not determine Γ uniquely. We may add a transverse part of vertex Γ_μ^T which satisfies $k_\mu \Gamma_\mu^T(q,p) = 0$. If one decompose the Γ into longitudinal part Γ^L and transverse one Γ^T, longitudinal part is determined by W-T identity as

$$\Gamma_\mu^L(q,p;k) = \frac{k_\mu}{k^2}[S(q)^{-1} - S(p)^{-1}], \quad (3)$$

whereas Γ^T is left arbitrary. What choice should be taken for the transvers vertex?
Kondo and Kikukawa proposed as the transverse vertex

$$\Gamma_\mu^T = \frac{k_\lambda}{k^2} \{S(q)^{-1}\sigma_{\lambda\mu} + \sigma_{\lambda\mu} S(p)^{-1}\} \quad (4)$$

and its some variations[5]. These are derived from Transeverse-Takahashi (T-T) identity with some truncation. T-T identity is a consequence of equation of motion, and relate Γ^T to two point (and higher) function[10].

Another interesting choice for Γ^T is

$$\Gamma^T_\mu = \{\delta_{\mu\nu} - k_\mu k_\nu/k^2\}\gamma_\nu F(\max(p^2, q^2)) \tag{5}$$

where F is a arbitrary function of p^2 or q^2. Though this vertex does not respect the Transverse-Takahashi identity, it is found to give weak gauge dependence[8].

As ansatz for transverse part of vertex, we take eq.(4) and its variation and eq.(5). Specifically we deal with following five cases.

	Γ^L	Γ^T
I	$\frac{k_\mu}{k^2}[S(q)^{-1} - S(p)^{-1}]$	$\frac{1}{k^2}\Big\{A(q^2)[(k\cdot q)\gamma_\mu - q_\mu \slashed{k}]$ $+A(p^2)[p_\mu\slashed{k} - (k\cdot p)\gamma_\mu]$ $+k_\lambda \sigma_{\lambda\mu}[B(q^2) + B(p^2)]\Big\}$
II	same as in I	$\frac{1}{k^2}\Big\{A(q^2)[(k\cdot q)\gamma_\mu - q_\mu \slashed{k}]$ $+A(p^2)[p_\mu\slashed{k} - (k\cdot p)\gamma_\mu]\Big\}$
III	$\frac{k_\mu}{k^2}[A(q)^{-1} - A(p)^{-1}]$	same as in II
IV	Bare γ	
V	same as in I	$\{\delta_{\mu\nu} - k_\mu k_\nu/k^2\}\gamma_\nu F(\max(p^2, q^2))$

where we use the notation $S(p) = [A(p^2)\slashed{p} + B(p^2)]^{-1}$

In case I,II and V, longitudinal part of vertex is determined so that W-T identity is satisfied. These three verteces differ in the form of transverse part. I and II respect the T-T identity, wheras vertex V does not. Vertex III, which is included for comparision, have different longitudinal part, which does not satisfy W-T identity when the dynamical mass is generated. Vertex IV, bare vertex, is also included for comparison.

With these ansatz, we can dictate S-D equation for fermion propagater in the form of coupled integral equation of A and B. Their explicit form is rather complicated, and is listed in Table I.

We numerically solved these S-D equation with ultra violet cut off, and the continuum limit (infinit cut off limit) was taken by extrapolation. Bifurcation technique[11] was also adopted to simplify the calculation. Critical values for

I and II

$$A(x) = 1 + \frac{1}{2}d_l g + d_l g A(x)\int_x^\Lambda dy\frac{A(y)}{yA^2(y)+B^2(y)} - d_l g\frac{B(x)}{x^2}\int_0^x dy\frac{y^2 B(y)}{yA^2(y)+B^2(y)}$$

$$+ 3g\int_0^\Lambda dy\frac{A(y)}{yA^2(y)+B^2(y)}\frac{1}{2}\{A(y)-A(x)\}\left[\theta(y-x) - \frac{y^2}{x^2}\theta(x-y)\right]$$

$$\left[-3g\frac{1}{x^2}\int_0^x dy\frac{y^2 B(y)}{yA^2(y)+B^2(y)}[B(x)+B(y)]\right]_{\text{for I only}}$$

$$B(x) = (3+d_l)g\frac{A(x)}{x}\int_0^x dy\frac{yB(y)}{yA^2(y)+B^2(y)}$$

$$+ d_l g B(x)\int_x^\Lambda dy\frac{A(y)}{yA^2(y)+B^2(y)} + 3g\int_x^\Lambda dy\frac{B(y)A(y)}{yA^2(y)+B^2(y)}$$

$$\left[+3g\int_x^\Lambda dy\frac{A(y)}{yA^2(y)+B^2(y)}[B(x)+B(y)]\right]_{\text{for I only}}$$

III

$$A(x) = 1 + d_l g\int_0^\Lambda dy\frac{A(y)}{yA^2(y)+B^2(y)}\left[\frac{y^2}{x^2}A(y)\theta(x-y) + A(x)\theta(y-x)\right]$$

$$+ 3g\int_0^\Lambda dy\frac{A(y)}{yA^2(y)+B^2(y)}\frac{1}{2}[A(x)-A(y)]\left[\frac{y^2}{x^2}\theta(x-y)-\theta(y-x)\right]$$

$$B(x) = (3+d_l)g\int_0^\Lambda dy\frac{B(y)}{yA^2(y)+B^2(y)}\left[\frac{y}{x}A(x)\theta(x-y)+A(y)\theta(y-x)\right]$$

IV

$$A(x) = 1 + d_l g\int_0^\Lambda dy\frac{A(y)}{yA^2(y)+B^2(y)}\left[\frac{y^2}{x^2}\theta(x-y)+\theta(y-x)\right]$$

$$B(x) = (3+d_l)g\int_0^\Lambda dy\frac{B(y)}{yA^2(y)+B^2(y)}\left[\frac{y}{x}\theta(x-y)+\theta(y-x)\right]$$

V

$$P(x)A(x) = 1 - d_l g\frac{B(x)}{x^2}\int_\mu^x dy\frac{yB(y)}{yA^2(y)+B^2(y)}$$

$$P(x)B(x) = d_l g\frac{A(x)}{x}\int_0^x dy\frac{yB(y)}{yA^2((y)+B^2(y)}$$

$$+ 3g\int_0^\Lambda dy\frac{yB(y)}{yA^2(y)+B^2(y)}\left[\frac{A(x)}{x}\theta(x-y)+\frac{A(y)}{y}\theta(y-x)\right]$$

$$P(x) = 1 - d_l g\int_x^\Lambda dy\frac{A(y)}{yA^2(y)+B^2(y)}$$

Table I: Coupled S-D equation for fermion propagater $S(p) = [A(p^2)\not{p} + B(p^2)]^{-1}$. Λ is ultra violet cut off, which is removed by extrapolation in our calculation.

vertex	I	II	III	IV(bare)	V
W-T id.	◯	◯	×	×	◯
T-T id.	◯	◯	–	–	–
gauge parameter					
$d_l = 3$			0.185	0.236	
$d_l = 2$	0	0	0.144	0.206	
$d_l = 1$			0.116	0.164	0.077
$d_l = 0$	0	$\frac{1}{12}$	$\frac{1}{12}$	$\frac{1}{12}$	$\frac{1}{12}$
$0 > d_l > -3$	0	?	?	0	
$d_l = -3$	no χSB	no χSB	no χSB	no χSB	
$d_l < -3$	?	?	no positive solution		

Table II : Critical values of gauge coupling $g \equiv e^2/(16\pi^2)$. Type I, II and V vertex respect the W-T identity. Transverse part of I and II are derived by T-T identity with some truncation.

gauge coupling constant $g \equiv e^2/(16\pi^2)$ are shown in Table II, which is the main result in this section.

Let us discuss the result of our improvement. We expected that W-T identity and T-T identity would restore the gauge invariance. Though the result of I, II and V, in which W-T id. and T-T id. are respected, have weaker gauge dependence than that of III and IV, critical coupling in case I and II become peculiar value, zero, which means that chiral symmetry breaking occoures for arbitrary small gauge coupling. Kondo and Kikukawa[5] showed that this situation is almost direct consequence of W-T identity imposed on vertex. Only when one choose transverse part specially, like as in case V, coupling constant become finite.

Though, at this stage, we can not say what is the systematic improvement, type V vertex is favorable in a sence that this vertex have weak gauge dependence and does not lead to unsound result.

INCLUSION OF VACUUM POLARIZATION

So far we neglected the vacuum polarization effect. In this section we take this into account and add to coupled S-D eq. a photon propagater part

$$D_{\mu\nu}^{-1}(k) = D_{\mu\nu}^{(0)-1}(k) + \Pi_{\mu\nu}(k)$$
$$\Pi_{\mu\nu}(k) = e^2 N_f \int \frac{d^4p}{(2\pi)^4} \text{tr}[\gamma_\mu S(p)\Gamma_\nu(p, p-k, k)S(p-k)], \qquad (6)$$

where N_f is the number of fermion flavor. One can, in priciple, solve the coupled S-D (1) and (6) provided that ansatz for vertex function is given. It is, however, practically difficult to solve this system for general vertex function. So we obliged to take type V vertex which make calculation practically possible. With this ansatz one can derive the integral equation for F from the S-D equation for vertex according to Landau's arguments[12].

$$F(\xi) = 1 + d_l g \int_\xi^L d\zeta \{F^3(\zeta) + F(\zeta)[A(\zeta) - F(\zeta)]^2\}/A(\zeta)^2 \qquad (7)$$

where $\xi = \ln p^2, \zeta = \ln q^2$ and $L = \ln \Lambda^2$ is an articficial cut-off which sould be removed finally. Moreover one can find by inspection a consistent solution for F and $D_{\mu\nu}$

$$F(p^2) = A(p^2)$$
$$d_t(k^2) = \left[1 + \frac{4}{3}gN_f \ln(\Lambda^2/k^2)\right]^{-1} \qquad (8)$$

d_t is the transverse part of $D_{\mu\nu}$ which defined by

$$D_{\mu\nu} = \frac{d_t(k^2)}{k^2}\{\delta_{\mu\nu} - k_\mu k_\nu/k^2\} + \frac{d_l}{k^2}\frac{k_\mu k_\nu}{k^2},$$

where d_l is gauge parameter. The reason why such simple solution is found is come from the special choice of vertex.

Fermion propagater part is reduced to

$$P(x)A(x) = 1 - d_l g \frac{B(x)}{x^2}\int_{\mu^2}^x dy \frac{yB(y)}{yA^2(y) + B^2(y)}$$
$$P(x)B(x) = d_l g \frac{A(x)}{x}\int_{\mu^2}^x dy \frac{yB(y)}{yA^2((y) + B^2(y)}$$
$$+ 3g\int_{\mu^2}^{\Lambda^2} dy \frac{yB(y)}{yA^2(y) + B^2(y)}\left[\frac{F(x)d_t(x)}{x}\theta(x-y) + \frac{F(y)d_t(y)}{y}\theta(y-x)\right] \qquad (9)$$

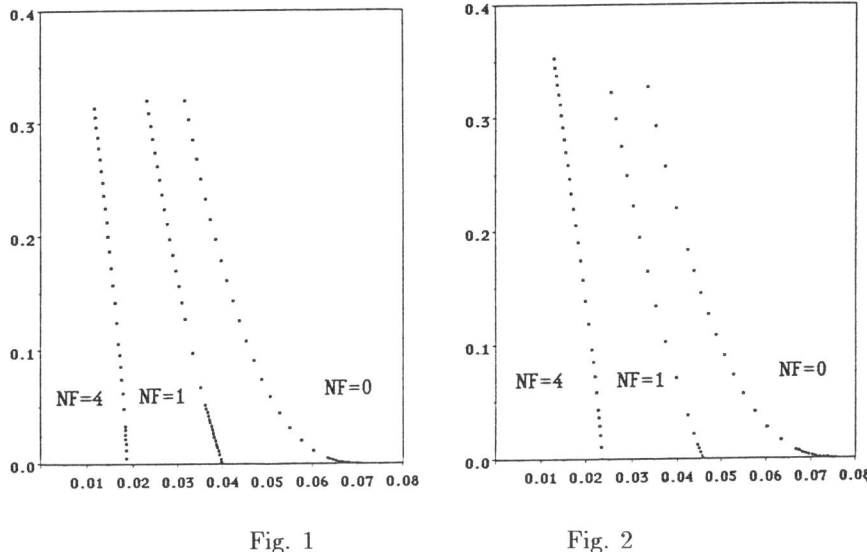

Fig. 1   Fig. 2

Critical behavior of orderparameter $\Sigma(0) \equiv B(0)/A(0)$, Fig.1 for Landau gauge and Fig.2 for Feynman gauge.

where
$$P(x) = 1 - d_l g \int_x^{\Lambda^2} dy \frac{A(y)}{yA^2(y) + B^2(y)},$$
and μ and Λ is infra red and ultra violet cut off. One can directly check that (8) is solution of (7) and (9).

We solved eq.(9) with (8) numerically and explored the chiral symmetry breaking solution in continuum limit by extrapolation. Our main result is that chiral symmetry breaking does occour even in the presence of vacuum polarization. This result, togather with the result of Monte Carlo study, would be a support for existence of ultra violet fixed point in QED, at which sensible continuum theory could be defined.

Phase transition point and critical behavior of Σ figured out in Fig 1. for Landau gauge and in Fig. 2 for Feynman gauge. Qualitatively, results in these gauge coincide.

Finally, there is a comment on comparison of our result to the one from

Monte Carlo study. Kogut et. al.[13] reported that phase transition of essential singularity type even in the presence of fermion loop as well as in quenched case. In contrast, our result suggest a power like behavior of order parameter. The origin of this discrepancy and its physical implication is not yet known.

SUMMARY

We stuied the spontaneous chiral symmetry breaking of QED beyond the ladder approximation. For various types of vertex function, we solve the S-D equation explicitly and obtain the critical value of gauge coupling. We found that transverse part of vertex is quite relevant to result. What is the systematic improvement for vertex function is not yet clear.

We also tried to include the vacuum polarization in S-D equation. Though it is performed only for special anzats for vertex, we found that chiral symmetry does occoures in the presence of vacuum polarization. Critical value and critical behavior of order parameter is also obtained. These results, togather with the one from Monte Carlo study, will be a significant information for deeper understanding of QED and of other strong coupling gauge theories.

References

1. T. Maskawa and H. Nakajima, Prog. Theor. Phys. **52** (1974) 1326; **54** (1975) 860.
   R. Fukuda and T. Kugo, Nucl. Phys. **B117** (1976) 250.
2. V. A. Miransky, Nouvo Cimento **90A** (1985) 149.
3. T. Nonoyama and M. Tanabashi, Nagoya Univ. Preprint (DPNU-88/22).
   M. Bando and K. Hasebe, private communication.
4. D. Atkinson and P. W. Johnson, J. Math. Phys. **28** (1987) 2488.
5. K.-I. Kondo and Y. Kikukawa, Nagoya Univ. Preprint (DPNU-88/20).
6. J. B. Kogut, E. Dagotto and A. Kocic, Phys. Rev. Lett. **60** (1988) 772.
7. M. Okawa, KEK preprint, KEK-TH-204 (1988).
8. D. Atkinson, P. W. Johnson and K. Stam, Phys. Lett. **B201** (1988) 105.
9. K.-I. Kondo, Y. Kikukawa and H. Mino, Nagoya Univ. Preprint (DPNU-88/21).
10. Y. Takahashi in "*Quantum Field Theory*" ed. by F. Mancin, (Elsevier Science Publishers, 1986).
11. D. Atkinsion, J. Math. Phys. **28** (1987) 2949.
12. L. D. Landau, in "*Niels Bohr and the Development of Physics*" ed. W. Pauli (Pergamon, London, 1955).
    L. D. Landau, A. Abrikosov and I. Khalatnikov, Nuovo Cimento, Supp. **3** (1956) 80.
13. J. B. Kogut, E. Dagotto and A. Kocic, ILL-(TH)-88-#14, 1988

Collapse of the Wavefunction and Catalyzed
Symmetry Breaking in QED4

J. B. Kogut

Department of Physics
University of Illinois at Urbana-Champaign
1110 West Green Street
Urbana, Illinois 61801
U.S.A.

ABSTRACT

We discuss "collapse of the wavefunction" as the phenomenon underlying chiral symmetry breaking in quenched QED4. The 1/r singularity in the "collapsed" $q\bar{q}$ wavefunction causes "catalyzed symmetry breaking" which is the field theoretic analog of "monopole induced proton decay."

1. INTRODUCTION: PHYSICAL PICTURES

In this short talk I want to emphasize the simple physical picture of "collapse of the wavefunction" which underlies chiral symmetry breaking and the interacting ultra-violet stable/infra-red unstable fixed point of quenched QED4.[1] Unlike chiral symmetry breaking in QCD which occurs at scales comparable to the sizes of meson bound states and is closely related to flux tube formation and the confinement mechanism in the quenched theory, chiral symmetry breaking in QED4 is a short distance phenomenon related to the strong 1/r attraction between constituents and the development of a 1/r singularity in the pair wavefunction.[2] This unique phenomenon is analogous to the singular wave functions which develop when a fermion binds to a monopole.[3] It causes the $\bar{q}q$ constituents to collide with probability $O(1)$ at r=0 and thus induces renormalizable four-fermi interactions in the fixed point Lagrangian. If quenched QED4 were unified with another theory characterized by a high energy scale Λ, then the 1/r singularities can amplify certain interaction terms of the high energy theory to $O(1)$ effects although naive dimensional analysis would have estimated them as $O(\Lambda^{-2})$. This phenomenon has been coined "catalyzed symmetry breaking" in the literature.[4] It represents a maximal violation of perturbative decoupling theorems[5] and can be understood in field theoretic language as the fact that various multi-fermion operators have large anomalous dimensions at the theory's fixed point.[1,6]

I will argue that all these bizarre phenomena can be anticipated and understood in elementary physics terms.

2. COLLAPSE OF THE WAVE FUNCTION

Consider a $q\bar{q}$ pair interacting through a Coulomb potential in a box of linear dimension L. The system has an expected potential energy $\sim -\alpha/L$ and an expected kinetic energy $\sim 1/L$. So, if $\alpha > L$ this system implodes.

To understand this phenomenon better, but still working in a two-body picture, take the Dirac equation with a fixed $-\alpha/r$ potential which is cutoff at length r_0. In the super-critical region $\alpha > 1$, for $r \geq r_0$,

$$-\left[\frac{d^2}{dr^2} + \frac{1}{r}\frac{d}{dr}\right](r\psi) + \frac{1-\alpha^2}{r^2}(r\psi) \approx 0 \tag{2.1}$$

Matching onto a plane wave for $r < r_0$ where the potential is flat $-\alpha/r_0$, we find the wavefunction[1],

$$\psi(r) \underset{r \to r_0^+}{\longrightarrow} \left(\frac{1}{r_0 \sqrt{\alpha^2 - 1}}\right) \sin\left(\sqrt{\alpha^2 - 1} \ln\left(1/r_0 |\varepsilon|\right)\right) \tag{2.2}$$

The collapse of the wave function is seen here -- the sin oscillates uncontrollably as $r_0 \to 0$. However, if we introduce a renormalization condition[1], that the s-wave bound state energy ε should be independent of r_0 as $r_0 \to 0$,

$$|\varepsilon| = \frac{1}{r_0}\exp\left(-\pi/\sqrt{\alpha-1}\right) \tag{2.3}$$

then the system will exist. Note, however, that this requires that α be a function of the cutoff r_0, $\alpha(r_0)$, and that $\alpha(r_0) \to 1$ as $r_0 \to 0$. Thus $\alpha_c = 1$ has become a ultra-violet stable/infrared-red unstable fixed point since Eq. (2.3) can be written[1],

$$\alpha^2(r_0) = \alpha_c^2 + \frac{\pi^2}{\ln^2(r_0|\varepsilon|)} \quad , \alpha_c = 1 \tag{2.4}$$

The wavefunction Eq. (2.2) now makes sense at short distances, $r \ll |\varepsilon|^{-1}$,

$$\psi(r) \sim \ln(1/r|\epsilon|)/r \qquad (2.5)$$

Recall that in the sub-critical region $\psi(r) \sim r^{-1+\sqrt{1-\alpha}}$, so the power-law in Eq. (2.5) is sensible.

There are several things which are very special about the fixed point $\alpha = \alpha_c$. Note that the probability to find the fermion near the origin $|\psi|^2 r^2 dr d\Omega$ behaves as $dr\, d\Omega$, up to an unimportant, non-universal logarithm-squared. Thus, the fermion passes through the origin with probability $O(1)$ and the phase space resembles 1+1 dimensions! We shall see below that this behavior generates renormalizable four-point contact interactions in a field theoretic discussion of this phenomenon, that it leads to large anomalous dimensions of multi-fermion operators, and that it acts as a field theoretic realization of monopole catalysis of proton decay!

3. WHY IS THE FIXED POINT INFRA-RED UNSTABLE?

Before discussing the implications of a $1/r$ wavefunction singularity, we should understand the coupling constant renormalization effect Eq. (2.4) which underlies it. Eq. (2.4) can be written as a Callan-Symanzik function[6],

$$\beta(\alpha) = \Lambda \frac{\partial}{\partial \Lambda} \alpha(\Lambda) = -\frac{2\alpha_c}{\pi}(\alpha/\alpha_c - 1)^{3/2}, \quad \Lambda = 1/r_0 \qquad (3.1)$$

for $\alpha > \alpha_c$ with $\beta(\alpha) = 0$ otherwise. The puzzling fact about Eq. (3.1) is the following: How can a potential model have a renormalization problem and why does the coupling grow large (the negative sign!) as the spatial cutoff grows? One finds Eq. (3.1) also for quenched QED4, so the two-body $q\bar{q}$ dynamics generates a negative β-function in the super-critical region even though the photon propagator is free. Since we are considering strong coupling, perturbative renormalization can not be expected to be a complete description. The instability in the Dirac equation for $\alpha > \alpha_c$ is a non-perturbative effect which produces a new source of coupling constant renormalization.[2] In particular, the $1/r$ singularity of $\psi(r)$ found as $\alpha \to \alpha_c^+$ is the crucial element here. Since $|\psi|^2 r^2 dr\, d\Omega \not\to 0$ as $r \to 0$, the low energy levels of the quantum system are sensitive to the behavior of the potential at the shortest distances. Typically in quantum problems $|\psi|^2 r^2 dr\, d\Omega \to 0$ as $r \to 0$ and the very short distance behavior of the potential $V(r)$ is irrelevant to the low energy features of the theory. But now we can also understand why $\beta(\alpha)$ is negative for $\alpha > \alpha_c$. Consider Eq. (2.1) with a cutoff $r'_0 > r_0$. We again want to calculate an energy level $|\epsilon|$ which is physical and should be independent or r'_0. But if the Coulomb potential is cutoff at $r'_0 > r_0$, then to <u>compensate</u> for some otherwise lost $q\bar{q}$ attraction, $\alpha(r'_0)$ must be chosen <u>greater</u> than $\alpha(r_0)$. In a renormalization group approach we would integrate out high frequency photon exchanges. Then the effects of the virtual photons would be incorporated into a new coupling. Clearly the output coupling is larger than the input coupling.

It is important to remember that $\psi(r) \sim 1/r$ is a crucial ingredient in finding $\beta(\alpha) < 0$ for $\alpha > \alpha_c$. If $\psi(r)$ were less singular, $\beta(\alpha)$ would vanish identically as in more typical potential problems.

4. CATALYZED SYMMETRY BREAKING

Let's make the analogy between collapse of the wavefunction in a super-critical field and monopole-induced proton decay[3] more explicit. Consider a charged Dirac fermion in the presence of a U(1) monopole. The orbital angular momentum is $\underline{L} = -i \underline{r} \times \underline{D} - g\hat{r}$ and the total angular momentum is $\underline{J} = \underline{L} + \underline{\sigma}$, where g is the magnetic charge of the monopole and $\underline{D}$ is the covariant derivative. We will write down the Dirac equation for energy eigenstates and will diagonalize $\underline{J}^2$ and $\underline{\sigma} \cdot \hat{r}$. The equation of motion for j=0 states reads[3]

$$\left[-\left(\frac{d^2}{dr^2} + \frac{2}{r}\frac{d}{dr}\right) + \frac{\frac{1}{2}\left(\frac{1}{2}+1\right) - g(g-\underline{\sigma}\cdot\hat{r})}{r^2} \right] \psi \approx 0 \qquad (4.1)$$

for small r. note that the centrifugal barrier term vanishes for $g = -1/2$, $\underline{\sigma}\cdot\hat{r} = 1$ or $g = 1/2$, $\underline{\sigma}\cdot\hat{r} = -1$. In this care $\psi(r) \sim 1/r$ and the fermion's probability density near the origin does not vanish. The fermion passes through the monopole core with probability O(1) and can experience the short distance Grand Unified (GUT) transitions characterized by distances $O(M_w^{-2})$ with no suppression. In this example we imagine that chiral symmetry is broken explicitly by the GUT so that the choice to diagonalize $\underline{\sigma}\cdot\hat{r}$ and make a "hedgehog" fermion field configuration is a physical choice. In a scattering picture, the bound fermion burrows through the monopole core and picks up the appropriate helicity so that $\underline{\sigma}\cdot\hat{r}$ is unchanged when it emerges. This one-body picture of the physics must be generalized to account for particle creation and describe the GUT dynamics appropriately.[3] When this is done one finds that the 1/r wavefunction singularity gives rise to four fermi interactions whose symmetries only respect those of the GUT and which generate transitions at large length scales unsuppressed by powers of M_W^{-2}.

Lets check that "collapse of the wavefunction" in QED4 amplifies short distance symmetry breaking in a similar fashion. Consider the Lagrangian,

$$L = L_{QED} + m_0(\Lambda)\,\bar\psi\psi + L_{pv} \tag{4.2a}$$

where $L_{pv} = g\bar\psi\psi\bar\psi i\gamma_5\psi$ which models short distance parity violation. The dimensional four fermi coupling g is characterized by a length scale Λ,

$$g = \bar g/\Lambda^2 \tag{4.2b}$$

where $\bar g$ is a dimensionless, fixed number. So, we think of Λ as a mass of a heavy boson which characterizes the energy scale where parity violation turns on perturbatively. L_{pv} can be thought of as the effective interaction resulting from a parity violating Yukawa interaction at high energy scales larger than Λ. Conventional wisdom would say that because g is $O(\Lambda^{-2})$ parity violation effects at low energies would be negligible. In fact, "collapse of the wavefunction" will amplify them to $O(1)$.[4]

This effect can be demonstrated straight-forwardly in the quenched ladder approximation to QED4. Consider the Schwinger-Dyson equation for the fermion propagator with the replacement,

$$\Sigma(p) \rightarrow \Sigma(p) + i\gamma_5\,\Sigma_5(p) \tag{4.3}$$

We then obtain the ordinary gap equations for Σ and Σ_5 with the replacements

$$m_0(\Lambda) \rightarrow m_0(\Lambda) - \frac{1}{2}g\,<\bar\psi i\gamma_5\psi>_\Lambda$$
$$m_0^{(5)}(\Lambda) \rightarrow -\frac{1}{2}g\,<\bar\psi\psi>_\Lambda \tag{4.4}$$

The crucial calculation is then to renormalize these bare quantities. In QED4 the wavefunction renormalization for $\bar\psi\psi$ and $\bar\psi\gamma_5\psi$ is,

$$Z_m = \begin{cases} (\mu/\Lambda)^{1-\gamma},\ \gamma = \sqrt{1-\alpha/\alpha_c} & \text{for } \alpha<\alpha_c \\ (\mu/\Lambda) & \text{for } \alpha>\alpha_c \end{cases} \tag{4.5}$$

Thus, the renormalized quantities are,

$$\langle \bar\psi \gamma_5 \psi \rangle_R = Z_m \langle \bar\psi \gamma_5 \psi \rangle_\Lambda$$

$$\langle \bar\psi \psi \rangle_R = Z_m \langle \bar\psi \psi \rangle_\Lambda \tag{4.6}$$

$$m_R = Z_m^{-1} m_o(\Lambda)$$

So, the shift in the bare mass due to L_{pv} in the gap equation is,

$$m_o(\Lambda) - \frac{1}{2} g \langle \bar\psi i \gamma_5 \psi \rangle_\Lambda = Z_m \left[m_R - \frac{1}{2} \left(g Z_m^{-2} \right) \langle \bar\psi i \gamma_5 \psi \rangle_R \right] \tag{4.7}$$

where we identify a renormalized four-fermi coupling characterizing the low energy effects of L_{pv},

$$g_R \equiv g Z_m^{-2} = g \left(\frac{\Lambda}{\mu} \right)^{2(1-\gamma)} = \frac{\bar g}{\mu^2} \left(\frac{\Lambda}{\mu} \right)^{2\gamma} \tag{4.8}$$

Several observations about Eq. (4.8) are in order. First, at the critical point $\alpha = \alpha_c$ the factor γ vanishes, so g_R is finite and <u>independent</u> of Λ! This is the amplification effect promised. The large anomalous dimension in Z_m and collapse of the wave function go hand-in-hand. Second, at weak coupling $\alpha \approx 0$ Eq. (4.8) becomes,

$$g_R = \frac{\bar g}{\Lambda^2} \left[1 + \left(\frac{\alpha}{\alpha_c} \right) \ln \left(\frac{\Lambda}{\mu} \right) + \frac{1}{2} \left(\frac{\alpha}{\alpha_c} \right)^2 \ln^2 \left(\frac{\Lambda}{\mu} \right) + \ldots \right] \tag{4.9}$$

where we have exposed the naive, free field suppression factor Λ^{-2} and the logarithms of the perturbative renormalization group. This formula illustrates how our calculation respects the perturbative decoupling theorems[5] at $\alpha \approx 0$ and yet gives O(1) effects at $\alpha = \alpha_c$.

5. OUTLOOK

There are many unanswered questions and problems hidden here. A partial list might be:

1. Are there realistic applications of "catalyzed symmetry breaking"? In particular models some symmetry breaking interactions will amplify and others will not. It depends on the anomalous dimension (and, therefore, conservation laws) of the operators involved.
2. What are the modifications to these considerations in strongly coupled but non-fixed point examples, which occur in Technicolor model building? The asymptotic freedom at the very highest scales in those models will cutoff our amplification effects at the appropriate scale.
3. Can "collapse of the wavefunction" be incorporated into field theoretic renormalization group calculations of coupling constant flow diagrams? Can Eq. (2.4) be obtained and assimilated in this way? The singular $1/r$ form of the fermion wavefunction will be crucial here.
4. Does any of this physics survive beyond the quenched "approximation" of strongly coupled QED4?
5. Are there interesting lower dimensional analogues of these effects, such as in RVB models of high T_c superconductivity?

ACKNOWLEDGEMENT

This work was done in collaboration with E. Dagotto and A. Kocic. JBK is supported in part by the National Science Foundation grant PHY87-01775.

REFERENCES

1. Miransky, V., Il. Nuovo CImento 90A, 149 (1985). Fomin, P., Gusynin, V., Miransky, V. and Sitenko, Yu., Riv. Nuovo Cimento 6, 1 (1983).
2. Kogut, J. B., Dagotto, E. and Kocic, A., Phys. Rev. Lett. 60, 772 (1988).
3. Rubakov, V., Nucl. Phys. B203, 311 (1982).
   Callan, C., Phys. Rev. D26, 2058 (1982).
4. Kogut, J.B., Dagotto, E. and Kocic, A., ILL-(TH)-88-#29, Aug. 1988.
5. Appelquist, T. and Carazzone, J., Phys. Rev. D11, 2856 (1975).
6. Bardeen, W., Leung, C. and Love, S., Nucl. Phys. B273, 649 (1986).

DYNAMICS OF SYMMETRY BREAKING IN STRONG COUPLED QED

William A. Bardeen
Fermi National Accelerator Laboratory
P.O. Box 500, Batavia, IL 60137

Dedicated to the memory of
Heinz Pagels
Physicist, Philospher, Friend

ABSTRACT

I review the dynamical structure of strong coupled QED in the quenched, planar limit. The symmetry structure of this theory is examined with reference to the nature of both chiral and scale symmetry breaking. The renormalization structure of the strong coupled phase is analysed. The compatibility of spontaneous scale and chiral symmetry breaking is studied using effective lagrangian methods.

1. MOTIVATIONS.

Gauge theories with slowly running coupling constants appear to have an approximate scale symmetry. Hence, the dynamical breaking of chiral symmetry in such models[1] should be associated with an approximate dynamical breaking of scale symmetry[2]. Quenched, planar QED has no perturbative running and provides an interesting laboratory for the study of dynamical symmetry breaking. The possible existence of a nonperturbative fixed point with large anomalous dimensions in this theory has led to new speculations about the role of technicolor models[3] and properties of fixed point gauge field theories.

2. DYNAMICS OF QUENCHED, PLANAR QED.

Quenched, planar QED contains the basic dynamical structure of gauge field theory and may represent an approximate treatment of QED and/or nonabelian gauge theories with slowly running couplings. Dynamical symmetry breaking is studied through the Schwinger-Dyson equations

$$S_F^{-1}(p) = p - \Sigma(p)$$
$$= p - m_0 - i(4\pi)^{-4}\int dk \{D_{\mu\nu}(k)\cdot e\gamma^\mu \cdot S_F(p-k)\cdot e\gamma^\nu\} \qquad (1)$$

Solutions to these equations have been extensive studied over the years[4,5,6]. There is a unique infrared solution[6,2] to the equations having a dynamical mass scale, $\Sigma(0) = \Sigma_0 \neq 0$,

$$\Sigma(p) = e^t \cdot u(t+t_0), \quad t = \log(p), \quad t_0 = -\log(\Sigma_0) \qquad (2)$$

where $u(x)$ satisfies $e^x \cdot u(x) \to 1$ as $x \to -\infty$. The ultraviolet behavior of this solution depends on the gauge coulping constant, α. For strong coupling, $\alpha > \alpha_c = \pi/3$, $u(x)$ is given by

$$u(x) \to A(\alpha) \cdot e^{-2\cdot x} \cdot \sin[\sqrt{\alpha/\alpha_c - 1} \cdot (x + \delta(\alpha))]$$
$$\approx \tilde{A}(\alpha) \cdot e^{-2\cdot x} \cdot \{\sin[\sqrt{\alpha/\alpha_c - 1} \cdot (x + \delta(\alpha))]/\sqrt{\alpha/\alpha_c - 1}\} \qquad (3)$$

where $\tilde{A}(\alpha) \approx 1.2$, $\delta(\alpha) \approx .55$ for $\alpha \approx \alpha_c$. For weak coupling, $\alpha < \alpha_c$,

$$u(x) \to \tilde{A}(\alpha) \cdot e^{-2\cdot x} \cdot \{\sinh[\sqrt{1-\alpha/\alpha_c} \cdot (x + \delta(\alpha))]/\sqrt{1-\alpha/\alpha_c}\} \qquad (4)$$

where $u(x,\alpha)$ is analytic in α for fixed x and $\alpha \approx \alpha_c$.

The weak coupling solution has pure power behavior

$$\Sigma(p) = p \cdot u(\log(p/\Sigma_0)) \to m/(p)^{\gamma_m} + \langle\bar{\Psi}\Psi\rangle/(p)^{2-\gamma_m} \qquad (5)$$

reflecting the scaling structure[7] of the operator product expansion of the fermion propagator and the finite anomalous dimension of the fermion mass operator, $\overline{\Psi}\Psi$, given by $\gamma_m = 1 - \sqrt{1-\alpha/\alpha_c}$. The presence of both terms in the expansion of Eq.(4) confirms the result[5] that there are no massive solutions in the chiral limit, m = 0, for weak coupling.

For strong coupling, we must use an ultraviolet cutoff, Λ, and examine the boundary condition for the bare mass[4,5]

$$m_0 = (\Lambda/2) \cdot [u' + 3 \cdot u](t_\Lambda + t_0), \quad t_\Lambda = \log(\Lambda) \tag{6}$$

In the chiral limit, $m_0 = 0$, there exists a massive solution with scale

$$\Sigma_0 = e^\delta \cdot \Lambda \cdot e^{-(\theta/\sqrt{\alpha/\alpha_c - 1})}, \quad 0 < \theta < \pi \tag{7}$$

This solution will correspond to dynamical chiral symmetry breaking with a finite fermion mass scale, Σ_0, only in the Miransky limit[8] with

$$\alpha = \alpha(\Lambda) \to \alpha_c + \alpha_c \cdot \theta^2/\log^2(\Lambda/\kappa), \quad \theta \to \pi, \quad \text{as } \Lambda \to \infty \tag{8}$$

However this solution is incomplete as it neglects four fermion operators which are generated from the gauge interactions. Some four fermion operators have large anomalous dimensions in the planar limit, $d_{(\overline{\Psi}\Psi)^2} = 6 - 2\cdot\gamma_m \to 4$, $\alpha \to \alpha_c$, and are relevant (or marginal) operators in the continuum limit. Hence we must consider instead a modified "Nambu-Jona-Lasinio" model[9] where the gauge interactions are included in the lagrangian

$$L_{MNJL} = \overline{\Psi}\{iD - \mu_0\}\Psi + (G_0/2) \cdot [(\overline{\Psi}\Psi)^2 + (\overline{\Psi}i\gamma_5\Psi)^2] \tag{9}$$

In planar approximation, the Schwinger-Dyson equations are simply modified by the inclusion of a fermion tadpole contribution to the bare mass term which changes the mass boundary condition of Eq.(6)

$$m_0 = \mu_0 - G_0 \cdot \langle \overline{\Psi}\Psi \rangle = (\Lambda/2) \cdot [u' + 3\cdot u](t_\Lambda + t_0) \tag{10}$$

and

$$\langle \overline{\Psi}\Psi \rangle = (1/2\pi^2)\cdot(\alpha_c/\alpha)\cdot e^{3\cdot t}\Lambda \cdot [u'+u](t_\Lambda + t_0) \tag{11}$$

The gap equation is modified to read, $G \equiv (G_0 \cdot \Lambda^2/\pi^2)\cdot(\alpha_c/\alpha)$,

$$\mu_0 \cdot \Lambda = (\Lambda^2/2) \cdot [(1+G)\cdot u' + (3+G)\cdot u](t_\Lambda + t_0)$$

$$= (\tilde{A}/2) \cdot \Sigma_0^2 \cdot [(1-G)\cdot \sin(\theta)/\sqrt{\alpha/\alpha_c - 1} + \cos(\theta)], \quad \alpha > \alpha_c \tag{12}$$

$$\theta = \sqrt{\alpha/\alpha_c - 1} \cdot (\log(\Lambda/\Sigma_0) + \delta)$$

$$= (\tilde{A}/2) \cdot \Sigma_0^2 \cdot [(1-G)\cdot \sinh(\tilde{\theta})/\sqrt{1-\alpha/\alpha_c} + \cosh(\tilde{\theta})], \quad \alpha < \alpha_c \tag{13}$$

$$\tilde{\theta} = \sqrt{1-\alpha/\alpha_c} \cdot (\log(\Lambda/\Sigma_0) + \delta)$$

For $\alpha > \alpha_c$, the vacuum solution requires $0 < \theta < \pi$ as before.

In our planar approximation, the full scattering amplitudes are modified by the four fermion interactions which generate the bubble diagrams of the NJL model dressed by the radiative corrections of the gauge interactions. For our calculation we need to know the full, dressed vertex functions as well as the bubble functions. Fortunately we are able to compute the exact solutions in terms of the asymptotic behavior of the self-energy function, $u(x)$ [2]. The results for the scalar and pseudoscalar vertex and bubble functions are

$$\Gamma_s^0(p,p) = \partial_{m_0}\Sigma(p) = e^t \Lambda \cdot u'(t_\Lambda + t_0)/(\partial m_0/\partial t_0)$$
$$= \Gamma_s^R(p,p)/Z_s, \quad Z_s = -e^{t_0} \cdot (\partial m_0/\partial t_0) \tag{14}$$

$$\Gamma_p{}^0(p,p) = \Sigma_0(p)/m_0 = e^{t_\Lambda} \cdot u(t_\Lambda + t_0)/m_0$$

$$= \Gamma_p{}^R(p,p)/Z_p, \quad Z_p = e^{t_0} \cdot m_0 \tag{15}$$

$$B_S(0) = \partial_{m_0}\langle\overline{\Psi}\Psi\rangle = \partial_{t_0}\langle\overline{\Psi}\Psi\rangle/(\partial m_0/\partial t_0)$$

$$= -(1/2\pi^2) \cdot (\alpha_c/\alpha) \cdot e^{(3 \cdot t_\Lambda + t_0)} \cdot [u'' + u'](t_\Lambda + t_0)/Z_S \tag{16}$$

$$B_p{}^0(0) = \langle\overline{\Psi}\Psi\rangle/m_0$$

$$= (1/2\pi^2) \cdot (\alpha_c/\alpha) \cdot e^{(3 \cdot t_\Lambda + t_0)} \cdot [u' + u](t_\Lambda + t_0)/Z_p \tag{17}$$

The gap equations, Eqs.(12,13) now have nontrivial solutions for all values of the gauge coupling constant, α: $G = G(\alpha, \Lambda/\Sigma_0)$. We may now search for scale invariant fixed points. We might expect to find a continuum limit for general values of the coupling, G_0, but only particular values on the induced four fermion interactions may preserve the scale invariance as was the case for scale invariant $\eta\varphi^6$ theory[10]. Actually, no scale invariant fixed point was found[2] and the apparent scale symmetry of quenched, planar QED is broken even when the induced four fermion interactions are incorporated.

However, the continuum limit is modified by the presence of the four fermion interactions. There is now a nontrivial continuum limit along a critical line, $G = G(\alpha) > 1$ and $\alpha < \alpha_c$, as emphasized by a number of authors[11]. The existence of this critical line at weak coupling is somewhat surprizing as the effective anomalous dimensions of the four fermion operators should make them irrelevant at weak coupling, $d(\overline{\Psi}\Psi)^2 = 6 - 2\cdot\sqrt{1-\alpha/\alpha_c} > 4$. This result for weak coupling may be an artifact of the factorization properties of the planar approximation, although it may also indicate an interesting renormalizable phase of the gauged, Nambu – Jona-Lasinio model.

3. DILATONS: FACT OR FANCY.

The dynamical generation of the fermion mass scale, Σ_0, breaks both chiral and scale symmetry. If these symmetries are not explicitly broken, then we expect corresponding Goldstone pole in the appropriate S-matrix elements. With the inclusion of the four fermion interactions, we expect the Goldstone poles to come from the bubble sums and not the ladder diagrams of the pure gauge theory. Therefore, we must examine the renormalized bubble denominators for zeros reflecting the existence of Goldstone poles in the full amplitudes. The renormalized scalar and pseudoscalar denominator functions are given by

$$D_p^R(0) = Z_p^2 \cdot (1/G_0 + B_p^0(0)) = (1/4\pi^2) \cdot \tilde{A}^2 \cdot \Sigma_0^2 \cdot (\alpha_c/\alpha) \cdot (1/G)$$

$$\cdot \{\sin(\theta)/\sqrt{\alpha/\alpha_c - 1} + \cos(\theta)\} \cdot \{(1-G)\cdot\sin(\theta)/\sqrt{\alpha/\alpha_c - 1} + (1+G)\cdot\cos(\theta)\}$$

$$= 0 \quad \text{(chiral limit)} \tag{18}$$

$$D_s^R(0) = Z_p^2 \cdot (1/G_0 + B_p^0(0)) = (1/4\pi^2) \cdot \tilde{A}^2 \cdot \Sigma_0^2 \cdot (\alpha_c/\alpha) \cdot (1/G)$$

$$\cdot \{(1+\alpha/\alpha_c)\cdot\sin(\theta)/\sqrt{\alpha/\alpha_c - 1} + \cos(\theta)\} \cdot \{[(2-G)+(1+G)\cdot(\alpha/\alpha_c - 1)]$$

$$\cdot \sin(\theta)/\sqrt{\alpha/\alpha_c - 1} + (1+3\cdot G)\cdot\cos(\theta)\}$$

$$= (1/4\pi^2) \cdot \tilde{A}^2 \cdot \Sigma_0^2 \cdot \{2\cdot\alpha_c/\alpha + 1 + 1/G\} \quad \text{(chiral limit)} \tag{19}$$

The above formula for the scalar denominator in the chiral limit is valid in both weak and strong coupling. Clearly, the scalar denominator does not vanish at strong or weak coupling confirming previous results[2,8,12], but contradicting recent claims of a fixed point along the critical line[13]. There is a spurious vanishing for repulsive coupling, G < 0, which is due to the vanishing of the scalar vertex renormalization factor, Z_s, but this pole in the bubble sum is exactly cancelled by a related pole in the ladder diagrams with no resulting singularity in the S-matrix elements.

There is, however, an interesting partial cancelation in the scalar denominator function. Dimensional analysis, with the known anomalous dimensions, would predict that the scalar denominator should diverge with the cutoff

$$D_S^R(0) \to \Lambda^{2\cdot\sqrt{1-\alpha/\alpha_c}} \to \infty \qquad (20)$$

as the four fermion operators should be irrelevant at weak coupling. Instead, the scalar denominator remains finite along the critical line[11] as previously discussed. This behavior may be an artifact of the factorization treatment of the four fermion interactions or may imply an interesting weak coupling, renormalizable phase of the full theory.

A final possibility would be that the observed behavior at zero momentum represents a decoupled dilation, much like the pseudoscalar Goldstone boson in the normal NJL model. In this situation we could have $F^2_{dilaton} \to \infty$ and $m^2_{dilaton} \to 0$ as $\Lambda \to \infty$. To rule out this possibility, we must compute the momentum derivative of the scalar denominator, $\partial_{p^2} D_S^R(p^2)_{p^2=0} \approx F^2_{dilaton}$, to determine whether it remains finite or becomes divergent. It is not possible to get an exact expression for the result, but an estimate of the diagrams indicates that it remains finite and the decoupled dilaton scenario is not viable.

By our analysis of the symmetry structure, the gauged NJL model the chiral symmetry is preserved and can be dynamically broken even at weak gauge coupling, $\alpha < \alpha_c$. The scale symmetry is explicitly broken, as in the pure gauge case, and there is no dilaton, or decoupled dilaton, in either theory. The four fermion operators seem to remain as relevant operators even though their physical dimension seems too large, as $d(\overline{\psi}\psi)^2 > 4$, although this feature may also be an artifact of the approximations. Finally, it is still possible that a scale invariant theory may exist beyond the quenched, planar theory with the presence of additonal relevant interactions.

4. RENORMALIZATION.

In this section we discuss the renormalization properties of the quenched, planar theory. The continuum limit seems to require a particular cutoff dependence of the bare coupling constants. This cutoff dependence would seem to imply nonperturbative β-functions describing the strong coupling phase of the the theory[8,2]. Normally these β-functions would be related to the physical behavior of the amplitudes of the theory through the application of the renormalization group equations. However, the quenched, planar theory does not seem to allow for a dynamical running of the gauge coupling constant and the normal renormalization properties of the theory are brought into question.

We could study the dynamical running of the coupling constants if we could explicitly integrate the high momentum behavior of the theory and study directly the renormalization flow of the coupling constants. Despite the great simplifications of the quenched, planar theory, it would seem to be quite difficult to integrate the high momentum parts of the various rungs of the ladder and self-energy diagrams as in Figure 1. From this diagram, it is clear that the four fermion interactons must be generated from the high momentum structure of the pure gauge theory and must be included in the renormalization flow if these interactions become relevant.

Figure 1: Renormalizaton Diagrams

Although we can not integrate the specific diagrams directly, we can compute their effect indirectly by using the fact that our theory is defined using a momentum cutoff on the fermion self-energy. By varying the cutoff, Λ, while holding the low energy physics constant, we can effectly study the renormalization flow of the coupling constants.

For the gauge coupling constant near the critical coupling, the fermion self-energy function has a nearly universal behavior. For moderate momentum, $\Sigma_0 < p \ll \Lambda$, we use Eqs.(2,3,4) to obtain,

$$\Sigma(p) \rightarrow \tilde{A}(\alpha) \cdot (\Sigma_0^2/p) \cdot (\log(p/\Sigma_0) + \delta(\alpha)) \qquad (21)$$

To hold low energy physics fixed we shall use the generated fermion mass scale, Σ_0, as an invariant quantity. An alternative would be the renormalized fermion condensate,

$$\langle \overline{\Psi}\Psi \rangle_R = Z_S \cdot \langle \overline{\Psi}\Psi \rangle = -(1/2\pi^2) \cdot \tilde{A}^2 \cdot (\alpha_c/\alpha) \cdot \Sigma_0^3 \qquad (22)$$

where we have used Eqs.(11,14,6) for the pure gauge theory result although exactly the same answer is obtained in the gauged, NJL model. We should be able to use Σ_0, $\langle \overline{\Psi}\Psi \rangle$, and $\Sigma(p)$ as the low energy parameters. This would seem to imply that α must be held fixed during the renormalization flow in agreement with the diagram structure.

The flow of the four fermion coupling constant, G, for constant Σ_0 can be computed directly from the gap equations Eqs.(12,13) in the chiral limit.

$$G = (\tan(\theta)/\sqrt{\alpha/\alpha_c - 1} + 1)/(\tan(\theta)/\sqrt{\alpha/\alpha_c - 1} - 1) \qquad (23)$$

$$\rightarrow 1 + 2/(\log(\Lambda/\Sigma_0) + \delta - 1), \quad \tan(\theta) \approx \theta$$

where this limit is valid for both the weak and strong coupling regions. The physical running of the four fermion coupling is required to keep the low energy physics fixed. This flow is shown in Fig.(2) where the bare coupling, $G_0 = G \cdot (\alpha/\alpha_c)$, is plotted as a function of the cutoff.

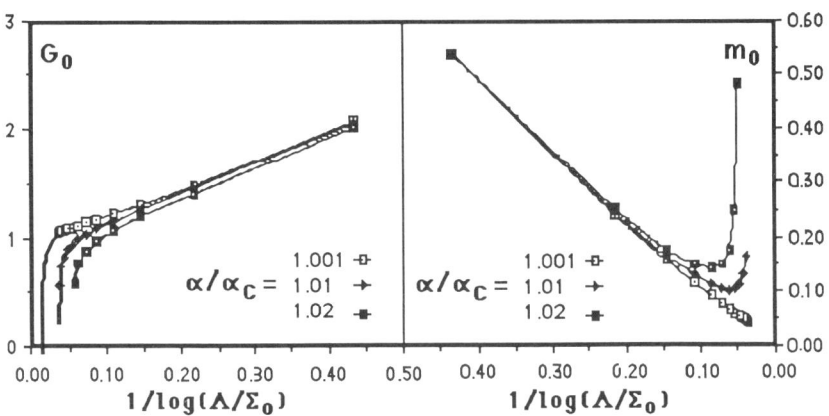

Figure 2: Renormalization Flows G_0 and m_0

The flow is given for various values of the gauge coupling, α, which does not flow with the cutoff in ladder approximation. The flow indicates the presence of an apparent ultraviolet fixed point[2] as $G \to 1$, $\alpha \to \alpha_C$. If we require that $G_0(\Lambda)$ be held fixed, for example at the pure gauge theory value: $G_0(\Lambda) = 0$, then a spurious renormalization of the gauge coupling constant would seem to be required to maintain a stable low energy theory. However, all relevant coupling constants must be included if the low energy theory is to be fully renormalized and the four fermion interactions are generated. Away from the chiral limit the bare mass parameter, m_0, will also flow with the cutoff and this dependence is also shown in Figure 2. Actually, the renormalization flow shown for $\alpha > \alpha_C$ can not be maintained to the continuum limit as the vacuum eventually becomes unstable due to short distance effects. However the continuum limit can be achieved for the theory in the range, $0 < \alpha \leq \alpha_C$. We have not investigated the stability of the full low energy effective theory and further running of other relevant coupling constants may be required to achieve a full renormalization of the complete theory.

We have examined the renormalization flow of the quenched, planar theory which requires the dynamical running of the four fermion coupling constants but not the gauge coupling constant. A smooth continuum limit seems to be associated with an ultraviolet fixed point where $G \to 1$ and $\alpha \to \alpha_c$ as well as the apparent critical line at weak coupling[11]. A more complete analysis of the renormalization properties of the full theory is needed to establish the complete continuum limit.

5. EFFECTIVE DYNAMICS OF SPONTANEOUSLY BROKEN SCALE AND CHIRAL SYMMETRIES.

Recently a NO-GO theorem has been proposed[14] which suggests a basic conflict between the low energy theorems for scale and chiral symmetry. We will show, by the explicit construction of an effective lagrangian, that both scale and chiral symmetry can be realized in the Goldstone model in the presence of explicit symmetry breaking of "fermion" mass term. The low energy theorems of both scale and chiral symmetry are shown to be satisfied by the the amplitudes generated by this effective lagrangian.

In a scale invariant gauge theory, only the fermion mass terms should explicitly break the scale symmetry. The divergence of the scale and axial currents should be given by

$$\partial^\mu D_\mu = \Theta_{\mu\mu} = (1+\gamma_m) \cdot \overline{\Psi} \, m \, \Psi$$

$$\partial^\mu A^a_\mu = \overline{\Psi} \, \{\lambda^a/2, m\} \cdot i \cdot \gamma_5 \, \Psi \tag{24}$$

where m is the fermion mass matrix and γ_m is the anomalous dimension for the fermion mass operator, $\overline{\Psi}\Psi$. Dynamical breaking of both scale and chiral symmetry would imply that Goldstone bosons carry the scale dimension and chirality. We can introduce the Goldstone fields, $\pi(x)$ and $D(x)$, by

$$U(\pi) = \exp\{ i\lambda \cdot \pi(x)/f_\pi \}$$

$$S(D) = \exp\{ D(x)/F_D \} \tag{25}$$

where $U(\pi)$ is a dimensionless matrix with $SU(N) \otimes SU(N)$ flavor symmetry and $S(D)$ is the dimension one, scale field. If the Goldstone bosons saturate the low energy theorems then all operators must have a Goldstone realization in terms of $\pi(x)$ and $D(x)$. The fermion bilinear operator will have the representation

$$\overline{\Psi}_{Rj}\Psi_{Li} = -r_0 \cdot F^2_\pi \cdot (S(D))^{3-\gamma_m} \cdot \{U(\pi)\}_{ij} \tag{26}$$

where the S factor generates the correct dimension and $U(\pi)$ the correct chirality.

The effective action is given in terms of the nonlinear lagrangian

$$L = (1/2) \cdot F^2_D \cdot (\partial_\mu S)^2 + (1/4) \cdot F^2_\pi \cdot S^2 \cdot \mathrm{tr}\{ \partial^\mu U^+ \partial_\mu U \}$$

$$+ r_0 \cdot F^2_\pi \cdot (S)^{3-\gamma_m} \cdot \mathrm{tr}\{ U^+ \cdot m + m \cdot U \} \tag{26}$$

$$- (1/2) \cdot r_0 \cdot F^2_\pi \cdot (3-\gamma_m) \cdot \mathrm{tr}\{ m \} \cdot S^4$$

where the coupling have been defined so that the classical vacuum state will have $\langle S \rangle_0 = 1$, $\langle \{U\}_{ij} \rangle_0 = \delta_{ij}$ or $\langle D \rangle_0 = \langle \pi \rangle_0 = 0$. The full energy momentum tensor is determined from the above lagrangian to be

$$\Theta_{\mu\nu} = F^2_D \cdot \{ \partial_\mu S \partial_\nu S - (1/2) \cdot g_{\mu\nu} \cdot (\partial_\alpha S)^2 \}$$

$$+ (1/4) \cdot F^2_\pi \cdot S^2 \cdot \mathrm{tr}\{ \partial_\mu U^+ \partial_\nu U + \partial_\nu U^+ \partial_\mu U - g_{\mu\nu} \cdot \partial^\alpha U^+ \partial_\alpha U \}$$

$$- g_{\mu\nu} \cdot r_0 \cdot F^2_\pi \cdot S^{3-\gamma_m} \cdot \mathrm{tr}\{ U^+ \cdot m + m \cdot U \} \tag{27}$$

$$+ g_{\mu\nu} \cdot (1/2) \cdot r_0 \cdot F^2_\pi \cdot (3-\gamma_m) \cdot tr\{m\} \cdot S^4$$

$$- F^2_D (1/6) \cdot \{\partial_\mu \partial_\nu - g_{\mu\nu} \cdot \partial^2\} S^2$$

where the final term is a necessary "improvement" term. Using the classical field equations derived from Eq.(26), the trace of the energy momentum tensor in Eq.(27) is given by

$$\Theta_{\mu\mu} = -(1+\gamma_m) \cdot r_0 \cdot F^2_\pi \cdot S^{3-\gamma_m} \cdot tr\{U^+ \cdot m + m \cdot U\}$$
$$= (1+\gamma_m) \, \overline{\Psi} \, m \, \Psi$$
(28)

in agreement with Eq.(24). Of course the axial current divergence is also correctly given by using the classical field equations.

We may now directly check the structure of the low energy theorems for scale and chiral symmetry. We can first determine the masses for the Goldstone particles

$$m^2_\pi = 2 \cdot r_0 \cdot (m_i + m_j) = -(1/F^2_\pi) \cdot (m_i + m_j) \cdot \langle \overline{\Psi}\Psi \rangle_0$$

$$m^2_D = 2 \cdot r_0 \cdot (F_\pi/F_D)^2 \cdot tr\{m\} \cdot (3-\gamma_m) \cdot (1+\gamma_m)$$
(29)

$$= - (1/F^2_D) \cdot (3-\gamma_m) \cdot (1+\gamma_m) \cdot tr\{m\} \cdot \langle \overline{\Psi}\Psi \rangle_0$$

If we use the divergence of the axial current for the interpolating field for the pseudoscalar Goldstone bosons as in Ref.(14), then we may directly evaluate the matrix elements for the trace of the energy momentum tensor. The pseudsoscalar field is given by

$$\phi^a \equiv \overline{\Psi}\{\lambda^a/2, m\} i\gamma_5 \Psi / F_\pi m^2_\pi \to \pi^a \cdot (1 + (3-\gamma_m) \cdot D/F_D + \cdots)$$
(30)

The appropriate diagrams are shown in Fig.(3) and yield

$$\Gamma = \langle \phi(p) \, \phi(p') \, \Theta_{\mu\mu}(p'-p) \rangle_0$$

$$= (p'^2-m^2_\pi)^{-1} \cdot (p^2-m^2_\pi)^{-1} \cdot \{ \, [-2p \cdot p' + 4m^2_\pi]$$

$$- F_D \cdot q^2 (q^2-m^2_D)^{-1} \cdot (1/F_D) \cdot [-2p \cdot p' + m^2_\pi \cdot (3-\gamma_m)] \, \} \quad (31)$$

$$- (p'^2-m^2_\pi)^{-1} \cdot F_D \cdot q^2 (q^2-m^2_D)^{-1} \cdot (1/F_D) \cdot (3-\gamma_m)$$

$$- (p^2-m^2_\pi)^{-1} \cdot F_D \cdot q^2 (q^2-m^2_D)^{-1} \cdot (1/F_D) \cdot (3-\gamma_m)$$

where it is essential to keep the contributions of the dilaton poles.

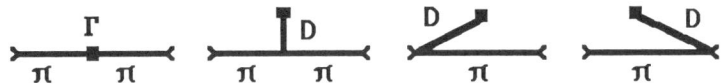

Figure 3: Diagrams for trace matrix elements

The matrix elements for the divergence of the scale current are given by

$$\Gamma_m = \langle \Phi(p) \, \Phi(p') \, (1+\gamma_m) \overline{\Psi} m \Psi (p'-p) \rangle_0$$

$$= (p'^2-m^2_\pi)^{-1} \cdot (p^2-m^2_\pi)^{-1} \cdot \{ \, [(1+\gamma_m) \cdot m^2_\pi]$$

$$- F_D \cdot m^2_D (q^2-m^2_D)^{-1} \cdot (1/F_D) \cdot [-2p \cdot p' + m^2_\pi \cdot (3-\gamma_m)] \, \} \quad (32)$$

$$- (p'^2-m^2_\pi)^{-1} \cdot F_D \cdot m^2_D (q^2-m^2_D)^{-1} \cdot (1/F_D) \cdot (3-\gamma_m)$$

$$- (p^2-m^2_\pi)^{-1} \cdot F_D \cdot m^2_D (q^2-m^2_D)^{-1} \cdot (1/F_D) \cdot (3-\gamma_m)$$

The scale identity is determined from Eqs.(31,32) and given as

$$\Gamma = \Gamma_m - (3-\gamma_m) \cdot (p'^2-m^2_\pi)^{-1} - (3-\gamma_m) \cdot (p^2-m^2_\pi)^{-1} \quad (33)$$

where $(3-\gamma_m)$ is the scale dimension of pseudoscalar field, Φ, used to compute the matrix element. The relation found above is precisely as expected, contrary to the result found in Ref.(14).

We can use these results to compute the on-shell relations for the meson matrix elements

$$\langle \pi(p')|\Theta_{\mu\mu}|\pi(p)\rangle = \lim (p'^2-m^2_\pi)\cdot(p^2-m^2_\pi)\cdot\Gamma$$

$$= [q^2+2\cdot m^2_\pi] - q^2\cdot(q^2-m^2_D)^{-1}\cdot[q^2+(1-\gamma_m)\cdot m^2_\pi] \qquad (34)$$

$$= \langle \pi(p')|(1+\gamma_m)\cdot\overline{\Psi}m\Psi|\pi(p)\rangle$$

It is clear that the presence of the dilaton pole is essential to a consistent evaluation of the low energy theorems for the meson matrix elements of the energy momentum tensor and the divergence of the scale current.

I conclude that there is no NO-GO theorem. There is a consistent low energy phenonemology for the Goldstone realization of scale and chiral symmetry. Further, there seems to be no constraint on the value of the anomalous dimension, γ_m. Hence, there is no consistency barrier to the simultaneous realization of scale and chiral symmetry.

Of course, it is still essential that the fundamental theory have the softly broken scale symmetry, as in Eq.(24), to apply the relations of scale current algebra. The results of the previous sections have shown that the nontrivial phase of quenched, planar QED is associated with a hard, explicit breaking of the scale symmetry and, hence, the scale current algebra can not be applied to this system.

ACKNOWLEDGEMENTS

I would like to thank the Japan Society for the Promotion of Science for its support and the Workshop organization for its efficiency and hospitality. This research was done in collaboration with Dr. C.N. Leung and Professor S.T. Love of Purdue University.

REFERENCES

1] W.J. Marciano, Phys. Rev. **D21**(1980)2425.

2] W.A. Bardeen, C.N. Leung and S.T. Love, Phys. Rev. Lett. **56**(1986)1230; C.N. Leung, S.T. Love and W.A. Bardeen, Nucl. Phys. **B273**(1986)649.

3] T. Akiba and T. Yanigida, Phys. Lett. **169B**(1986)432; T. Appelquist, D. Karabali and L.C.R. Wijewardhana, Phys. Rev. Lett. **57**(1986)957, T. Appelquist and L.C.R. Wijewardhana, Phys. Rev. **D35**(1987)774; ibid, Phys. Rev. **D36**(1987)568; K. Yamawaki, M. Bando and K. Matumoto, Phys. Rev. Lett. **56**(1986)1335; K. Aoki, M. Bando, H. Mino, T. Nonoyama, H. So and K. Yamawaki, Nagoya Preprint, DPNU-88-7; B. Holdom, Phys. Rev. Lett. **60**(1988)1233.

4] K. Johnson, M. Baker, R. Willey, Phys. Rev. **136**(1964)B1111; ibid, Phys. Rev. **163**(1967)1699.

5] T. Maskawa and H. Nakajima, Prog. Theor. Phys **52**(1974)1326; ibid, **54**(1975)860.

6] R. Fukuda and T. Kugo, Nucl. Phys. **B117**(1976)250.

7] S.L. Adler and W.A. Bardeen, Phys. Rev. **D4**(1971)3045; R. Jackiw and K. Johnson, Phys. Rev. **D8**(1973)2386; H. Pagels and S. Stokar, Phys. Rev. **D20**(1979)2947.

8] P.I. Fomin, V.P. Gusyinin, V.A. Miransky and Yu. Sitenko, Riv. Nuovo Cimento **6**(1983)1; P.I. Fomin, V.P. Gusyinin, V.A. Miransky, Phys. Lett. **78B**(1978)136; V.P. Gusynin and V.A. Miransky, Phys Lett. **B198**(1987)79; ibid, Phys. Lett. **B198**(1987)362.

9] Y. Nambu and G. Jona-Lasinio, Phys. Rev. **122**(1961)345.

10] M. Bander, W. Bardeen and M. Moshe, Phys. Rev. Lett. 52(1984)1188;

11] K. Aoki, Japan Physical Society Meeting (April, 1988); K. Kondo, H. Mino, K. Yamawaki, Nagoya Preprint, DPNU-88-18 T. Appelquist, M. Soldate, T. Takeuchi and L.C.R. Wijewardhana, Yale Preprint, YCTP-P19-88.

12] B. Holdom, Phys. Lett. **200B**(1988)338; ibid., Phys. Lett. 187B(1987)357.

13] K. Kondo, H. Mino and K. Yamawaki, Nagoya Preprint DPNU-88-18; K. Yamawaki, College Park Superstrings, 1987:514; K. Yamawaki, M. Bando and K. Matumoto, Phys. Rev. Lett. **56**(1986)1335.

14] V.P. Gusynin, V.A. Kushnir and V.A. Miransky, Kiev Preprint ITP-88-84E.

PHASE STRUCTURE OF QUANTUM ELECTRODYNAMICS IN THE FRAMEWORK OF THE SCHWINGER-DYSON EQUATION

Kei-ichi Kondo[†]
Department of Physics, Nagoya University,
Chikusa-ku, Nagoya 464,
JAPAN

ABSTRACTS

Based on the chiral-symmetry-breaking solution of coupled SD (Schwinger-Dyson) equation, we study the phase structure and critical behaviors of massless QED beyond the quenched ladder approximation.

1. INTRODUCTION

In this talk we report some recent progress on the study of phase structure and continuum limit of massless QED (Quantum Electrodynamics) in the framework of the SD equation.

1.1. Notation

Of course, the Lagrangian of QED is given by
$$\mathcal{L} = -\tfrac{1}{4}F_{\mu\nu}F^{\mu\nu} + \overline{\psi}i\gamma^\mu\partial_\mu\psi - m\overline{\psi}\psi + e\overline{\psi}\gamma^\mu\psi A_\mu. \tag{1}$$
The full set of the SD equation for QED with the lagrangian (1) is given as follows. For the fermion propagator
$$S^{-1}(p) = S^{(0)-1}(p) + \Sigma(p), \tag{2}$$
for the photon propagator
$$D_{\mu\nu}^{-1}(k) = D_{\mu\nu}^{(0)-1}(k) + \Pi_{\mu\nu}(k), \tag{3}$$
for the vertex function
$$\Gamma_\mu(p, p-k; k) = \gamma_\mu + \Lambda_\mu(p, p-k; k), \tag{4}$$
Here $\Sigma(p)$ is the fermion self-energy function
$$\Sigma(p) \equiv e^2 \int d^4k/(2\pi)^4 \, \gamma_\mu S(p-k)\Gamma_\nu(p, p-k; k)D_{\mu\nu}(k), \tag{5}$$

the vacuum polarization tensor $\Pi_{\mu\nu}(k)$ is given by

$$\Pi_{\mu\nu}(k) \equiv e^2 \int d^4p/(2\pi)^4 \, \text{tr}[\gamma_\mu S(p)\Gamma_\nu(p, p-k; k)S(p-k)], \tag{6}$$

the vertex correction function $\Lambda_\mu(p, p-k; k)$ is given by

$$\Lambda_\mu(p, p-k; k)_{\delta\gamma} \equiv \int d^4p/(2\pi)^4 \, [S(p+q)\Gamma_\mu(p+q, p'+q; p-p')S(p'+q)]_{\beta\alpha}$$
$$\cdot K(p'+q, p+q, q)_{\alpha\beta,\gamma\delta} \tag{7}$$

where $K(p'+q, p+q, q)$ is an electron–positron scattering kernel.

The solutions for the fermion propagator and photon one are obtained in the following form

$$S(p) = [A(p^2)\slashed{p} + B(p^2)]^{-1}, \tag{8}$$
$$D_{\mu\nu}(k) = (d_t(k^2)/k^2)(\delta_{\mu\nu} - k_\mu k_\nu/k^2) + (d_\ell/k^2)(k_\mu k_\nu/k^2), \tag{9}$$

where d_ℓ is the gauge parameter. The bare propagators $S^{(0)}(p)$ and $D_{\mu\nu}^{(0)}(k)$ correspond to the choices of $A(p^2) \equiv 1$, $B(p^2) \equiv m$, $d_t(k^2) \equiv 1$:

$$S^{(0)}(p) = [\slashed{p} + m]^{-1}, \tag{10}$$
$$D_{\mu\nu}^{(0)}(k) = (1/k^2)(\delta_{\mu\nu} - k_\mu k_\nu/k^2) + (d_\ell/k^2)(k_\mu k_\nu/k^2). \tag{11}$$

In our notation

$$\Sigma(p) \equiv [A(p^2)-1]\slashed{p} + B(p^2). \tag{12}$$

We use another notation

$$S(p) = [A(p^2)\slashed{p} + B(p^2)]^{-1} = Z(p^2)[\slashed{p} + M(p^2)]^{-1}, \tag{13}$$
$$Z(p^2) \equiv 1/A(p^2), \quad M(p^2) \equiv B(p^2)/A(p^2). \tag{14}$$

1.2. Motivation

Recently, in the framework of the SD (Schwinger–Dyson) equation, Miransky[1] has shown within the quenched ladder approximation in the Landau gauge that the continuum limit of massless QED with cutoff may be nontrivial by approaching the critical point from the strong coupling phase where the chiral symmetry is spontaneously broken. In the strong coupling phase we have a Nambu–Goldstone particle in the spectrum as a result of spontaneous breaking of the chiral symmetry. Even in the Landau–Pomeranchuk zero-charge situation[2,3], the continuum QED is an interacting field theory with the Yukawa type coupling of fermion, antifermion and pseudoscalar composed Nambu–Goldstone bosons.

In order to control the continuum limit, we need to know the critical behavior of the cutoff theory near the critical point. In massless QED with cutoff Λ, the critical point is given by the chiral transition point which separates the weak coupling phase from the strong coupling phase where the chiral symmetry is spontaneously broken. In the framework of SD equation, the occurrence of spontaneous breaking of chiral symmetry is expressed by the existence of nontrivial solutions for the dynamical mass function $M(p^2)$ in the absence of the bare fermion mass[4].

In the scaling region the dynamical mass $M(p^2)$ is governed by the scaling function $f(g)$ of the bare coupling (charge) g as

$$M(0)/\Lambda = f(g), \tag{1}$$

since the cutoff Λ is the only dimensionful parameter of the theory. In order to have a meaningful continuum limit, the scaling function $f(g)$ must have a zero at the critical value g_c, which implies continuous transition.

If we require $M(0)$ does not change when the cutoff is varied, i.e.

$$dM(0)/d\Lambda = 0, \tag{2}$$

then the cutoff dependence of the bare coupling $g = g(\Lambda)$ is governed by the bare β-function $\beta(g)$ as

$$\beta(g) = \Lambda(dg/d\Lambda), \tag{3}$$

where $\beta(g)$ is given by

$$\beta(g) \equiv -f(g)/f'(g). \tag{4}$$

In the quenched ladder approximation[1], it is shown that for the bare coupling

$$g \equiv e^2/(16\pi^2) \tag{5}$$

sufficiently close to the critical one $g_c = 1/12$,

$$f(g) \propto \exp[-\theta/(g/g_c-1)^\rho], \rho = 1/2, \tag{6}$$

where θ is a constant. Corresponding to this scaling, we obtain

$$\beta(g) = -(g_c/\rho\theta)(g/g_c-1)^{1+\rho}, \tag{7}$$

which is tangent to the g-axis at the critical point g_c.

If the scaling function exhibits the power-law behavior

$$f(g) \propto (g-g_c)^\nu \quad (\nu > 0), \tag{8}$$

with critical exponent ν, then the corresponding bare β-function reads

$$\beta(g) = -(g_c/\nu)(g/g_c-1). \tag{9}$$

In any case, the continuum limit of cutoff QED is reached at the continuous phase transition point where the bare β-function vanishes:

$\beta(g_c) = 0.$ (10)

It is quite interesting to know to what extent the improvement of the quenched ladder approximation, e.g. the effect of the vacuum polarization, can influence the Miransky's scenario to obtain nontrivial QED.

2. QUENCHED LADDER APPROXIMATION

The simplest approximation is the quenched ladder approximation: $\Pi_{\mu\nu}(k) \equiv 0$ and $\Lambda_\mu(p, p-k; k) \equiv 0$, i.e.

$$D_{\mu\nu}(k) = D_{\mu\nu}{}^{(0)}(k), \quad \Gamma_\mu(p, q; k) = \gamma_\mu. \tag{1}$$

Then the SD equation for the fermion propagator reads

$$S^{-1}(p) = S^{(0)-1}(p) + \Sigma(p), \tag{2}$$

where the fermion self-energy function $\Sigma(p)$ is given by

$$\Sigma(p) \equiv e^2 \int d^4k/(2\pi)^4 \, \gamma_\mu S(p-k) \gamma_\nu D_{\mu\nu}{}^{(0)}(k). \tag{3}$$

In what follows we study the dynamics of spontaneous chiral symmetry breaking in massless QED. In the classical work[4] on the SD equation in the quenched ladder approximation, in order to determine whether any solution obtained by solving the SD equation without cutoff corresponds to explicit breaking or spontaneous breaking, we must introduce the cutoff Λ in the SD equation and subsequently take the continuum limit of removing the cutoff. Therefore the SD equation without cutoff must be obtained as a limit of cutoff SD equation. In the SD equation we introduce the UV (ultraviolet) cutoff Λ and IR (infrared) cutoff ε. It is shown that, in the quenched ladder approximation, the SD equation has nontrivial solution corresponding to the spontaneous breaking of chiral symmetry in the Landau gauge[4].

3. QED FROM THE RENORMALIZATION GROUP POINT OF VIEW

3.1. The SD Equation

In this section we investigate the phase structure of QED from the renormalization group point of view. From the RG viewpoint, we can add more interactions to the system to investigate the phase structure and renormalization flow. In view of this, the four fermion interaction has primary importance, since photon exchange diagrams generate it. Furthermore, the dynamical mass behaves

$\Sigma(p) \to (p^2)^{-\frac{1}{2}+\sigma} \equiv (p^2)^{-\frac{1}{2}(3-\dim[\bar{\psi}\psi])}$ as $p \to \infty$, $\sigma \equiv \frac{1}{2}\sqrt{1-\lambda/\lambda_c}$.
Then the dimension of mass operator is given by $\dim[\bar{\psi}\psi] = 2+\sqrt{1-\lambda/\lambda_c} \to 2$ ($\neq 3$: canonical dimension), and hence $\dim[(\bar{\psi}\psi)^2] = 4 + 2\sqrt{1-\lambda/\lambda_c} \to 4$ as $\lambda \to \lambda_c$. Hence the four-fermi interaction is relevant to electromagnetic interaction, since $\dim[\bar{\psi}\gamma^\mu\psi A_\mu] = 4$. Therefore the four-fermion interaction must be included to study the phase structure of QED from the renormalization group point of view.

Thus we consider the system with the lagrangian

$$\mathcal{L} = -\tfrac{1}{4}F_{\mu\nu}F^{\mu\nu} + \bar{\psi}i\gamma^\mu\partial_\mu\psi - \mu_0\bar{\psi}\psi + e_0\bar{\psi}\gamma^\mu\psi A_\mu$$
$$+ \tfrac{1}{2}G_0[(\bar{\psi}\psi)^2 + (\bar{\psi}i\gamma_5\psi)^2]. \quad (1)$$

Bardeen et al.[5] have considered the following SD equation for the fermion propagator

$$S^{-1}(p) = \slashed{p} + m_0 + \Sigma(p) = A(p^2)\slashed{p} + B(p^2), \quad (2)$$

where m_0 is determined by the relativistic Hartree-Fock-Bogoliubov approximation

$$m_0 = \mu_0 - G_0\langle\bar{\psi}\psi\rangle. \quad (3)$$

If we adopt the quenched ladder approximation for $\Sigma(p)$, then after integrating over angles

$$A(x) = 1 + d_\ell\lambda \int_{\varepsilon^2}^{\Lambda^2} dy \frac{A(y)}{yA^2(y) + B^2(y)} \left[\frac{y^2}{x^2}\theta(x-y) + \theta(y-x)\right], \quad (4)$$

$$B(x) = m_0 + (d_\ell+3)\lambda \int_{\varepsilon^2}^{\Lambda^2} dy \frac{B(y)}{yA^2(y) + B^2(y)} \left[\frac{y}{x}\theta(x-y) + \theta(y-x)\right], \quad (5)$$

$$m_0 = \mu_0 + \frac{g}{\Lambda^2}\int_{\varepsilon^2}^{\Lambda^2} dy \frac{yB(y)}{yA^2(y) + B^2(y)}, \quad (6)$$

where the coupling constants are redefined as follows.

$$g_0 = \tfrac{1}{2}G_0, \quad g \equiv g_0\Lambda^2/(2\pi^2), \quad \lambda \equiv e_0^2/(16\pi^2). \quad (7)$$

and the Euclidean distance in momentum space is defined by

$x \equiv p^2 = p \cdot p$, $y \equiv q^2 = q \cdot q$. (8)

In the limit $\lambda = 0$, the model reduces to the Nambu and Jona-Lasinio model[6]. In this model, $A(x) \equiv 1$, and $B(x) = m_0$ ($\mu_0 = 0$). In this model the solution does not depend on the external momentum $x \equiv p^2$, and hence we can put $B(x) \equiv M$ for some constant M. Explicit calculation of the integral leads to (in the limit $\varepsilon \to 0$)

$$1/g = 1 - (M^2/\Lambda^2) \ln(\Lambda^2/M^2 + 1). \qquad (9)$$

3.2. The Critical Line[7,8]

In the Landau gauge $d_\ell = 0$, $A(x) \equiv 1$. Then the bifurcation solution[9] obey the integral equation:

$$\delta B(x) = \frac{g}{\Lambda^2} \int_{\varepsilon^2}^{\Lambda^2} dy\, \delta B(y) + 3\lambda \int_{\varepsilon^2}^{\Lambda^2} dy\, \frac{\delta B(y)}{y}\left[\frac{y}{x}\theta(x-y) + \theta(y-x)\right]. \qquad (1)$$

Defining $M(x) \equiv \delta B(x)$, this is equivalent to the following differential equation

$$x^2 M''(x) + 2x M'(x) + 3\lambda M(x) = 0, \qquad (2)$$

together with the boundary conditions:

$$\text{UV b.c.} \quad [xM(x)]'\Big|_{x=\Lambda^2} = \frac{g}{\Lambda^2} \int_{\varepsilon^2}^{\Lambda^2} dy\, M(y), \qquad (3)$$

i.e.
$$M(\Lambda^2) = -[1 + g/(3\lambda)]\Lambda^2 M'(\Lambda^2),$$

$$\text{IR b.c.} \quad x^2 M'(x)\Big|_{x=\varepsilon^2} = 0, \quad \text{i.e.} \quad M'(\varepsilon^2) = 0. \qquad (4)$$

The general solutions of the above differential equation are given by

$$M(x) = \begin{cases} A\, x^{-1/2+\sigma} + B\, x^{-1/2-\sigma}, & \text{for } 0 < \lambda < \lambda_c, \\ x^{-1/2}(C + D \ln x), & \text{for } \lambda = \lambda_c, \\ E\, x^{-1/2+i\rho} + F\, x^{-1/2-i\rho}, & \text{for } \lambda > \lambda_c, \end{cases} \qquad (5)$$

where

$$\sigma \equiv \tfrac{1}{2}\sqrt{1-\lambda/\lambda_c}, \qquad \rho \equiv \tfrac{1}{2}\sqrt{\lambda/\lambda_c - 1}, \qquad \lambda_c \equiv 1/12. \tag{6}$$

In order for the general solution (5) of the differential equation (2) to be the solution of the original integral equation, they must satisfy both IR b.c. and UV b.c. Equating the ratios, say B/A, obtained from each boundary condition, we obtain the equation of the line, (we call this line the critical line at cutoff Λ, by abuse of language) on which the nontrivial (spontaneous XSB) solutions for the SD equation exist for some fixed values of ε and Λ.

i) in the strong coupling region $(0 < \lambda < \lambda_c)$, we have

$$\left(\frac{\Lambda^2}{\varepsilon^2}\right)^{2\sigma} = \frac{g - (\tfrac{1}{2}-\sigma)^2}{g - (\tfrac{1}{2}+\sigma)^2}, \quad \text{or} \quad g = \tfrac{1}{4} + \sigma^2 + \frac{\sigma}{\tanh[\sigma \ln(\Lambda^2/\varepsilon^2)]}. \tag{7}$$

In particular, without four-fermi interaction $(g = 0)$, this relation can not be satisfied for $\sigma > 0$, and hence there is no nontrivial solution for $0 < \lambda < \lambda_c$.

ii) for $\lambda = \lambda_c$,

$$\ln\frac{\Lambda^2}{\varepsilon^2} = \frac{1}{g - \tfrac{1}{4}}, \qquad \text{or} \quad g = \tfrac{1}{4} + [\ln(\Lambda^2/\varepsilon^2)]^{-1}. \tag{8}$$

In the absence of the four-fermi interaction, there is no nontrivial solution for $\lambda = \lambda_c$.

iii) In the strong coupling region $(\lambda > \lambda_c)$,

$$\ln\frac{\Lambda^2}{\varepsilon^2} = \frac{n\pi + \tan^{-1}[\rho/(g-\tfrac{1}{4}+\rho^2)]}{\rho}, \quad (n: \text{integer}), \tag{9a}$$

i.e.

$$g = \tfrac{1}{4} - \rho^2 + \frac{\rho}{\tan[\rho \ln(\Lambda^2/\varepsilon^2)]}. \tag{9b}$$

In the case of pure gauge $(g = 0)$, eq. (9a) with $n = 1, 2 \cdots$ reduces to the scaling form obtained by Miransky[1]

$$\Lambda/\varepsilon = e^\delta \exp[n\pi/\sqrt{\lambda/\lambda_c - 1}], \qquad \text{for } |g - g_c| \ll 1. \tag{10}$$

Thus the critical line is given by[7]

$$g = \begin{cases} \frac{1}{4} + \sigma^2 + \sigma/\tanh[\sigma \ln(\Lambda^2/\varepsilon^2)], & \text{for } 0 < \lambda < \lambda_c, \\ \frac{1}{4} + 1/[\ln(\Lambda^2/\varepsilon^2)], & \text{for } \lambda = \lambda_c, \\ \frac{1}{4} - \rho^2 + \rho/\tan[\rho \ln(\Lambda^2/\varepsilon^2)], & \text{for } \lambda > \lambda_c. \end{cases} \tag{11}$$

In the continuum limit $\Lambda/\varepsilon \to \infty$, we obtain[7]

$$g = (\tfrac{1}{2} + \sigma)^2 = \tfrac{1}{4}(1 + \sqrt{1 - \lambda/\lambda_c})^2 \qquad \text{for } \lambda \leq \lambda_c. \tag{12}$$

This expression for the critical line is exact and is connected with the N & J–L point $(\lambda, g) = (0, 1)$ in the limit $\lambda \to 0$. Bardeen et al.[5] studied this system, but they missed to get a complete set of solutions.

Taking into account the IR b.c., the coefficients A, D, E are determined. Then the general bifurcation solution is rewritten as follows.

i) For $0 < \lambda < \lambda_c$,

$$M(x) = \frac{\tfrac{1}{2} + \sigma}{2\sigma} M(\varepsilon^2) \left(\frac{x}{\varepsilon^2}\right)^{-\tfrac{1}{2}} \left[\left(\frac{x}{\varepsilon^2}\right)^\sigma - \frac{\tfrac{1}{2} - \sigma}{\tfrac{1}{2} + \sigma}\left(\frac{x}{\varepsilon^2}\right)^{-\sigma}\right]. \tag{13}$$

ii) For $\lambda = \lambda_c$,

$$M(x) = \tfrac{1}{2} M(\varepsilon^2) \left(\frac{x}{\varepsilon^2}\right)^{-\tfrac{1}{2}} \left[\ln \frac{x}{\varepsilon^2} + 2\right]. \tag{14}$$

iii) For $\lambda > \lambda_c$,

$$M(x) = \frac{\sqrt{1 + (2\rho)^2}}{2\rho} M(\varepsilon^2) \left(\frac{x}{\varepsilon^2}\right)^{-\tfrac{1}{2}} \sin\left[\rho \ln \frac{x}{\varepsilon^2} + \theta\right], \tag{15}$$

where $\theta \equiv \tan^{-1}(2\rho)$. The boundary value of $M(x)$ is given as follows.

$$M(\Lambda^2) = \frac{\sqrt{1 + (2\rho)^2}}{2\rho} \frac{\varepsilon M(\varepsilon^2)}{\Lambda} \sin\left[\tan^{-1}\left(2\rho \frac{g + 3\lambda}{g - 3\lambda}\right)\right]. \tag{16}$$

In particular, for $g = 0$ (no four-fermi interaction),
$$M(\Lambda^2) \simeq \varepsilon M(\varepsilon^2)/\Lambda \quad \text{for } \rho \ll 1. \tag{17}$$

3.3. Renormalization Group Flow and Renormalized Trajectories[8]

Now we have obtained the critical line in the bare parameter space with two bare parameters (λ, g). Next we try to obtain the RG (renormalization group) flow in the framework of the SD equation. In order to obtain the RG flow, we must find RG invariant under change of the cutoff: $\Lambda \to \Lambda'$. To find renormalized trajectory in the framework of the SD equation, it will be natural to take the renormalized condensation $\langle \bar{\psi}\psi \rangle_R$ of the bare one $\langle \bar{\psi}\psi \rangle_\Lambda$ as the renormalization group invariant, since bare condensation $\langle \bar{\psi}\psi \rangle_\Lambda$ is divergent in the limit $\Lambda \to \infty$. In any case we need to know the renormalization constant $Z_2(\Lambda)$ defined by
$$\langle \bar{\psi}\psi \rangle_R \equiv Z_2(\Lambda)^{-1} \langle \bar{\psi}\psi \rangle_\Lambda, \tag{1}$$
such that
$$0 < \langle \bar{\psi}\psi \rangle_R < \infty \quad \text{in the limit } \Lambda \to \infty. \tag{2}$$
However, the rate of divergence of $\langle \bar{\psi}\psi \rangle_\Lambda$ may be different depending on what critical point we may choose and how the cutoff theory is approached to the critical point. Hence, without appointing the critical point at which the the continuum limit is taken and way of approaching the critical point, the renormalization constant can not be determined uniquely. This situation is in sharp contrast with the case of pure gauge interaction where the theory is specified by only one (gauge) coupling constant.

In the Landau gauge, for $x \equiv p^2$
$$\langle \bar{\psi}\psi \rangle = \frac{1}{4\pi^2} \int_{\varepsilon^2}^{\Lambda^2} dy \, \frac{yM(y)}{y+M^2(y)}. \tag{3}$$
From the differential equation and IR b.c.: $x^2 M'(x) = 0$ at $x=\varepsilon^2$, $\langle \bar{\psi}\psi \rangle$ is written by the boundary value of $M'(\Lambda^2)$:
$$\langle \bar{\psi}\psi \rangle = -\frac{1}{\pi^2} \frac{\lambda_c}{\lambda} \Lambda^4 M'(\Lambda^2). \tag{4}$$
i) For $0 < \lambda < \lambda_c$,

$$\langle\bar\psi\psi\rangle = \varepsilon^{-2\sigma} \frac{\varepsilon M(\varepsilon^2)}{4\pi^2} \frac{\Lambda}{\sigma} \sinh\left[\sigma \ln \frac{\Lambda^2}{\varepsilon^2}\right]. \tag{5}$$

ii) For $\lambda = \lambda_c$,

$$\langle\bar\psi\psi\rangle = \frac{\varepsilon M(\varepsilon^2)}{4\pi^2} \Lambda \ln \frac{\Lambda^2}{\varepsilon^2} . \tag{6}$$

iii) For $\lambda > \lambda_c$,

$$\langle\bar\psi\psi\rangle = \frac{\varepsilon M(\varepsilon^2)}{4\pi^2} \left(\frac{\lambda_c}{\lambda}\right)^{\frac{1}{2}} \frac{\Lambda}{\rho} \sin\left[\rho \ln \frac{\Lambda^2}{\varepsilon^2}\right] \tag{7}$$

In any case, $\langle\bar\psi\psi\rangle \to \infty$ as $\Lambda \to \infty$.

We consider the renormalization flow which lies in the strong coupling region $\lambda > \lambda_c$. In this case, by using the scaling relation (3.2.9) and the formula $\sin \tan^{-1} y = y/\sqrt{1+y^2}$, we have

$$\langle\bar\psi\psi\rangle = \frac{\varepsilon M(\varepsilon^2)}{4\pi^2} \left(\frac{\lambda_c}{\lambda}\right)^{\frac{1}{2}} \frac{\Lambda}{\rho} \frac{1}{\sqrt{1 + [(g-\frac{1}{4})/\rho+\rho]^2}}. \tag{8}$$

However, the way to renormalize $\langle\bar\psi\psi\rangle$ is not unique. Note that, in what follows, $\varepsilon M(\varepsilon^2)$ ($\Rightarrow \Sigma(0)^2$) is held fixed in the limit $\Lambda \to \infty$. We consider two typical cases.

Renormalization condition (I):

$$Z(\Lambda) \propto \rho/\Lambda. \tag{9}$$

Under this condition, the RG flows are obtained

$$g - \tfrac{1}{4} \propto \rho^m \text{ (for all } m \geq 1\text{), and } g = \tfrac{1}{4} \text{ in the region } \lambda \geq \lambda_c. \tag{10}$$

In this case, the critical point (1/12, 1/4) is also a fixed point and this system seems to have two relevant operators. (See Fig. 1)

Renormalization condition (II):

$$Z(\Lambda) \propto \rho^n/\Lambda \ (0 < n < 1). \tag{11}$$

Under this condition, the RG flows are obtained

$$g - \tfrac{1}{4} \propto \rho^n \text{ in the region } \lambda \geq \lambda_c. \tag{12}$$

In this case, the theory has one marginal operator.

In the pure gauge case, the renormalization constant is uniquely determined up to the numerical factor as follows.

$$Z(\Lambda) \propto \Lambda^{-1}, \qquad \text{for } \lambda > \lambda_c. \tag{13}$$

If we start from the weak coupling region, the point (1/12, 1/4) is not a special point. Any point on the critical line $g = (\frac{1}{2}+\sigma)^2$ may become a fixed point depending on the renormalization condition. Thus the requirement (2) does not determine the renormalized trajectories uniquely.

3.4. Critical Exponent and Order of the Chiral Transition[8]

Since the theory has two bare parameters, the bare β-function should be defined on the renormalized trajectories. First of all, we consider the scaling function near the critical point. [Similar observation was done by Appelquist et al. (YCTP-P19-88).]

i) For $0 < \lambda < \lambda_c$: For example, consider the continuum limit approaching a critical point on the critical line $g = (\frac{1}{2}+\sigma)^2$ along the (vertical) line $\sigma = \sigma_0$ ($0 < \sigma_0 < \frac{1}{2}$), the scaling function is of power-law type:

$$f(g) \propto [1/(2\sigma_0)]^{1/(4\sigma_0)}[g-(\frac{1}{2}+\sigma_0)^2]^{1/(4\sigma_0)}, \tag{1}$$

and hence order of the transition is finite. Then the critical exponent ν is given by $\nu = (4\sigma_0)^{-1} \in (\frac{1}{2}, \infty)$, which changes continuously, while ν takes the mean-field value $\nu = \frac{1}{2}$ in the Nambu and Jona-Lasinio limit $\lambda \searrow 0$.

ii) On the line $\lambda = \lambda_c$, the scaling function has essential singularity at $g = \frac{1}{4}$, and hence the transition is infinite order, i.e.

$$f(g) \propto \exp[-\frac{1}{2}/(g-\frac{1}{4})]. \tag{2}$$

iii) For $\lambda > \lambda_c$, we can parametrize the renormalized trajectories starting at the point (1/12, 1/4) by the parameter s ($s = 0 \iff (1/12, 1/4)$), such that

$$f(s) \propto \exp(-\Theta/s^n) \text{ for some real } n > 0, \tag{3}$$

in the neighborhood of (1/12, 1/4). In this case, $f(s)$ has essential singularity at (1/12, 1/4), which is infinite-order phase transition point.

Corresponding bare β-functions are obtained from eq. (1.2.4). In the presence of vacuum polarization, we again encounter such cross-over phenomena between exponential behavior to power-law one, see §4.

3.5. Scale Invariance and Dilaton[7,8]

The SD equation in the quenched ladder approximation preserves an exact scale invariance in the sense that the SD equation (3.1.5) (in the Landau gauge) with $\varepsilon = 0$ is invariant under the transformation

$$\Sigma(x) \to \kappa\Sigma(x/\kappa^2), \quad \Lambda \to \Lambda/\kappa, \tag{1}$$

provided that the couplings λ and g are held fixed. The scale symmetry is usually broken explicitly by quantum effects, but it may be recovered at the fixed point. In view of this, we consider the (induced) fermion-antifermion scattering amplitude

$$S_{ff}(p, p; p', p') = S_{ff}(\text{ladder}) - [\Gamma^R_p(p, p)][\Gamma^R_p(p', p')]/D^R_p(0)$$
$$- [\Gamma^R_s(p, p)][\Gamma^R_s(p', p')]/D^R_s(0). \tag{2}$$

The bare scalar denominator $D^0_s(0)$ is obtained by

$$D^0_s(0) \equiv [1 + G_0 B^0_s(0)]/G_0, \tag{3}$$

where $B^0_s(0)$ is the scalar bubble function at zero momentum given by

$$B^0_s(0) \equiv -\int d^4p/(2\pi)^4 \ \text{tr} \ [S(p)\Gamma^0_s(p, p)S(p)], \tag{4}$$

and $\Gamma^0_s(p, p)$ is the bare scalar vertex given by $\Gamma^0_s(p, p) \equiv \partial_{m_0}\Sigma(p)$. The renormalized scalar vertex function $\Gamma^R_s(p, p) \equiv Z_s\Gamma^0_s(p, p)$ is defined by requiring that $\Gamma^R_s(p, p)$ is proportional to $\Gamma^0_s(p, p)$ and $\Gamma^R_s(p, p) \to 1$ as $p \to 0$. The renormalized scalar denominator is defined by

$$D^R_s(0) \equiv Z_s^2 D^0_s(0) = [1 + G_0 B^0_s(0)]Z_s^2/G_0, \tag{5}$$

If $D^R_s(0) \to 0$ as $\Lambda \to \infty$, this shows existence of the Goldstone pole in the scalar channel, dilaton.

We have discovered the following exact formula[7,8]

$$D^R_s(0) = \frac{1}{2\pi^2\Sigma(0)^2}\left[1 + \frac{6\lambda}{g}\frac{g+3\lambda}{1+[M(\Lambda^2)/\Lambda]^2}\right]S(\lambda, g; \Lambda)[\Lambda M(\Lambda^2)]^2,$$

where

$$S(\lambda, g; \Lambda) \equiv \frac{g}{(g+3\lambda)^2} - \frac{1}{1+[M(\Lambda^2)/\Lambda]^2}. \tag{6}$$

Note that the following identity holds ($G \equiv g/(3\lambda)$):

$$\Lambda M(\Lambda^2) = -(1+G)\Lambda^3 M'(\Lambda^2) = \pi^2(1+G)(\lambda/\lambda c)(\langle\bar{\psi}\psi\rangle/\Lambda). \tag{7}$$

Then we have

$$D^R_s(0) = \frac{\pi^2}{2\Sigma(0)^2}(1+G)^2\left(\frac{\lambda}{\lambda_c}\right)^2\left(1 + \frac{6\lambda}{g}\frac{g+3\lambda}{1+[M(\Lambda^2)/\Lambda]^2}\right)$$

$$\cdot\left[\frac{g}{(g+3\lambda)^2} - \frac{1}{1+[M(\Lambda^2)/\Lambda]^2}\right]\left[\frac{Z_2(\Lambda)}{\Lambda}\right]^2 \langle\bar\psi\psi\rangle_R^2. \quad (8)$$

By substituting the (bifurcation) solution of the SD equation into the above formula, we can calculate the renormalized scalar denominator $D^R_s(0)$. We determine the renormalization constant $Z_2(\Lambda)$ such that $\langle\bar\psi\psi\rangle_R$ remains finite in the limit $\Lambda \to \infty$. Here note that $S(\lambda, g; \Lambda) \simeq g/(g+3\lambda)^2 - 1 \to 0$ as $\Lambda \to \infty$. On the other hand, for $g \neq 0$, $Z_2(\Lambda)/\Lambda \to \infty$ as $\Lambda \to \infty$. Other factors are finite for arbitrary Λ.

For example, in the strong coupling region: $\lambda > \lambda_c$ (i.e., $\rho > 0$), the most general renormalization constant is given by

$Z_2(\Lambda)/\Lambda \propto \sin(\rho \ln \kappa^2)/\rho, \quad \kappa \equiv \Lambda/\varepsilon$.

Substituting the equation for the critical line, we have

$$S(\lambda, g; \Lambda) \simeq \frac{-4[\rho/\sin(\rho \ln \kappa^2)]^2}{[1 + 2\rho/\tan(\rho \ln \kappa^2)]^2}.$$

Since $\rho/\tan(\rho \ln \kappa^2) = \sqrt{\rho^2 + (g-\frac{1}{4}+\rho^2)^2} \to 0$ as $\Lambda \to \infty$, we can conclude that $D^R_s(0) \to$ finite constant $\neq 0$ as $\Lambda \to \infty$.

In order to study whether the Goldstone pole in the scalar channel exists or not, we have calculated the renormalized scalar denominator $D^R_s(0)$ at zero momentum. We take the renormalization such that $0 < \langle\bar\psi\psi\rangle_R < \infty$ in the continuum limit $\Lambda \to \infty$. Under this renormalization, we have exhausted all the possible cases of approaching the critical point. As a result,

$0 < D^R_s(0) < \infty$ in the continuum limit $\Lambda \to \infty$.

This implies absence of the Goldstone pole in the scalar channel, i.e. no dilaton. Then there is explicit breaking of scale invariance.

4. EFFECTS OF VACUUM POLARIZATION ON THE PHASE STRUCTURE

4.1. The Coupled SD equation[10,11]

For the vertex function, we consider the following correction[2]

$$\Lambda_\mu(p, p-\ell; \ell) = e^2 \int d^4k/(2\pi)^4 \Gamma_\rho(p, p-k; k) S(p-k) \Gamma_\mu(p-k, p-k-\ell; \ell)$$

$$\cdot S(p-k-\ell) \Gamma_\sigma(p-k-\ell, p-\ell; -k) D_{\rho\sigma}(k). \quad (1)$$

The longitudinal part of the full vertex $\Gamma_\mu^L(p, q; k) \equiv (k_\mu k_\nu/k^2) \Gamma_\nu(p, q; k)$ is uniquely determined by the Ward–Takahashi identity as

$$\Gamma_\mu^L(p, q; k) = (k_\mu/k^2)[S^{-1}(p) - S^{-1}(q)], \quad k_\mu \equiv p_\mu - q_\mu. \quad (2)$$

For the transverse part $\Gamma_\mu^T(p, q; k) \equiv (\delta_{\mu\nu} - k_\mu k_\nu/k^2) \Gamma_\nu(p, q; k)$, we take the ansatz:

$$\Gamma_\mu^T(p, q; k) = (\delta_{\mu\nu} - k_\mu k_\nu/k^2) \gamma_\nu F(f^2). \quad (3)$$

We assume that $F(f^2)$ is a (slowly changing) function of the argument f^2 where f^2 represents any of the quantities p^2, q^2 and k^2, when these are of the same order of magnitude. Landau et. al[2] choose the largest f^2, when one of these quantities is small compared with the two others. More specifically we take the approximation[11]:

$$F(f^2) = F(p^2)\theta(p^2-q^2) + F(q^2)\theta(q^2-p^2). \quad (4)$$

Analogous analysis as Landau[1] leads to the following integral equation

$$F(\xi) = 1 + d_\ell g \int_\xi^L d\zeta \, \{F^3(\zeta) + F(\zeta)[A(\zeta)-F(\zeta)]^2\}/A(\zeta)^2, \quad (5)$$

where $\xi \equiv \ln p^2$, $\zeta \equiv \ln q^2$ and $L \equiv \ln \Lambda^2$. Moreover, defining $x \equiv p^2$, $y \equiv q^2$, we obtain the following coupled integral equations for $A(x)$, $B(x)$ and $F(z)$ after angular integral:

$$P(x)A(x) = 1 - d_\ell g \frac{B(x)}{x^2} \int_{\mu^2}^x dy \frac{yB(y)}{yA^2(y) + B^2(y)}, \quad (6)$$

$$P(x)B(x) = d_\ell g \frac{A(x)}{x} \int_{\mu^2}^{x} dy \frac{yB(y)}{yA^2(y) + B^2(y)}$$

$$+ 3g \int_{\mu^2}^{\Lambda^2} dy \frac{yB(y)}{yA^2(y) + B^2(y)} \left[\frac{F(x)d_t(x)}{x} \theta(x-y) + \right.$$

$$\left. + \frac{F(y)d_t(y)}{y} \theta(y-x) \right], \quad (7)$$

where

$$P(x) \equiv 1 - d_\ell g \int_{x}^{\Lambda^2} dy \frac{A(y)}{yA^2(y) + B^2(y)}. \quad (8)$$

It should be remarked that eq. (6) is obtained only by assuming $F(\ell^2) = F(p^2)$, $F(q^2)$ or $F(k^2)$ without any further assumption, e.g. (4). We can find a solution for eq. (5) such that

$$F(p^2) = A(p^2), \quad (9)$$

which is consistent with multiplicative renormalizability, i.e. $\Gamma_\mu(p,p; 0) = A(m^2)\gamma_\mu$ at $p^2 = m^2$ and with the above approximation for the vertex (2). Then eq. (5) is simplified as

$$F(\xi) = 1 + d_\ell g \int_{\xi}^{L} d\zeta\, F(\zeta), \quad (10)$$

whose solution is given by

$$F(p^2) = (p^2/\Lambda^2)^{-d_\ell g}. \quad (11)$$

Note that, in the chiral symmetric phase where $B(x) \equiv 0$, eq. (6) is solved to give

$$A(p^2) = (p^2/\Lambda^2)^{-d_\ell g}, \quad (12)$$

which is consistent with the solution (11) under the above approximation. Unless the exact expression for the longitudinal vertex is taken into account in the SD equation for the fermion propagator, such a set of consistent solutions can not be obtained, see Ref. 9). The effect of the vacuum polarization is introduced through $d_t(k^2)$. Its explicit form is obtained in a similar way:

$$d_t(k^2) = 1/[1 + (4/3)gN_f \ln(\Lambda^2/k^2)], \tag{13}$$

which is also the leading term in the $1/N_f$ expansion where gN_f is held fixed.

In the Landau gauge $d_\ell = 0$, $A(p^2) \equiv 1$ and $\Sigma(p) = B(p^2)$. In the quenched ladder approximation, the SD equation has no consistent solution, except the Landau gauge, even in the chiral symmetric phase[10,13]. In order to go beyond the ladder approximation, we must consider, first of all, the correction of the vertex. It should be remarked that the longitudinal vertex is uniquely determined by the W-T identity which is a necessary condition for the gauge invariance, see (2). (In the quenched ladder approximation, the W-T identity is not satisfied) Furthermore, so as to maintain gauge independence, the choice of the transverse vertex is essential[10]. In fact, some choices of the transverse vertex may lead to quite different conclusion for respective gauge[10]. Note that the function $P(x)$ is coming from the longitudinal vertex obeying the Ward-Takahashi identity. Note that the function $P(x)$ may violate the positivity of the dynamical mass function $M(x)$, unless the transverse vertex is appropriately chosen[10]. The choice of the transverse vertex in this paper does not suffer from this type of difficulty, since for the solution (12)

$P(x) = 1/A(x) > 0$,

if $B(x) \equiv 0$. Detailed analysis of the SD equation will be given elsewhere[12].

In the limit $N_f = \infty$, it is proved[12] that the spontaneous breaking of the chiral symmetry does not occur for all gauges even for finite Λ (which is trivial in the Landau gauge where $\Sigma(p) \equiv 0$), provided that $d_t(k^2) \equiv 0$ in the limit $N_f = \infty$, consistently with (13). In the quenched limit $N_f = 0$, It can be shown analytically that the scaling function $f(g)$ is of the essential singularity type and hence the order of the transition is infinite, irrespective of the gauge choice, see e.g. Ref. 16).

4.2. Numerical Results[11,12]

We have solved numerically the coupled integral equations (6)-(9). In addition we have paid special attention to the influence of the number of flavors N_f on the value of the critical coupling and the order of the transition. The limit $N_f = 0$ corresponds to the quenched limit.

Our data show that, even in the presence of the vacuum polarization, there exists a transition exhibiting spontaneous chiral symmetry breaking in QED for sufficiently large coupling, and the transition is continuous for $0 < N_f < \infty$. The critical coupling grows towards the strong coupling, as N_f is increased, see Fig. 2.

Our numerical and analytical analyses show that the chiral transition does not disappears for sufficiently large $N_f < \infty$, in contrast with the result of Ref. 14). Therefore, if we define the critical value N_f^c below which the chiral transition does occur at some critical coupling $g_c < \infty$, then the above result suggests for finite cutoff $\Lambda < \infty$

$N_f^c = \infty$ in $(QED)_4$,

in sharp contrast with $(QED)_3$: $0 < N_f^c < \infty$, see Ref. 15).

Numerical data for the scaling function is plotted in Fig. 3 for the number of flavors $N_f = 0, 1, 4$. <u>In the quenched limit $N_f = 0$, the scaling function $f(g)$ is of the essential singularity type and the order of the transition is infinite irrespective of the gauge choice</u>, in agreement with the analytical result. If the vacuum polarization effect is included, however, our numerical solution suggests the power-law behavior for the scaling function

$f(g) \propto (g-g_c)^\nu, \nu > 0$,

in any gauge and hence <u>the order of the transition is finite for large N_f</u>, even for $N_f = 1$. Our result should be compared with Monte Carlo result[17]. Then, in the presence of vacuum polarization, we have the bare β-function as (1.2.9).

5. ZERO CHARGE PROBLEM OF QED

5.1. Landau Ghost and Triviality of QED

In spite of the great success of perturbative QED in describing low energy phenomena, it has been pointed out that QED is logically incomplete theory. This is represented by the notorious Landau ghost problem, or zero charge problem. Landau et al.[2] have obtained the relation between the bare coupling constant e and the true charge of an electron e_R by considering the scattering of an electron by an electron within $e_0 \ll 1$:

$$e_R^2 \cong e_0^2 d_t(m^2) = \frac{e_0^2}{1 + (e_0^2/12\pi^2) \ln(\Lambda^2/m^2)}. \tag{1}$$

If the cutoff Λ is removed, i.e. $\Lambda \to \infty$, the physical (renormalized) charge e_R (describing strength of the interaction among photon, fermion and antifermion) goes to zero and the continuum theory is free, as long as the bare charge $e_0(\Lambda)$ is real. In other word, to maintain the physical charge nonzero in the continuum limit, the bare charge must take pure imaginary value, contradictory to the selfadjointness of the Hamiltonian.

Thus, as $\Lambda \uparrow \infty$,

$$0 \leqslant e_R^2 \leqslant \frac{2\pi^2}{\ln(\Lambda/m)} \downarrow 0, \tag{2}$$

irrespective of the choice of the bare charge $e_0 = e_0(\Lambda)$ as a function of Λ, provided only that $e_0^2(\Lambda) \geqslant 0$. Thus, in this approximation, the continuum limit of the theory is inevitably a (generalized) free field. Of course, we must answer the question whether it is a property of the full theory, or merely an artifact of the above approximation. Recent progress of constructive field theory suggests and partially proves that it is the former for the φ^4 theory even in the broken-symmetry phase, see e.g. Ref. 18, 19). For Monte Carlo simulation to the zero charge problem, see Ref. 20).

5.2. Bare and Renormalized β-functions

The <u>renormalized β-function</u> $\beta_R = \beta_R(g_R)$ ($g_R \equiv e_R^2$) can be explicitly calculated from the above relation (5.1.1) as

$$\beta_R(g_R) \equiv \Lambda \frac{dg_R}{d\Lambda} = \frac{1}{12\pi^2} g_R^2. \tag{1}$$

On the other hand, <u>the bare β-function $\beta(g)$ as a function of the bare charge g determines how the bare charge $g = g(\Lambda)$ must approach to the critical point g_c to take the continuum limit.</u> The renormalized β-function above has a zero only at $g_R = 0$ (trivial fixed point), while the bare β-function has a zero at the critical point $g = g_c$.

Note that the bare β-function obtained does not necessarily coincide with the one obtained in perturbation theory, in contrast with the case of QCD. In QCD the critical point $g_c = 0$ is also the fixed point g_R^* and hence the perturbation theory is applicable in the neighborhood of this point, which enables us to replace the bare charge e_0 with the renormalized charge e_R.

Here it should be remarked that in the above treatment, the scattering amplitude is obtained by considering the contribution from the lowest order graph with the scattering kernel $K(p'+q, p+q, q)_{\alpha\beta,\delta\gamma} = e_0^2 \Gamma_\rho(p,p+q; -q)_{\delta\beta} D_{\rho\sigma}(q)\Gamma_\sigma(p'+q,p'; q)_{\alpha\gamma}$, which is consistent with (4.1.1). However the SD equation for the vertex function is not closed and we can consider higher order graph with dressed propagator. By this modification, it is highly nontrivial how the solutions of the coupled SD equation with modified kernel may alter the relation between the bare and renormalized charges. If e_R is related to e_0 in quite different manner for $e_0 \gg 1$, QED may have a nontrivial fixed point g_R^*:

$\beta_R(g_R^*) = 0$ at $g_R = g_R^* > 0$?

6. DISCUSSION

The triviality or nontriviality of the continuum theory can be judged, if we know sufficient set of critical exponents describing critical behavior near the transition point. Our result reported in this talk is the first step to obtain such critical exponents. It is worthwhile to remark that, even if the critical theory exactly obey the mean field theory, the continuum limit in four dimension is not necessarily trivial. In fact, in four dimensions, the scalar φ^4 theory would be nontrivial if the critical behavior exactly obeys the mean field theory. It is the log correction to the mean field theory that leads to the triviality of $(\varphi^4)_4$ theory, which is in sharp contrast with the triviality for $d > 4$. See, e.g. Ref. 18, 19).

In the case of QED, we expect similar situation, but it is not so clear in the present stage of our investigation. It is quite difficult but interesting problem to answer whether QED may become an example of nontrivial and non-asymptotically free quantum field theory in 4 dimension beyond the quenched ladder approximation.

†Address after October 1, 1988: Department of Physics, Chiba University, 1-33 Yayoi-cho, Chiba-shi, Chiba 260, Japan

REFERENCES

1. Miransky, V. A., Nuovo Cimento 90A, 149 (1985).
2. Landau, L. D., In Niels Bohr and the Development of Physics, ed. W. Pauli (Pergamon, London,1955). Landau, L. D., Abrikosov, A. and KHalatnikov, I., Nuovo Cimento, Supplement, 3, 80 (1956).
3. Landau, L. D. and Pomeranchuk, I., JETP 29, 89 (1955). Pomeranchuk, I., DAN. 103, 1005 (1955).
4. Maskawa, T. and Nakajima, H., Prog. Theor. Phys. 52, 1326 (1974); 54, 860 (1975). Fukuda, R. and Kugo, T., Nucl. Phys. B117, 250 (1976).
5. Bardeen, W. A., Leung, C. N. and Love, S. T., Phys. Rev. Lett. 56, 1230 (1986). Leung, C. N., Love, S. T. and Bardeen, W. A., Nucl. Phys. B273, 649 (1986).
6. Nambu, Y. and Jona-Lasinio, G., Phys. Rev. 122, 345 (1961).
7. Kondo, K.-I., Mino, H. and Yamawaki, K., (DPNU-88-18).
8. Kondo, K.-I., Mino, H., Suzuki, S. T., Nonoyama, T. and Yamawaki, K., in preparation.
9. Atkinson, D., J. Math. Phys. 28, 2494 (1987).
10. Kondo, K.-I. and Kikukawa, Y., Nagoya Univ. Preprint (DPNU-88-20).
11. Kondo, K.-I., Kikukawa, Y. and Mino, H., (DPNU-88-21).
12. Kondo, K.-I., Kikukawa, Y. and Mino, H., papers in preparation.
13. Nonoyama, T. and Tanabashi, M., (DPNU-88-22).
14. Dagotto, E., Nucl. Phys. B (Proc. Suppl.) 4, 607 (1988).
15. Appelquist, T. W., Bowick, M., Karabari, D. and Wijiwardhana, C. R., Phys. Rev. D33, 3704 (1986). Appelquist, T. W., Nash, D. and Wijewardhana, L. C. R.: YCTP-P2-88 (March 1988).
16. Atkinson, D., Johnson, P. W. and Stam, K., Phys. Lett. B201, 105 (1988).
17. Kogut, J. B., Dagotto, E. and Kocić, A., ILL-(TH)-88-#14, 1988.
18. Fröhlich, J., In: Progress in gauge field theory, eds. by G.'t Hooft et al. (Plenum Press, New York, 1984).
19. Kondo, K.-I., Unpublished manuscript (1986); Prog. Theor. Phys. 79, 1217 (1988); J. Stat. Phys. 46, 35 (1987).
20. Kogut, J. B., Dagotto, E. and Kocić, A., Phys. Rev. Lett. 60, 772 (1988).

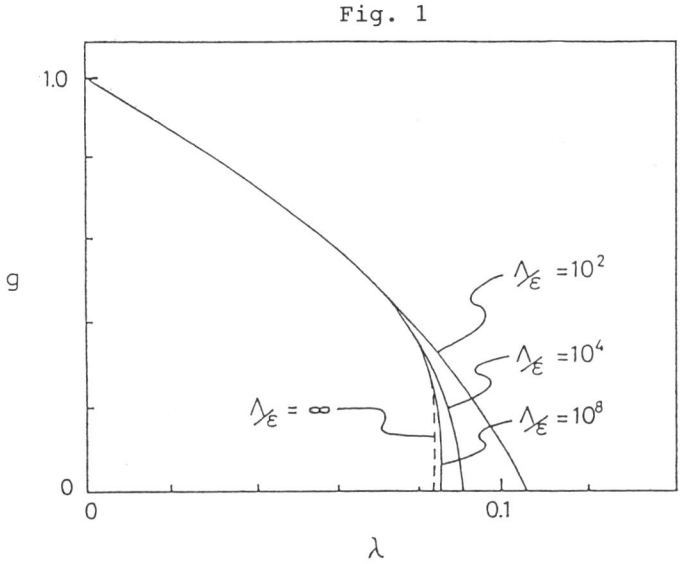

Fig. 1

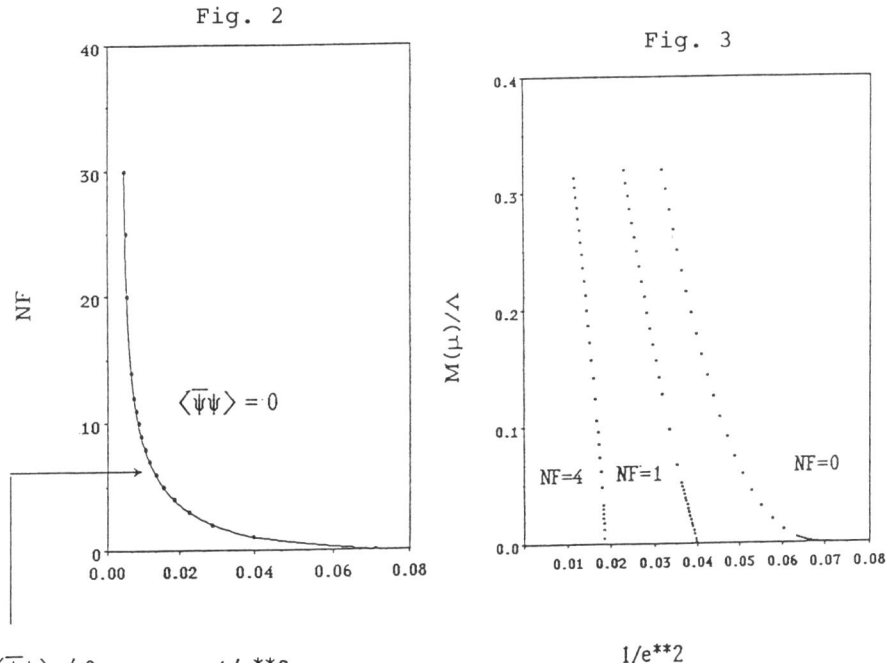

Fig. 2

Fig. 3

RENORMALIZATION GROUP ANALYSIS OF STRONG COUPLING PHASE IN QED

Ken-Ichi AOKI
Research Institute for Fundamental Physics
Kyoto University, Kyoto 606
JAPAN

ABSTRACT

A brief review of recent approaches to a possible strong coupling phase in QED is given. We stress significance of the renormalization group method for the problem. General concept of the renormalization flow and fixed point structure are introduced, and then we set up our strategy. We sketch a coupled Schwinger Dyson Equation approach. A non-perturbative renormalization group equation in QED is formulated including dimension-6 operators, which is able to give a non-trivial fixed point.

1. Modern Renormalization Theory Framework for Investigating Phase Structure of the Field Theory

We start with the modern renormalization theory framework initiated by Wilson and Kogut.[1] This is because it is the best tool and the most convenient framework to work with multi-phase structure in the field theory. The keywords of framework are listed: Renormalization group flow, fixed points, phase structure, critical surface, physical cutoff *etc.*. For detailed introduction to this subject, refer to a review by the author.[2] Here we just recapitulate the basic notions.

In our modern framework, renormalization group equation (RGE) comes first. Now we define a set of effective Lagrangians. Take a theory, which is defined by a bare Lagrangian $\mathcal{L}_0$ with momentum cut off Λ_0. The quantum theory with $\mathcal{L}_0$ is given by the effective action, $\Gamma_{\Lambda_0}(\mathcal{L}_{\Lambda_0})$. We then define an effective Lagrangian $\mathcal{L}_\Lambda$ by

$$\Gamma_\Lambda[\mathcal{L}_\Lambda] = \Gamma_{\Lambda_0}[\mathcal{L}_{\Lambda_0}] \;, \tag{1}$$

where the suffix Λ of Γ represents the cutoff for the loop momentum integration when evaluating the quantum corrections. The effective Lagrangian $\mathcal{L}_\Lambda$ gives the same (low energy) physics with lower cutoff Λ ($< \Lambda_0$) as the original system. We may understand that $\mathcal{L}_\Lambda$ is the effective action for the high frequency modes while it is the effective quantum Lagrangian for the low frequency modes. $\mathcal{L}_\Lambda$ satisfies the following differential equation by definition,

$$\delta \Gamma_\Lambda[\mathcal{L}_\Lambda] = 0 \ , \tag{2}$$

where δ is the total derivative with respect to Λ. This equation defines a "flow" of the effective Lagrangian in the space of possible interactions which is infinite dimensional. Every flow line defines a theory with a bare Lagrangian at scale Λ_0 which is nothing but an initial condition of the flow. Every points ($\mathcal{L}_\Lambda$ at Λ) on a flow gives the same physics ($\Gamma_\Lambda(\mathcal{L}_\Lambda)$). Flow is determined locally by Eq.(2) and the coefficients of the differential equation, *i.e.*, β functions, are well defined without suffering the divergences. This equation is nothing but our RGE.

The main and important feature of the flow is "Strong convergence to an infrared fixed point sub-manifold in the space of possible interactions". This means that almost any change of the initial condition of the flow does not essentially affect the final (sufficiently low energy) behavior of the flow.

It is easy to see the convergence in the perturbation theory. Take a Lagrangian at scale Λ_0 with two generic interactions,

$$\mathcal{L}_{\Lambda_0} = g \bar{\psi} \psi \phi + \frac{G}{\Lambda_0^2} \bar{\psi} \psi \bar{\psi} \psi + \text{(kinetic terms)} \ , \tag{3}$$

that is, g represents the so-called renormalizable interactions and G represents higher dimensional operators. The four fermion interaction G in the quantum Lagrangian does not affect terms whose operator dimensions are greater than four in the effective action, since such terms must contain negative powers of Λ_0 in their coefficients and must be suppressed by μ/Λ_0 where μ is a low energy scale. On the other hand, for operators whose dimensions are less than or equal to four, the G interactions do affect the effective action. Now the point becomes clear. Infrared effective action has just one degree of freedom. In this case we can take the renormalized Yukawa interaction g_R (which is nothing but g_μ) as a parameter which defines a renormalized theory. In this renormalization procedure, the effect of G to the low energy Yukawa interaction is irrelevant, since anyway there is a counter term δg with which we will fix the renormalized g_μ to be a finite and experimentally consistent value. In other words, G effects are absorbed into δg. With this g_μ, the rest of the effective action (including the four fermion term) is uniquely determined, *i.e.*, the dependence of the (rest of) effective action on G will vanish in the limit of large Λ (or equivalently in the limit of the infrared observables). Note that this statement is exactly equivalent to the usual renormalizability. We now understand that the initial two dimensional space (parametrized by g and G) converges into one dimensional subspace (parametrized by g_μ), which is called 'relevant operator'. The above argument is based on the fact that the naive dimensional analysis of the Feynman integrals does hold. Thus using the usual renormalizability proof, we can prove the strong convergence structure of the flow of effective Lagrangians (within the perturbation theory). Furthermore, starting with this RGE, the renormalizability itself is proved strictly and more easily than the usual graphical proof.[3]

Thus we may imagine flow lines in the possible interaction space which are going together to form a strongly converged lower dimensional submanifold. Let us see the convergence of flow in Fig. 1. Take point A as an initial point. Then at scale μ, the effective Lagrangian arrives at point B. Add G interaction to the initial Lagrangian and the starting point is now C. Then the final point moves to point D. Note that the final points B and D are separated even in g direction. This is the inevitable effect to the low energy Yukawa interaction by the four fermion interactions. When we compare these two theories (two flows), we take the same renormalized g (g_μ), that is, we compare point B and E instead of point D. The difference between these two points B and E decreases in the infrared limit of the flow. Thus the initial G interaction becomes irrelevant in the infrared physics.

On the other hand, the usual renormalization procedure goes as follows. The bare coupling constant g_Λ is to be adjusted when increasing Λ, while G_Λ is ignored from the beginning. This ignorance doesn't matter due to the above convergence. Take a renormalization condition, $g_\mu = g_B$. Increasing Λ, we adjust g_Λ from point A to point F, so as to get a fixed g_μ. Then the end point B is moved to E, and is finally converged to H on the renormalized trajectory in the infinite cutoff limit.

Now the second notion of the framework comes in: Fixed points of flow. Figure 2 shows a typical structure of the flow. The first fixed point is the infrared (IR) fixed point A. Every flow goes into the fixed point A in the infrared limit. Then one may expect that there is a region in the total space in which all points are attracted to point A. The boundary of the attractive region is the shaded surface in Fig. 2, i.e., x-z plane. The points on the boundary move only on the boundary by definition. Then for these on-boundary points there is the second IR fixed point on the boundary, which we plot as point B. This point B is IR unstable only in the

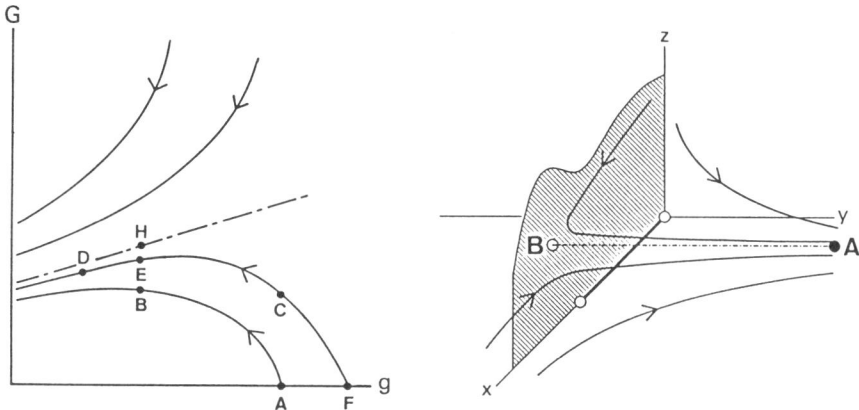

Fig. 1 Renormalization Flow   Fig. 2 Fixed Points

direction normal to the surface and is IR stable in any other directions.

Now consider a renormalized theory characterized by these fixed points A and B. In the limit of infinite cutoff, the effective Lagrangian have to spend an infinitely 'long time' until it arrives at a point at scale μ. The only way to spend infinitely long time is to pass very near to a fixed point. Consider a initial point which is close to the surface. It starts to move almost parallel to the surface, since attraction to point A is weak in this region of the space. The flow goes towards point B, takes a long rest near B, and then moves to point A. Hence infinite cutoff limit with a fixed finite renormalized interaction enforces that the corresponding flow is almost on the 'renormalized trajectory' (dot-dashed line in Fig. 2). For this renormalized theory, point B is the UV fixed point and point A is the IR fixed point. The set of fixed points, A and B, gives a renormalized theory with one relevant operator. The number of relevant operators is equal to the number of IR unstable directions of the corresponding UV fixed point. We can proceed to classify fixed points and corresponding field theories step by step, and obtain n-relevant operator renormalized theory. Then a hierarchical structure of fixed points will show up.

Here we introduce a completely different notion, phase and phase boundary. If a system under consideration brings about a second order phase transition, there is a phase boundary (critical hyper-surface). Near the critical surface, the correlation length (inverse of the relevant mass parameter) diverges when approaching to the surface. We will see that the critical surface of the second order phase transition becomes a boundary of the renormalization flow discussed in the above. Remember that flow diagrams are drawn on the possible interaction space whose axes represent dimensionless coupling constants scaled by the corresponding cutoff Λ (lattice spacing a^{-1} in case of lattice theories). The inverse correlation length ℓ^{-1} is given by

$$\ell^{-1} = f(C_i)\Lambda \ , \tag{4}$$

where C_i are dimensionless coupling constants normalized by Λ. The above equation is trivial, since there is only one dimensionful parameter Λ in the theory, and the theory is well defined (free of divergence) so that the naive dimensional analysis holds. Then, the contour for ℓ on C_i–space does not depend on Λ. When the cutoff Λ is decreased, the C_i's must move so as to satisfy

$$\delta \ell^{-1} = \delta \{f(C_i(\Lambda))\Lambda\} = 0 \ , \tag{5}$$

since all points on a flow line must be equivalent for infrared physical observables. Hence if the cutoff becomes a half, the function $f(C_i)$ have to be doubled, which means that the flow go away from the critical surface making a step to a point on the doubled mass (half correlation length) contour. This means that the phase transition critical surface plays a role of a boundary surface for the renormalization flow.

Table 1   Flow of Basic Notions

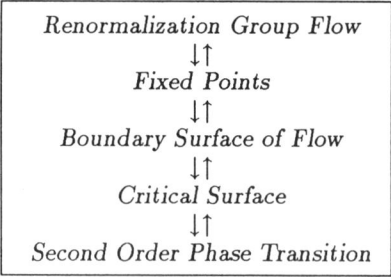

The basic notions introduced so far are tabulated in Table 1. We have two directions of studying a new phase.

1. Upward Approach: Starting with a nonperturbative analysis giving a phase structure, then investigate the renormalization flow on that phase diagram.
2. Downward Approach: Starting with a nonperturbative renormalization flow, then find the phase structure of the system.

In any case, the renormalization flow is essentially important, because it determines the basic characteristics of the new phase; the number of relevant operators,

components of the relevant operators, anomalous dimensions near the UV fixed point
etc.. In the previous literature, using Schwinger-Dyson equation or numerical simulations on the lattice, most efforts are devoted to find out a new phase, but not for RGE properties.

2. Analyses with a Coupled Schwinger-Dyson Equation
— Upward Approach —

Here we mention our recent results with a coupled Schwinger Dyson equation.[2,4] (We do not describe the basic formulation and results.[5-11] Also refer to many talks in this proceedings on the Schwinger Dyson Equation approach.) We work with a coupled system:

$$\Sigma = C_1 \;\;\raisebox{-2pt}{\rule{0pt}{0pt}}\!\!\!\!\! + C_2 \;\;\raisebox{-2pt}{\rule{0pt}{0pt}}\!\!\!\!\! (M^2), \qquad (6)$$

where the second term is due to a chiral invariant four fermion interactions with a form factor, for which we take a usual massive exchange form $1/(M^2 + q^2)$. By varying M, we get a total picture of the system.

Why did we add the four fermion interactions to the usual ladder Schwinger Dyson Equation ? There may be three (not exclusive) answers.

1. To trigger strong phase fixed point dynamics: Note that we do *not* have to include strong operators at near fixed point, unless they are really relevant. However, at any rate, we are working with some crude approximation (ladder in this case), and adding other operators may help the essential lack due to the approximation.

2. Since 4-fermion interaction becomes relevant: This is the original reason when Bardeen et. al.[8] first investigated the coupled system. However note that *relevant or irrelevant* needs independent argument to confirm it.

3. Since physical particles mediate it: This is a trivial answer. In usual type of Extended Technicolor theories, there are such particles indeed. In this case, we may calculate more physical parameters like F_π, m_{quark}, *etc.*. Note that this standing point is completely different from the above two.

We calculated solutions $\Sigma(p)$ of the integral equation by modifying it into second rank differential equation. Fig.3 shows the critical surface on the 3-dimensional space of C_1-C_2-M. At $M = 0$, our system is just a sum of two independent gauge interactions, and the criticality lies on a straight line of $C_1 + C_2 = 1$. Switching on M, first we look at a sudden change of criticality for negative C_2, which is seen in Fig. 4. That is, for negative C_2, the *decoupling* does hold completely: C_2 interaction does not affect the criticality, $C_1 = 1$.

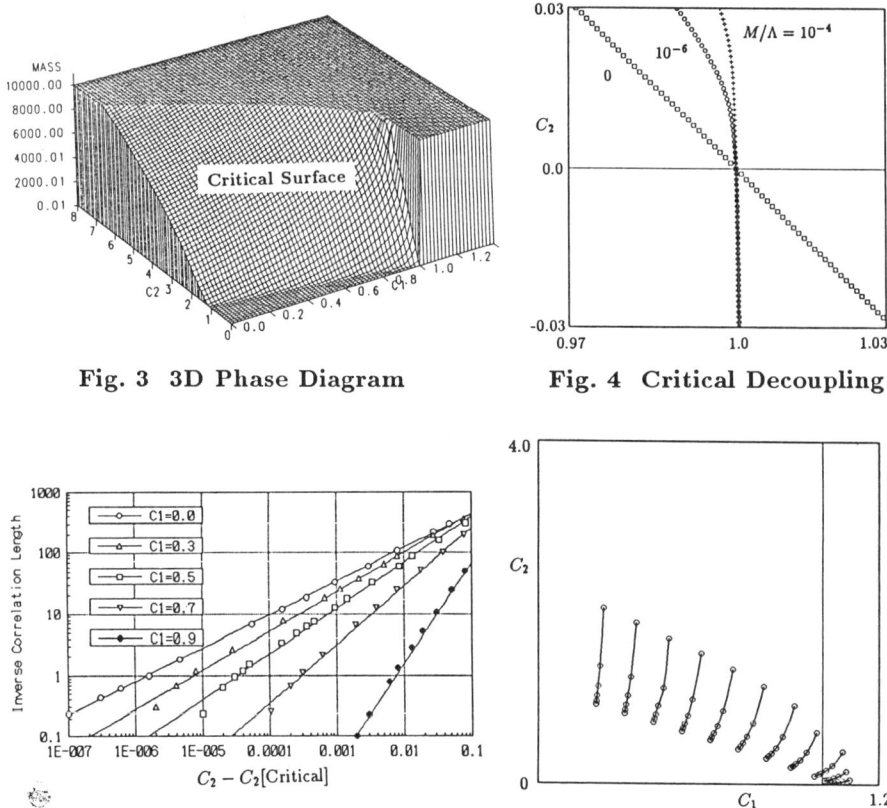

Fig. 3 3D Phase Diagram

Fig. 4 Critical Decoupling

Fig. 5 Critical Exponent: $\Sigma(0)$

Fig. 6 RGE Flow: m_q Fixed

Critical exponent for $\Sigma(0)$ varies smoothly along the critical surface. For example along the line for the period of $C_1 = [0,1]$, we plot in Fig. 5 the log–log plot of $\Sigma(0)$ vs. $C_2 - C_2[\text{critical}]$ (M=Λ). Also the critical form of $\Sigma(p)$ varies along the criticality. Using $\Sigma(0)$ and other physical quantities (F_π, m_{quark}) depending on high momentum part of Σ function, we may argue renormalization flows on this three parameter plane. For example, as fixed infrared physical parameters, we take M, $\Sigma(0)$ and m_{quark} (generic mass integral of $\Sigma(p)$), and we can define flows of lowering Λ (see Fig. 6). Note that only with a local four fermion interactions, infrared structure of $\Sigma(p)$ does *not* depend on the four fermion coupling constant (this is because local four fermion interactions cannot modify the differential equation structure and appear only in the ultra-violet boundary condition), and we cannot naturally define flows of traversing the gauge coupling. On the other hand, in Fig. 6, we get such a traversing flow. These results show, however, that our system has three relevant parameters (as naively expected). Hence only with these results, we may not argue about the fixed point structure. We need more investigation so that we might extract information about fixed points from Schwinger Dyson solutions, if any. (Our analysis on the Schwinger Dyson Equation with the gauge parameter α gives an interesting result in relation to the above point.[12])

3. A Non-Perturbative Renormalization Group Equation and Non-Trivial Fixed Point Solutions
   — Downward Approach —

In this section we briefly look at how to set up a non-perturbative Renormalization group of QED appropriate for investigating its strong coupling region.[13] Here we stress rather conceptual aspects and the basic framework.

We start with the RGE defined in Section 1,

$$\delta \Gamma_\Lambda[\mathcal{L}_\Lambda] = 0 \ . \tag{7}$$

We use the cutoff method *a la* Polchinsky,[3] that is, every propagators are multiplied by a factor K,

$$\frac{1}{m^2 + p^2} K(p^2/\Lambda^2) \ , \tag{8}$$

where we take the theta function for K: $K(x) = \theta(1-x)$. Actually, the theta function cutoff causes a spurious singularity in the RGE, which we do not care here.

Theory is defined through the generating functional of Green functions,

$$Z[J, \mathcal{L}_\Lambda, \Lambda] = \int \mathcal{D}\phi \exp\left\{-\mathcal{L}_\text{kinetic}(\phi, \Lambda) - \mathcal{L}_\text{int}(\phi, \Lambda) - J(p)\phi(-p)\right\} \ , \tag{9}$$

$$\mathcal{L}_\text{kinetic}(\phi, \Lambda) = \int \frac{d^4p}{(2\pi)^4} \left\{\frac{1}{2}\phi(p)\phi(-p)(p^2 + m^2)K^{-1}(p^2/\Lambda^2)\right\} \ . \tag{10}$$

The renormalization group equation(7) gives the derivative of $\mathcal{L}_\text{int}(\phi, \Lambda)$ with respect to Λ:

$$\Lambda\frac{\partial \mathcal{L}}{\partial \Lambda} \equiv -\frac{1}{2}\int d^4p \frac{(2\pi)^4}{p^2+m^2} \Lambda\frac{\partial K(p^2/\Lambda^2)}{\partial \Lambda} \left\{\frac{\partial \mathcal{L}}{\partial \phi(p)} \cdot \frac{\partial \mathcal{L}}{\partial \phi(-p)} + \frac{\partial^2 \mathcal{L}}{\partial \phi(p)\partial \phi(-p)}\right\} \ . \tag{11}$$

Terms in the above equation have very intuitive meanings. When lowering the cutoff Λ, propagators in each diagram are modified, *i.e.*, a *part* of virtual particles does no longer propagate. The part is just a particle with momentum of as much as Λ. Hence we have to compensate this loss due to non-propagation of particles with adding appropriate terms to the interaction Lagrangian so that the resulting effective action is not changed. Thus, the first term is a compensating interaction term for a propagator connecting two interaction vertices originally, and the second is that for self-looping propagator:

$$\frac{\partial \mathcal{L}}{\partial \phi(p)} \cdot \frac{\partial \mathcal{L}}{\partial \phi(-p)} : \qquad\qquad \frac{\partial^2 \mathcal{L}}{\partial \phi(p) \partial \phi(-p)} :$$

In diagrams above, the thick lines represent a propagator on which momentum Λ flows. These diagrams are nothing but the graphical representation for calculating the coefficients of our renormalization group equation. Note that there is effectively no loop integration at all in this evaluation, since the momentum of the thick lines are on Λ-shell.

To set up a renormalization group equation for QED, we first pick up some definite number of operators. (Certainly we cannot work with the infinite-dimensional space of all possible interactions.) Then, within the set of operators, we may write down the RGE explicitly, search for fixed points of it, and classify operators around each fixed point. In this way, our approach is completely different from the usual RGE with *perturbative* β functions. We are free of perturbation with respect to the *loop number*, since there is absolutely no loop expansion in our framework. Instead, we are limited by the fact that *we have to pick up only a tiny part of operators* in the infinite dimensional possible interaction space. Hence, if the operator set is appropriate for some fixed point we are looking for, *i.e.*, if our subspace is very near to the real place of the fixed point, then our method might work, and work well *non-perturbatively*.

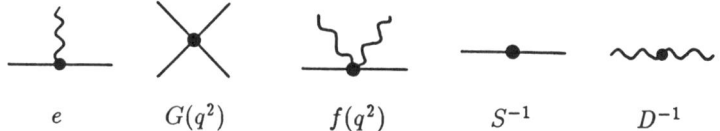

$$e \qquad G(q^2) \qquad f(q^2) \qquad S^{-1} \qquad D^{-1}$$

Fig. 7 Our Set of Operators

Here we take the following set of operators: The original gauge coupling, 4-fermion interactions, 2-fermion 2-photon vertices, and self energies of course (Fig. 7). We may see, diagrammatically, which part of the total effective action is included through an RGE with a specific set of operators: A diagram (by which we mean that also the inner momentum configuration is specified) is included, if and only if

1. by contracting one propagator at each step which has the highest momentum in the diagram,
2. generating a diagram which can be constructed only with the specific set of operators,
3. finally reduce it to a vertex (without loop) which is also a member of the set of operators.

When the above condition is satisfied for all possible sequences of contraction (for all possible loop momentum configuration), then the whole diagram is contained in the RGE. When some paths satisfy the conditions but others do not, then the diagram is partially contained: The part with appropriate loop momentum distribution corresponding to the *right* path is contained.

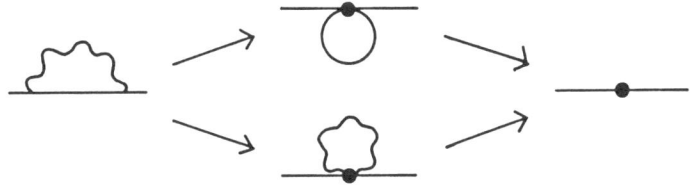

Fig. 8 Fermion Self Energy (one-loop)

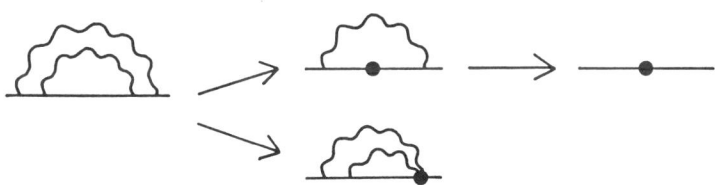

Fig. 9 Fermion Self Energy (planar)

Let us see some diagrams whether they are contained in our RGE or not. As for the one-loop fermion self energy, it is completely contained (see Fig. 8). As for the planar fermion self-energy (Fig. 9), it is partially contained. The *right* path is, so to speak, an *ordered* contraction. For example, the part in which the loop momenta are ordered from the innermost to the outermost is included. Note that this type of diagram gives the leading part in the usual *perturbative* RGE calculation. On the other hand, for a non-planar diagram, even 2-loop diagram is not contained at all. For the photon self energy, fermion one-loop diagram is included.

Some comments follow. We will evaluate RGE coefficients using the massless fermion propagator. Hence our strategy of investigating the phase structure is the following: We start with a chiral invariant Lagrangian $\mathcal{L}_0$, and by integrating out high frequency modes, we get the chiral invariant effective Lagrangian $\mathcal{L}_\Lambda$; After obtaining $\mathcal{L}_\Lambda$, we may calculate the effective potential for Σ and it might possess a chiral non-invariant vacuum solution. This means that, for instance, the corresponding Schwinger-Dyson kernel for the effective potential $V(\Sigma)$ obtained as above does not contain high frequency components $(\Sigma(p); p > \Lambda)$. Naively this negligence of the high frequency part does not modify the basic structure of the spontaneous symmetry breaking. If we work with a source term from the beginning, *i.e.*, calculate RGE coefficients using $\Sigma(p, J(x,y))$ propagator, then we may get full effective potential $V(\Sigma)$ from $\mathcal{L}_\Lambda$ thus obtained.

As above we add higher dimensional operators in addition to the original gauge interaction. The role of these operators is, first of all, to trigger a possible strong phase dynamics. Note that naively at near the strong coupling phase UV fixed point (if any), the gauge interaction is strong and other interactions including higher dimensional operators are also to be strong. Thus unless the fixed point sits on very near to the gauge coupling axis (this is very unlikely anyway), we have to include other strong operators in order to investigate the near fixed point structure with our type of RGE. These higher dimensional operators may be compared with source terms to trigger the spontaneous symmetry breakdown. Inclusion of some operators is *by no means* related to that they are *relevant* operators. We even do not expect that. Such properties, *relevant* or *irrelevant*, are determined after we expand the theory around a fixed point.

We lower the cutoff as

$$\Lambda_0 \to \Lambda \equiv \lambda \Lambda_0 \equiv \exp(-t)\Lambda_0 . \qquad (12)$$

All dimensionful quantities are to be scaled by Λ at that time. For example, the momentum itself is measured by a dimensionless parameter p/Λ which takes a value between 0 and 1. We perform the wave function renormalization at each step of lowering the cutoff. Wave function renormalization is nothing but a rescaling of fields, *i.e.*, just a transformation of variables, and certainly does not affect physical observables. Thus, with this un-physical degrees of freedom of field rescaling, we may define an equivalence class of theories in the possible interaction space. We take an ordinary condition to pick up a representative of this equivalence class: The kinetic terms are normalized to give the free propagators on the mass-shell.

Let us first consider the basic scaling of operators. Take a term in the effective Lagrangian $\mathcal{L}_\Lambda$,

$$G_\Lambda(p_i)\,(\psi_\Lambda)^{n_\psi}\left(A_\Lambda^\mu\right)^{n_A} p^{n_p}\,. \tag{13}$$

Then a change of cutoff modifies the coefficient function G as,

$$G_{\lambda\Lambda}(p_i) = G_\Lambda(\lambda p_i)\zeta_\psi^{n_\psi}(\Lambda)\zeta_A^{n_A}(\Lambda)\lambda^{n_p}\lambda^{(n_\psi+n_A-1)d}\,, \tag{14}$$

where d is the space-time dimension and ζ's are the stepwise wave function renormalization constants defined by

$$\zeta_\psi(\Lambda)\psi_{\lambda\Lambda} = \psi_\Lambda\,,\quad \zeta_A(\Lambda)A_{\lambda\Lambda}^\mu = A_\Lambda^\mu\,. \tag{15}$$

In case of the free theory, these constants take

$$\zeta_\psi(\Lambda) = \lambda^{-\frac{d+1}{2}},\quad \zeta_A(\Lambda) = \lambda^{-\frac{d+2}{2}}\,. \tag{16}$$

Of course we add contributions by interactions to these free formulas (Eq.(15) and (16)). For our operator set, we tabulate in Fig. 10 the interactions contributing the RGE coefficients.

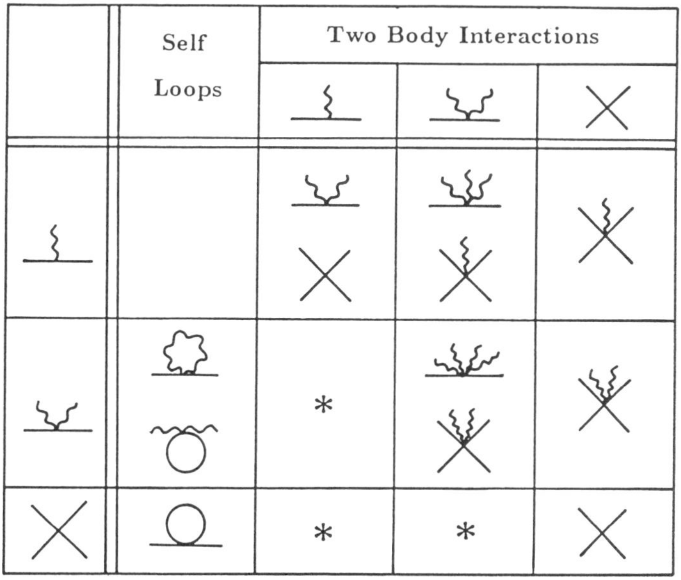

Fig. 10 Interactions Contributing to RGE Coefficients.

We first calculate self-energy corrections. Our operator set in Fig. 7 looks just 5 operators. This is not true, since we take account of the momentum dependence of the amplitudes (form factor), and it means we are actually working with infinite dimensional space of operators. Although our RGE formulation in the infinitesimal form is not involved in the loop integration itself, we need to integrate *operators*, which gives a set of integral-differential equations.

$$dS^{-1} = \{2d\zeta_\psi - (d+1)dt\}S^{-1} + + , \tag{17}$$

$$dD^{-1} = \{2d\zeta_A - (d+2)dt\}D^{-1} + , \tag{18}$$

$$dG(x) = (4d\zeta_\psi - 3d \cdot dt)G(x) - 2xG'(x)dt + , \tag{19}$$

$$df(x) = \{2d\zeta_\psi + 2d\zeta_A - (3d+1)dt\}f(x) - 2xf'(x)dt + , \tag{20}$$

$$de = (2d\zeta_\psi + d\zeta_A - 2d \cdot dt)e , \tag{21}$$

where x is the rescaled momentum variable: $x = p^2/\Lambda$, and $'$ represents derivative with respect to x.

Interaction part for the self energies do include *operator integration*. Here we omit form factors for these S and D amplitudes and take $x \to 0$ limit ($\sim x$-independent part for the infrared amplitudes) of the right hand side. This reduces the operator integration to local terms around $x = 1$. That is, loop diagrams are expressed by a linear sum of $G(1)$, $G'(1)$, $f(1)$, $f'(1)$. On the other hand, interaction part for G, f are evaluated as $2e^2\delta(x-1)dt$. The wave function renormalization constants ζ_ψ, ζ_A are determined as usual by $dS^{-1} = dD^{-1} = 0$.

Then our RGE becomes a set of partial differential equations for $G(x)$, $f(x)$ with respect to x and t, containing local source terms ($\delta(x-1)$), self local source terms ($G(1)$, $G'(1)$, $f(1)$, $f'(1)$), and with a *coupling constant* (e). Thus our system is something like a 2-dimensional field theory with two fields ($G(x,t)$, $f(x,t)$):

$$\frac{de}{dt} = \left\{-(C_1 G'(1) + C_2 f(1) + C_3 f'(1)) - \frac{1}{2}C_4 f'(1)\right\} e , \tag{22}$$

$$\frac{dG(x)}{dt} = -2G(x) + 2e^2\delta(x-1) - 2xG'(x)$$
$$- 2\left\{C_1 G'(1) + C_2 f(1) + C_3 f'(1)\right\} G(x) , \tag{23}$$

$$\frac{df(x)}{dt} = -2f(x) + 2e^2\delta(x-1) - 2xf'(x)$$
$$- \left\{C_1 G'(1) + C_2 f(1) + C_3 f'(1) + C_4 f'(1)\right\} f(x) , \tag{24}$$

where C_i's are numerical constants. Let us see some limiting solutions of the above system. First, omitting interactions for $G(x,t)(f(x,t))$, we have

$$\dot{G}(x,t) = -2G(x,t) - 2xG'(x) , \qquad (25)$$

and it has a set of solutions:

$$G(x,t) = \exp\{-2(n+1)t\}x^n . \qquad (26)$$

At the infinite *time* limit ($t \to \infty$), only the $n = -1$ solution gives a fixed point solution $1/x$, while otherwise $G(x,\infty)$ will diverges or vanishes. The fixed point solution is a scaling solution corresponding to a scale invariant vertex:

$$\frac{1}{x} \cdot \frac{1}{\Lambda^2}\bar{\psi}\psi\bar{\psi}\psi = \frac{1}{q^2}\bar{\psi}\psi\bar{\psi}\psi . \qquad (27)$$

Second, omitting x-dependence and f-interactions, we may set an equation:

$$\dot{G}(t) = -2G(t) + 2e^2 , \qquad (28)$$

which has a solution:

$$G(t) = (G(0) - e^2)\exp(-2t) + e^2 . \qquad (29)$$

This solution indicates that G operator is irrelevant because infinite t limit will wipe out initial value (G(0)) dependence and $G(\infty)$ is uniquely determined by $e(\infty)$.

Obviously there are some singularities in our system due to the δ-function source terms. We can handle these singularities to give more regular expression. Then we look for fixed point solution, *i.e.*, t-invariant solutions. Fixed point solutions [$G_c(x)$, $f_c(x)$, e_c] satisfy a first rank differential equation with respect to x. After imposing boundary conditions at $x = 0$, we get a set of fixed point solutions parametrized by a discrete number, which correspond to various values of e_c. We make eigen-operator analysis around these fixed point solutions, and for some fixed points, the local four-fermion operator (x-independent G mode) does become a relevant operator.[13]

In this talk, we emphasized the role of non-perturbative renormalization group analysis to investigate our new strong phase. The RGE analysis (*Downward approach*) when coupled with other non-perturbative methods (*Upward approach*), may give a physical picture of the strong phase.[2] In this way of thought, the following subjects are highly desired to perform.

 i) RGE approach with more operators: In our version of the operator set introduced here, we cannot include the usual vertex correction diagram.
 ii) RGE with Σ-source terms: Inclusion of source terms may help us to compare

RGE results with those from the effective potential or from the Schwinger Dyson Equation.

iii) Get $V(\Sigma)$ from $\mathcal{L}_\Lambda$: Inclusion of higher dimensional operators may modify the phase structure.

The author would like to express his sincere thanks for valuable discussions and comments by T. Maskawa, C. DeTar, B. Holdom, T. Kugo, M. Bando, K. Hasebe, H. Nakatani and M. Imachi.

References

1. K.G. Wilson, Rev. Mod. Phys. 47 (1975) 773; K.G. Wilson and J. Kogut, Phys. Rep. 12C (1974) 75.
2. K-I. Aoki, preprint, RIFP–758, to appear in the Proceedings "Physics at TeV Scale" published by KEK.
3. J. Polchinski, Nucl. Phys. B231 (1984) 269; B.J. Warr, preprint CALT-68-1334, 1393.
4. K-I. Aoki, in preparation; K-I. Aoki, M. Bando, T. Kugo, K. Hasebe and H. Nakatani, in preparation.
5. T. Maskawa and H. Nakajima, Prog. Theor. Phys. 52 (1974) 1326; 54 (1975) 860.
6. R. Fukuda and T. Kugo, Nucl. Phys. B117 (1976) 250.
7. P.I. Fomin, V.P. Gusynin, V.A. Miransky and Yu.A. Sitenko, Rivista Del Nuovo Cimento 6 (1983) 1; V.A. Miransky, Phys. Lett. 91B (1980) 421; P.I. Fomin, V.P. Gusynin and V.A. Miransky, Phys. Lett. 78B (1978) 136.
8. W.A. Bardeen, C.N. Leung and S.T. Love, Phys. Rev. Lett. 56 (1986) 1230; C.N. Leung, S.T. Love and W.A. Bardeen, Nucl. Phys. B273 (1986) 649.
9. K. Yamawaki, Preprint DPNU-88-34.
10. T. Appelquist, M. Soldate, T. Takeuchi and L.C.R. Wijewardhana, Preprint YCTP-P19-88.
11. K-I. Kondo, H. Mino and K. Yamawaki, preprint DPNU-88-18, DPNU-88-18-Rev.
12. K-I. Aoki, M. Bando, T. Kugo, K. Hasebe and H. Nakatani, preprint, RIFP–769.
13. K-I. Aoki, preprint, RIFP–779.

A New Expansion for Quantum Field Theories[*]

Moshe Moshe

Physics Department
Technion - Israel Inst. of Technology , Haifa 32000 , Israel

ABSTRACT

This is a summary of several recent results in developing a novel expansion for quantum field theory. The expansion reveals, already at low orders, the nonperturbative features of the theory at strong coupling. For scalar theories this method consists of an expansion in the parameter δ that appears in the self interaction of the scalar field $\lambda \phi^{2(1+\delta)}$; it interpolates between the free field theory at $\delta = 0$ and $\lambda \phi^4$ theory at $\delta = 1$. Encouraging results were obtained in d=0,1,2,3 and four dimensions for various theories and several physical problems were studied. For an action S of fermionic or gauge theory the interpolating action is linear, of the form $\delta S + (1-\delta)S_0$. Here the nonperturbative features of the expansion in δ are due to an optimization procedure which chooses the best S_0 action.

---

[*] *Work supported in part by the Israel-U.S Bi-National Science Foundation, by the fund for the promotion of research at the Technion and the Lawrence Deutsch Research Fund.*

1. Introduction

This talk will summarize the work on a new computational scheme for quantum field theories that has been recently proposed[1] [2] and which presents a novel analytic technique for studying nonperturbative features. It is an expansion in an artificially introduced parameter, called generically "δ", which interpolates between a solvable model at $\delta=0$ and the desired theory at $\delta=1$. In some sense, the expansion parameter is of the type of the artificially introduced $\epsilon = d - 4$ parameter in the ϵ expansion rather than of the type of an expansion parameter that appears naturally in the lagrangian like in the customary weak or strong coupling expansion. The technique has been first introduced for self interacting scalar field theories. Its extension to fermions and gauge fields[3] differs in details but is similar in spirit.

Conventional, commonly used expansions have numerous disadvantages and limitations. Most expansions in quantum field theory are divergent and some even do not give an asymptotic expansion; for many, nonperturbative results are inaccessible. Semiclassical, loop expansions and large N expansions are very difficult to compute beyond leading order. Strong coupling expansions often need a great number of terms in the expansion, being very slow to reach a meaningful result.

The studies done by now on the δ expansion show that this technique has several advantages:

(a) There is good evidence that the expansion has a finite radius of convergence.

(b) The coefficients in the expansion depend in a nontrivial way on the coupling constant, revealing nonperturbative features of the theory.

(c) Though the procedure for computing the δ expansion gives divergent coefficients in the limit where the ultraviolet cutoff $\Lambda \to \infty$, one finds that the divergence is weaker here than in a typical weak coupling expansion.

(d) One finds considerable improvements over strong coupling expansion in several cases, where the δ expansion can be compared to the former, as can be seen in Z(2) and U(1) gauge theories.

(e) When combined with an optimization procedure, the δ expansion approaches the exact solution (Gross Neveu model at $N \to \infty$) or Monte Carlo data (U(1) and Z(2) gauge theories) already at low order in δ.

2. δ Expansion for Scalar Theories

The δ expansion will be first demonstrated for $\lambda\phi^4$ in d space-time dimensions. The procedure consists of several steps. First consider an interpolating lagrangian

$$\mathcal{L} = \tfrac{1}{2}(\partial_\nu \phi)^2 + \tfrac{1}{2}\mu^2\phi^2 + \lambda M^2 \phi^2 (M^{2-d}\phi^2)^\delta \tag{1}$$

where μ is the bare mass, M is a scale parameter so that the bare coupling λ is dimensionless in any dimension d. In Eq. (1) $\mathcal{L}$ interpolates between a $\lambda\phi^4$ theory at $\delta=1$ and a free field theory with a mass squared $\mu^2 + 2\lambda M^2$ at $\delta=0$ about which we expand. The aim is to produce a set of computational rules that will enable one to expand the Green's function of the theory in δ

$$G^{(n)}(x_1, x_2, ..., x_n) = \sum_{k=0}^{\infty} \delta^k g_k^{(n)}(x_1, x_2, ..., x_n) \tag{2}$$

A straightforward expansion of the lagrangian $\mathcal{L}$ will give an interaction consisting of powers of logarithms of the field ϕ, which is certainly far from trivial to use in a conventional perturbation theory. Namely;

$$\mathcal{L} = \tfrac{1}{2}(\partial_\nu \phi)^2 + \tfrac{1}{2}(\mu^2 + 2\lambda M^2)\phi^2 + \lambda M^2 \phi^2 \sum_{k=1}^{\infty} \tfrac{\delta^k}{k!}[\ln(\phi^2 M^{2-d})]^k \quad . \tag{3}$$

Our solution[1] [2] to this is as follows: The procedure to compute the Green's functions in Eq. (2) to order K consists of two steps. In step one a conventional perturbation expansion is computed for a Green's functions $\tilde{G}^{(n)}$ of a theory given by a lagrangian $\tilde{\mathcal{L}}$. In the second step a differential operator D_K is applied to the Green's functions $\tilde{G}^{(n)}$. This gives $G^{(n)}$ of Eq. (2). Thus, other than to performing a conventional perturbative calculation one has also to compute $\tilde{\mathcal{L}}$ and the differential operator D_K. They are given by:

$$\tilde{\mathcal{L}} = \tfrac{1}{2}(\partial_\nu \phi)^2 + \tfrac{1}{2}(\mu^2 + 2\lambda M^2)\phi^2 + \lambda M^d \sum_{k=1}^{K} [\phi^2 M^{(2-d)}]^{\alpha_k + 1} P_k \quad . \quad (4)$$

where P_k are polynomials in $\alpha_1, \alpha_2, ..., \alpha_K$ (in Ref. 1 we calculated these polynomials to order K=4). We initially regard $\alpha_1, \alpha_2, ..., \alpha_K$ as integers and thus, it is straightforward to calculate the Green's function $\tilde{G}^{(n)}$ from Eq. (4). The operator D_K is given by[4] :

$$D_K = \frac{1}{K} \sum_{j=1}^{K} \sum_{k=1}^{K} \frac{exp[2\pi i j(1-k)/K]}{j!} [\frac{\partial}{\partial \alpha_k}]^j \quad . \quad (5)$$

Finally, $G^{(n)}(x_1, x_2..., x_n)$ of Eq. (2) is given by $G^{(n)}(x_1, x_2..., x_n) = D_K \tilde{G}^{(n)}(x_1, x_2..., x_n)$ evaluated at $\alpha_1 = \alpha_2 = ... = \alpha_K = 0$.

In ref. 2 we used this procedure for scalar theories in dimensions d=0,1 and 4. When comparison is possible with exact results we find that this expansion converges in a finite radius of convergence and it is very useful in rapidly revealing the true nonperturbative features of the theory. For example, the trivial zero dimensional case (simply, a one dimensional integral) shows an accumulation of singularities in δ at $\delta_m = -(2m+3)(2m+2)^{-1}$, where m=0,1,2...which determines the radius of convergence of the expansion. In the case of a linear interpolating lagrangian $\tilde{\mathcal{L}}$ in ref. 3, maximizing the radius of convergence of the δ expansion has been used to optimize the expansion in the solvable case of the large N Gross-Neveu model. Quantum mechanical calculations (d = 1 case) were done and the results compared to the conventional perturbation expansion. In four dimensions the δ expansion has been compared with the results of the solvable large N $\lambda\phi^4$

theory. More recent developments on δ expansion in scalar field theory δ were recently published: A set of diagrammatic rules which avoid the direct use of the lagrangian $\tilde{\mathcal{L}}$ can be formulated[5]. Techniques for evaluating the integrals that appear in $O(\delta^2)$ and their necessary regularization in dimensions $d \geq 2$ were proposed[6]. This has been used[7] in the discussion of the triviality of $\lambda\phi^4_4$ where the preliminary results suggest that the theory is free at $d \geq 4$. The δ expansion has been applied to supersymmetric quantum mechanics and two dimensional supersymmetric quantum field theory[8]. The expected zero ground state energy in global supersymmetric theories is clearly reproduced in the δ expansion. Improved accuracy of the δ expansion can be obtained by exploiting the freedom of choosing the unphysical parameters in the interpolating lagrangian $\tilde{\mathcal{L}}$ as has been shown in ref. 9 and will be explained below in the case of fermionic and gauge theories.

3. δ expansion for Fermionic and Gauge Theories

For fermionic field theories one finds that an interpolating lagrangian which contains powers of fields can be used with the help of auxiliary fields[8]. We found, however, that a linear interpolating lagrangian[3] gives a larger radius of convergence and thus a faster approach to the true answer. This has been examined on the large N, solvable, Gross-Neveu model[10]. We also applied a δ expansion using a linear interpolating lagrangian in U(1) and Z(2) gauge theories in three dimensions.

For the case of Gross-Neveu model the interpolating lagrangian was chosen as

$$\mathcal{L}_\delta = \bar{\psi}\partial\!\!\!/\psi + \tfrac{1}{2}\sigma^2 + g\sigma\delta\bar{\psi}\psi + g\mu\Lambda(1-\delta)\bar{\psi}\psi \tag{6}$$

which interpolates between the GN lagrangian at $\delta=1$ and a massive free field theory at $\delta=0$. Thus, the expansion is around a field theory which roughly reflects the true physics of dynamical chiral symmetry breaking in this model. In Eq. (6) σ is an auxiliary field and Λ is the UV cutoff.

One notes that at $\delta=1$ the theory does not depend on the mass parameter μ. To finite order in δ this is certainly not true and thus we exploit this to demand that the arbitrary parameter μ be chosen so that the radius of convergence of the expansion in δ be maximal. This optimization procedure has two important implications: (1) It gives an expansion which already at low orders in δ reproduces very well[3] the correct answer (the effective potential has been calculated to order δ^4). (2) The dependence of μ on the coupling constant g is far from trivial and in fact produces the important nonperturbative information into the physical results. The optimization procedure gives the typical essential singularity dependence [$exp(\frac{-2\pi}{g^2 N})$] known also from the large N exact answer.

The general approach to an optimized δ expansion has been applied[3] in order to calculate the plaquette energy of gauge theories. Here again, one has to choose the $\delta=0$ action (S_0) so that (a) The calculations should be doable with a reasonable effort and thus S_0 should not be too complicated and (b) the $\delta=0$ action S_0 should contain some of the physics one expects to find in these theories and thus cannot be a too trivial action. A possible interpolating action is:

$$S = \beta\delta \sum_p s_p + \beta'(1-\delta) \sum_p{}' s_p \qquad (7)$$

where s_p denotes $\prod_{\ell \epsilon p} s_\ell$ and $s_\ell = \pm 1$ for each link in plaquette p for Z_2 gauge theory or $cos(\sum_{\ell \epsilon p} \theta_\ell)$ for U(1) theory (and the usual $tr(\Pi_{\ell \epsilon p} U_\ell)$ for SU(N) theory). The $\sum_p'$ in the second term runs over a subset of plaquettes (a "maximal tree") with the property that the addition of any single plaquette to the set gives rise to a closed surface in the lattice tiled by plaquettes in the new set. For example, we took a maximal tree in three space-time euclidean dimension in which all (xz) and (yz) plaquettes and also the z=0 (xy) plaquettes are included. The exactly solvable $\delta=0$ action has a number of nice properties which makes it a good candidate around which one would like to formulate the δ expansion.

Eq. (7) can be rewritten in the form

$$S = \beta\delta \sum_p{}'' s_p + (\beta' + \delta\beta_t) \sum_p{}' s_p \tag{8}$$

where $\sum_p''$ runs over all plaquettes which are not included in the maximal tree (the $\sum'$ term), $\beta_t \equiv \beta - \beta'$ and the expansion can be made now with respect to the first term $(\beta\delta \sum_p'')$. Clearly, the "zero" order term and higher order terms are now expressed in terms of Bessel functions. Later, the Bessel functions of argument $\beta' + \delta\beta_t$ can be further expanded in δ to give the final series. For example, a typical contribution to the free energy per plaquette from two nontree plaquettes (second order in contributions from $\sum_p''$) is given by:

$$W_2 = \frac{\beta^2\delta^2}{12} \frac{1 + c_1{}^4}{1 - c_1{}^4} \tag{9}$$

where $c_1 = \frac{I_1(\beta'+\delta\beta_t)}{I_0(\beta'+\delta\beta_t)}$. There is an infinite sum over graphs at each finite order in δ. Clearly, from Eq. (7), the physical results at $\delta=1$ cannot depend on β'. This, of course, is not true at any finite order in the δ expansion but can be used for choosing the best optimized value of β'. Indeed, we find that as the order of the expansion in δ increases, the independence of the physical results on β' is impressively improved and it extends to a larger region of β. In Fig.1 below one sees that already at order δ^4 the plaquette energy $\left(\frac{\partial W}{\partial \beta}\right)$ is independent of β' for a wide range of β. At large β, the optimal value of β' (which is around $\beta' = 0.8\beta$) gives reliable values for the plaquette energy up to values of β which are out of reach to strong coupling expansion to order β^7. This is seen in Fig.2, where we compare our results[3] with the strong coupling expansion in ref.11 and also with the Monte Carlo data of ref. 12. Further work on four dimensional U(1) gauge theory is in progress[13].

ACKNOWLEDGMENTS

I would like to thank the organizers of the workshop for their warm hospitality in Nagoya and for an interesting and stimulating meeting. I wish to thank many people with whom I discussed the issues described in this talk. Most of all many thanks to Carl Bender and Tony Duncan for a pleasant collaboration. I also thank the theory group at Fermilab for its support and hospitality this summer.

REFERENCES

1. C.M. Bender, K.A. Milton, M. Moshe, S.S. Pinsky and L.M. Simmons,Jr., Phys. Rev. Lett. 58, 2615 (1987).

2. C.M. Bender, K.A. Milton, M. Moshe, S.S. Pinsky and L.M. Simmons,Jr., Phys. Rev. D37 1472 (1988).

3. A. Duncan and M. Moshe Fermilab preprint 88/99 - T , August 1988

4. Clearly, the method proposed is based on the simple fact that one can write the logarithms as derivatives of exponents. Namely, $\frac{da^x}{dx} = a^x \ln a$.

5. S.S. Pinsky and L.M. Simmons Jr., Los Alamos - June 88 preprint. For a rather different point of view and very different conclusions see: N. Brown, Brookhaven preprint - 1988

6. C.M. Bender and H.F. Jones, Imperial College preprint TP - 88/11.

7. C.M. Bender and H.F. Jones, Imperial College preprint TP - 88/28.

8. C.M. Bender, K. Milton, S.S. Pinsky and L.M. Simmons Jr. - Los Alamos preprint DOE-ER-01545-402.-1988.

9. H.F. Jones and M. Monoyios, Imperial College preprint TH-88/21.

10. D.J. Gross and A. Neveu, Phys. Rev. D10, 3235 (1974).

11. See, for example, R. Balian, J.M. Drouffe and C. Itzykson, Phys. Rev. D11 2104 (1975) and (E) D19 2514 (1979).

12. G. Bhanot and M. Creutz, Phys. Rev. D21, 2892 (1980).

13. A. Duncan and H.F. Jones, Imperial College preprint to appear soon

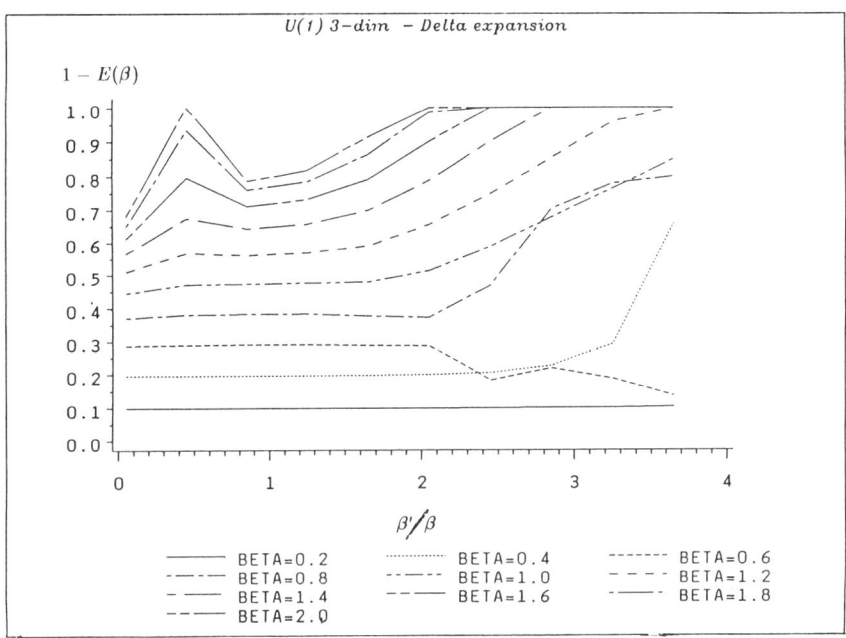

Fig. 1 The dependence on β' of the plaquette energy in U(1) gauge theory in three dimensions calculated to order δ^4. Choosing the optimal value around $\frac{\beta'}{\beta} = 0.8$ gave the result shown in Fig.2.

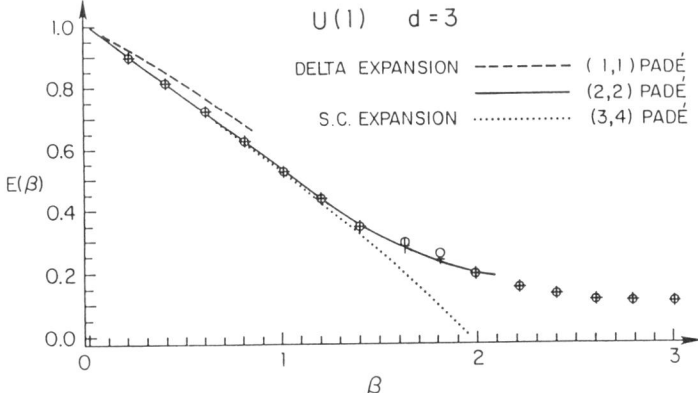

Fig. 2 The plaquette energy for U(1) gauge theory in three dimensions calculated to order δ^4 and compared to strong coupling expansion[11] to order β^7 and Monte Carlo data[12].

THE RENORMALIZATION GROUP STUDY OF THE EFFECTIVE THEORY OF LATTICE QED

YŪKI SUGIYAMA

Department of Physics, Nagoya University,

Nagoya 464-01, Japan

ABSTRACT

The compact U(1) lattice gauge theory with massless fermions (Lattice QED) is studied through the effective model analytically, using the renormalization group method. The obtained effective model is the local boson field system with non-local interactions. We study the exsistence of non-trivial fixed point and its scaling behavior. This fixed point seems to be 'tri-critical'. Such fixed point is interpreted in terms of the original Lattice QED model, and my rusults are consistent with the Monte Calro study.

1. INTRODUCTION

Recently, we are interested in the possibility of the theory of Strong Coupling QED.[1] We suppose that Strong Coupling QED is constructed at a non-trivial fixed point in the strong coupling regions. In order to find such fixed points, we must use some non-perturbative methods. One well-known method is the Lattice formulation of the field theory.[2]

The starting point is the Lattice QED model. The Lattice QED model is defined on a d-dimensional hypercubic lattice by the euclidean U(1) gauge action with fermions. The continuum theory is defined as the critical phenomena at each fixed point. One powerful method of such studies is the renormalization group (RG) method.[3] But it is hard to apply this method to the original lattice QED model directly. There are several difficulties, especially for analytic studies. So, we derive the effective model first and next apply RG method to this model.

2. THE EFFECTIVE MODEL OF LATTICE QED

2.1. Original Model

Now, we study the 4-dimensional Compact U(1)-Gauge theory with fermions.[4,5] We choose the gauge action for the mixed fundamental and adjoint representation action and the fermion charge is one.

$$S = -\frac{\beta_F}{2} \sum_{x,\mu>\nu} U_\mu(x) U_\nu(x+\hat{\mu}) U_\mu^\dagger(x+\hat{\nu}) U_\nu^\dagger(x) + \text{h.c.}$$
$$-\frac{\beta_A}{2} \sum_{x,\mu>\nu} [U_\mu(x) U_\nu(x+\hat{\mu}) U_\mu^\dagger(x+\hat{\nu}) U_\nu^\dagger(x)]^2 \qquad (1)$$
$$-K \sum_{x,\mu} \frac{1}{2} [\bar{\psi}(x)\gamma_\mu U_\mu(x)\psi(x+\hat{\mu}) - \bar{\psi}(x+\hat{\mu})\gamma_\mu U_\mu^\dagger(x)\psi(x)]$$
$$+m \sum_x \bar{\psi}(x)\psi(x)$$

First two terms denote gauge action. β_F is the gauge coupling of the fundamental action, and β_A is that of the adjoint one. $\psi(x)$ denotes the naive Wilson fermion.

2.2 The Effective Model

Now we derive the effective QED model. Several steps are needed.

1) At the first step, we make β-expansion, formally the strong coupling expansion, and integrate out the gauge variables. After that, we obtain the sequence of the interaction terms. They consist of bilocal composite fermion fields, such as $\bar{\psi}(x)\gamma_\mu\psi(x+\hat{\mu})$ denoted by ⟶$_\mu$. The obtained action is shown graphically as follows.

$$S = -\frac{K^2}{2} \sum \rightrightarrows \quad +\frac{K^4}{2}\beta \sum \square$$

$+ o(\beta^2):$ ▱ + ▱ + ▱ + ▱

+ higher order terms

We note all these interaction terms keep gauge invariant.

2) At the second step, we perform Fiertz rearrangement[6] of all interaction terms, and we get the action rewritten with the local composite fields such as $\bar{\psi}(x)\Gamma\psi(x)$, where Γ denotes the 16-Dirac matrices.

3) Finally, using the auxiliary field method we integrate out the fermion fields. Then we obtain the local bosonic system with non local interactions. Such non local interactions are induced by the plaquette gauge interaction.

$$\text{scalar}: \bar{\psi}(x)\psi(x) \longrightarrow \sigma(x)$$
$$\text{vector}: \bar{\psi}(x)\gamma_\mu\psi(x) \longrightarrow V_\mu(x)$$
$$\vdots$$

Writing up to the vector channel, the obtained effective model is given as follows.

$$\begin{aligned} S = & -\frac{K^2}{2}\sum_{x,\mu}\sigma(x)\sigma(x+\hat{\mu}) \\ & -\frac{K^2}{8}\sum_{x,\mu,\nu}V_\mu(x)(1-2\delta_{\mu\nu})V_\mu(x+\hat{\nu}) + [A,T] \\ & +\frac{d}{2}\sum_x \ln\Big\{\Big[K\sum_\mu \sigma(x+\hat{\mu}) + 4m\Big]^2 \\ & \qquad\qquad + K\sum_\mu\Big[\sum_\nu(1-2\delta_{\mu\nu})V_\mu(x)\Big]^2 + ...\Big\} \\ & -(\frac{1}{4})^3 \beta K^4 V_4[\sigma,V] \\ & + o(\beta^2)\, K^6 V_6[\sigma,V] + \text{ higher order terms....} \\ & + \text{ other interaction terms} \end{aligned} \quad (2)$$

where V_4 and V_6 are non-local fourth power bosonic interactions and that of sixth power's, respectivelly.

2.3. The Effective Model in Momentum Space

Now, this effective action has infinite series of non-local interactions, and they are very complicated. But fortunately, not all the interactions are important for the study of the critical behavior. For the viewpoint of the renormalization group method, almost all the terms must be 'transient' (which means ; during the iteration of the RG steps, many interaction terms disappear.) In our experience of the statistical mechanics higher power polynomial interactions are such irrelevant terms.[3]

So, the lower power interaction terms would be considered first, and we take only scalar fields as dynamical variables. Then, the effective model reduces to the scalar model with non local polynomial interactions. The expectation value of the scalar field $<\sigma>$ is corresponding to $<\bar{\psi}\psi>$ - condensate. So, we can study the chiral transition of the original QED model by the effective scalar model. The remaining scalar non-local interactions are given in Fig.1.

```
σ(x+ν̂) ──── σ(x+μ̂+ν̂)
  │              │
σ(x)  ────────  σ(x+μ̂)


σ(x+ν̂)²         σ(x+ν̂)²
  │               └──── σ(x+μ̂)²
σ(x)²
```

Fig.1.

We are interested in the long wavelength behavior for studying the critical phenomena. Then we rewrite the model in momentum space for small p^2. So, any kind of non-local polynomial interactions reduce to the following much simpler form.

$$S' = d\sigma(0) \times \tilde{m}$$
$$-\frac{1}{2}\int dp(p^2+r)\sigma(p)\sigma(-p)$$
$$-\int \prod_{i=1}^{3} dp_i \delta(\sum_{i=1}^{3} p_i)\sigma(p_1)\sigma(p_2)\sigma(p_3)\{\alpha_1 + \alpha_2 \sum_{i=1}^{3} p_i^2\} \quad (3)$$
$$-\int \prod_{i=1}^{4} dp_i \delta(\sum_{i=1}^{4} p_i)\sigma(p_1)\sigma(p_2)\sigma(p_3)\sigma(p_4)\{\beta_1 + \beta_2 \sum_{i=1}^{4} p_i^2\}$$

− higher power polynomial interactions....

We remark interaction kernel has momentum dependence. That shows the interactions are non-local. The contribution of the bare mass of fermions corresponds to the external field. And, when we take the bare mass vanishing, the odd polynomial coupling terms are dropped. We consider the massless fermions in the later sections.

This is the final form to apply the RG method.

3. THE RENORMALIZATION GROUP STUDY OF THE EFFECTIVE MODEL

3.1. Renormalization Group Procedure in Momentum Space

The renormalization group equations are derived by diagrammatic expansion.[3] RG procedure in momentum space consists of following two steps.

1) Integration out of high frequency mode

2) Rescales of momentum and the fields

The exact RG-equations are the sets of non linear functional equations. But they are too difficult to be solved. Usually, they are replaced by algebraic equations of the coupling parameters.

Up to the fourth order couplings the RG-equations are as follows.

$$
\begin{aligned}
r' &= \frac{4}{1+\frac{6c\beta_2}{1+r}}\{r + \frac{3c\beta_1}{1+r} + \frac{6c\beta_2}{1+r} + ...\} \\
\beta'_1 &= \frac{2^{4-d}}{(1+\frac{6c\beta_2}{1+r})^2}\{\beta_1 - \frac{9c\beta_1^2}{(1+r)^2} - \frac{36c\beta_1\beta_2}{(1+r)^2} + ...\} \\
\beta'_2 &= \frac{2^{2-d}}{(1+\frac{6c\beta_2}{1+r})^2}\{\beta_2 - \frac{9c\beta_1\beta_2}{(1+r)^2} + ...\}
\end{aligned}
\quad (4)
$$

The later investigations in this section are based on these equations.

Three cases can be considered in dimension $d = 4$.

1) $\beta_2 = 0$. The equations become that of the ordinal φ^4-model. In dimension $d = 4$, the model has only trivial fixed point. $r^* = \beta_1^* = \beta_2^* = 0$

2) $\beta_2 > 0$. In this case nothing change. Only trivial fixed point exists.

3) $\beta_2 < 0$. When β_2 takes negative value the situation is changed, and this is the only possibility of existence of non-trivial fixed point. If we can expect the small negative value of β_2^* at the non-trivial fixed point, we can choose β_2 as the expansion parameter near that fixed point for analytic study.

3.2. Analytic Study and 'τ-expansion' Method

Actually Fig.2 indicates such fixed point solution of the RG-equations (4). So we can chose $\beta_2 \sim -\tau$ for the expansion parameter near that fixed point. This expansion method is similar to ε-expansion method of scalar models in dimension $d = 4 - \varepsilon$.[3] This τ acts the same role as ε. Using this expansion method, we drive the non-trivial fixed point of order τ, and the results are consistent with the perturbative study of the non-local interaction terms. Moreover, higher power terms can be studied systematically, and sixth power or higher power interaction terms don't disturb the following results. Now we show some analytic results.

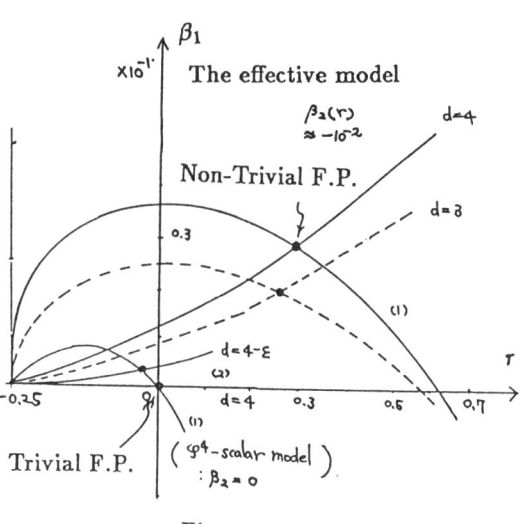

Fig.2

For the purpose of studying the scaling behavior, we linearize the RG-equations near the fixed points, and study the transition matrix, which derives the eigenvalues of the scaling operators.

1) At the trivial fixed point, the transition matrix and the eigevalues are as follows.

$$\tau^*, \beta_1^*, \beta_2^* = 0$$

$$M = \begin{pmatrix} 4 & 3c & 24c \\ 0 & 1 & 0 \\ 0 & 0 & \frac{1}{4} \end{pmatrix}$$

eigenvalues : $\lambda_1 = 4, \lambda_2 = 1, \lambda_3 = \frac{1}{4}$

1-relevant, 1-marginal, 1-irrelevant

The results show that the trivial fixed point is Gaussian Fixed point as well as the case of φ^4-model.

2) At the non trivial fixed point, the transition matrix and the eigenvalues are as follows.

$$r^* = -\frac{160}{3}\tau, \ \beta_1^* = \frac{58}{9c}\tau, \ \beta_2^* \sim -\frac{1}{c}\tau$$

$$M = \begin{pmatrix} 4(1-\frac{22}{3}\tau) & 12c(1+\frac{198}{3}\tau) & * \\ o(\tau^2) & 1-68\tau & * \\ o(\tau) & o(\tau) & * \end{pmatrix}$$

eigenvalues : $\lambda_1 = 4(1-\frac{22}{3}\tau)$, $\lambda_2 = 1-68\tau$, $\lambda_3 = *$

1-relevant, 1-irrelevant, 'not irrelevant'

The eigenvalues show the scaling operators are 1-relevant and 1-irrelevant. The third eigenvalue is unknown in the framework of this trial study. But considering the behavior of the RG-equations and the property of the trivial fixed point, the RG-flow near the trivial fixed point goes into the trivial fixed point along the β_2 direction. Then we can suppose the third scaling operator would be 'not irrelevant'.

Combining these results, we can draw the RG flow diagram roughly in Fig.3a

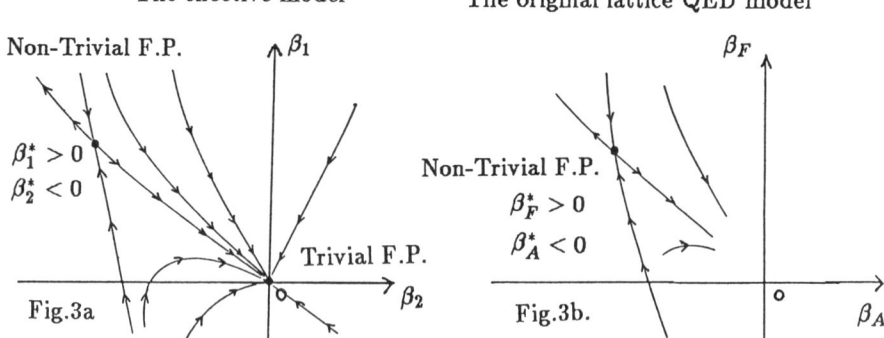

Fig.3a,b The left is the RG-flow diagram of the effective model and the right is the translated diagram near non-trivial fixed point into the parameter space of the original QED model.

Actually the flow diagram is drawn in the three dimensional parameter space, but the figure is the projected one in the (β_1, β_2)-plane. There is one more axis, and there is 1-relevant scaling operator along this direction.

This flow diagram is transformed into the (β_F, β_A)-plane of the original QED model, which diagram is drawn in Fig.3b. With some considerations,* non trivial fixed point exist in the region $\beta_F^* > 0$, $\beta_A^* < 0$. This causes the results of 2); $r^* < 0$, $\beta_1^* > 0$, $\beta_2^* < 0$.

4. SUMMARY AND DISCUSSION

1. We drive the effective model of Lattice QED model. The obtained effective model is the local bosonic system with non local interactions. This model keeps gauge invariant.

Including non single plaquette gauge action term, four fermi coupling interaction and Higgs scalars, the similar effective model can be derived. So, the wide class of gauge models on lattice can be described by this type of the effective models.

2. Using the renormalization group method we investigate the exsistence of the non-trivial fixed point and its critical behavior for studying the continuum theory defined at this fixed point.

We choose the original gauge model for the mixed fundamental and adjoint representation of single plaquette action. The non trivial fixed point seems to exist in the coupling region of $\beta_1^* > 0$, $\beta_2^* < 0$, which corresponds to $\beta_F^* > 0$, $\beta_A^* < 0$ of the original QED model. This fixed point would have two relevant operators, and seems to be somewhat 'tri-critical' point. These rusults are consistent with the Monte Calro study of the same Lattice QED model with compact U(1) mixed gauge action.[5] This tri-critical point is some different kind of the popular one. (which is in the $\lambda\varphi^4 + \mu\varphi^6$-model for $\lambda < 0, \mu > 0$.) [7]

The analysis of this work is preleminary. The estimation of the vector chanel etc. contributions must be investigated. Such contributions make the analysis fully systematic.* Nevertheless, the results show the basic property of the non-trivial fixed point, which is the first step of constructimg the consistent non-trivial continuum theory.

---

* In order to take away the ambiguity of the reparamerlization of the couplings to the original model ,which causes the only scalar fields to be taken for dynamical variables, we remove the contribution from the trivial gauge couplings.

REFERENCES

1. Miransky, V. A., Report in this proceedings.

2. Wilson, K. G., Phys. Rev. $\underline{B4}$, 3174 (1971)

3. Wilson, K.G. and Kogut, J. B., Phys. Rep. $\underline{12C}$, 75 (1974)

4. Kogut, J. and Dagotto, E., Phys. Rev. Lett. $\underline{59}$, 617 (1987); Dagotto, E. and Kogut, J., Nucl. Phys. $\underline{B295}$ [FS21], 123 (1988).

5. Okawa, M., KEK-TH-204

6. Greensite, J. and Primack, J., Nucl. Phys. $\underline{B180}$ [FS2], 170 (1981)

7. Riedel, E.K. and Wegner, F.J., Phys. Rev. Lett. $\underline{29}$, 349 (1972)

EFFECTIVE THEORY OF STRONG COUPLING QED

Takuya Morozumi[†], Hiroto So[*] and Misako Suwa[**]

National Laboratory for High Energy Physics (KEK)
Oho, Tsukuba, Ibaragi 305, Japan

[*] Department of Physics, Niigata University,
Ikarashi, Niigata 950-21, Japan

[**] Graduate School of Science and Technology, Niigata University,
Ikarashi, Niigata 950-21, Japan

ABSTRACT

Using the effective theory of the strong coupling QED(SCQED), we have investigated the symmetries which the theory contains and checked the Miransky Limit for physical quantities. The mass spectra of fermion bound states have been also calculated numerically.

1. INTRODUCTION

The strong coupling QED(SCQED) has two aspects: one is finite QED or an Ultraviolet Fixed Point Theory and the other is chiral symmetry breaking and continuum limit of its breaking scale[1]. In 1978 - 1985, Miransky proposed the connection between them based on Landau gauge - ladder approximation, that is Miransky Limit[2]. His proposal is, however, necessary for the check of some approximations beyond the ladder aproximation and the gauge invariance.

Our purposes in this talk are the following three points along its Miransky's interpretation; the first point is to investigate the symmetry of the effective theory of SCQED constructed by bilocal auxiliary fields method[3],[4],[5]. The second is to

---

[†] On leave of absence from Kyoto University, Kyoto 606, Japan

check the consistency among various quantities for the Miransky Limit in SCQED. The third is to calculate the mass spectra of bound states. All calculations are done within Landau gauge - ladder approximation. It becomes important when we consider the next order calculation such as loop effects of composite fields.

2. EFFECTIVE THEORY OF STRONG COUPLING QED

As previous Dr. Morozumi's talk, in order to construct the effective theory, we take the following four procedures from 1) to 4);

1) Firstly, adopt the Euclidean QED action which is fixed to Landau gauge, and integrate out the photon field.

2) Perform Fierz transformation of induced interactions $J_\mu(x)J_\nu(y)$.

3) Introduce bilocal auxiliary fields; $S(x,y)$, $P(x,y)$, $V_\mu(x,y)$, $AV_\mu(x,y)$ and $AT_{\mu\nu}(x,y)$ in order to eliminate the four-fermi interactions.

4) Integrate out the fermion field.

After we carry out them, we can obtain the following effective action which consists of the scalar kinetic term, the pseudo-scalar kinetic term, other kinetic terms and interaction terms,

$$S_{eff} = \frac{1}{2} \iint dx\,dy (S(x,y)D_s^{-1}S(y,x) + P(x,y)D_p^{-1}P(y,x) + \ldots)$$

$$+ \text{interaction terms}, \qquad (1)$$

and the effective potential for vacuum expectation value (B) of scalar field is written by,

$$V_{eff} = \frac{1}{3e^2} \int_0^{\Lambda^2} dq^2 \left(\frac{d}{dq^2}(q^2 B(q^2))\right)^2 - \frac{1}{8\pi^2} \int_0^{\Lambda^2} dq^2 q^2 ln(q^2 + B^2(q^2)), \qquad (2)$$

where q_μ is a space-like relative momentum[5]. The variation of the effective potential leads us to the Schwinger-Dyson equation in the Landau gauge - ladder approximation. This is a differential equation with two boundary conditions, i.e., ultraviolet one and infrared one.

After the second order variation of V_{eff}, we can obtain the stability operator, which is the second order differential one,

$$\delta^2 V_{eff} = -\frac{2}{3C}\left(4x\frac{d^2}{dx^2} + 8\frac{d}{dx}\right) + \frac{2(-x+B^2)}{(x+B^2)^2}, \tag{3}$$

where x is the squre of a relative momentum, $q_\mu q_\mu$, and C means $\frac{e^2}{4\pi^2}$. The stability operator, $\delta^2 V_{eff}$, is divided into two parts; positive-semidefinite operator, $-\frac{2}{3C}(4x\frac{d^2}{dx^2} + 8\frac{d}{dx}) - \frac{2}{x+B^2}$, in the conditiion that C is larger than 1/3 and a positive function, $\frac{4B^2}{(x+B^2)^2}$. The semidefiniteness of the former operator is due to gap equation. Therefore, its operator is positive definite [5].

If B goes to zero, then the lowest eigenvalue of stability operator goes to zero. Accordingly, this property seems to imply a smooth restoration of chiral symmetry. However, it needs more carefull argument. In section 3, we will return to this subject.

The present effective action has not only the well-known chiral symmetry but also the hidden symmetry. The chiral symmetry is scalar-pseudo scalar symmetry and the hidden symmetry is among scalar, pseudo scalar, vector, and axial vector symmetry in our action. Contrary to Nambu-Jona-Lasino model analysis by Kugo et al.[6], the symmetry among those fields is only 'partially' realized in our action. It means that just only restricted configurations have not only the same vacuum structure and the same chiral symmetry as the original effective action, but also extra-symmetry, i.e. "hidden local symmetry". Because of bilocal fields, it is not so easy to calcurate the vector mass pole which means that the dynamical generation of the kinetic term.

3. MIRANSKY LIMIT OF VARIOUS QUANTITIES

I shall examine the Miransky Limit of various quantities including stability operator eigenvalue and the bound state mass spectra. Firstly, we can understand the Miransky Limit as a substitution rule of cutoff Λ into dynamical scale B_0. The explicit form of this was found by Fukuda-Kugo [7] and the cutoff dependence of coupling constant C_Λ found by Miransky [2],

$$B_0 = \Sigma(0) = f(C_\Lambda)\Lambda\exp\left(-\frac{\pi}{\sqrt{3C_\Lambda - 1}}\right), \tag{4}$$

where $f(C)$ is a smooth function of coupling constant C. Let's first examine fermionic quantities; dynamical fermion has no real mass pole (firstly found by Fukuda and Kugo) and the existence of complex singularities was found by Atkinson and Blatt [8]. The complex singularity means the fermion inverse propagator $(S_F^{-1}(q))$ vanishes at the square momentum q_μ^2 which is naively proportional to the square of the cutoff, by Miransky's substitution, however, it becomes finite; $(1.4 \pm 0.7i)B_0^2$. On the other hand, the chiral order parameter, $<\bar{\Psi}\Psi>$, is linearly divergent even after the Miransky Limit; $-0.1B_0^2\Lambda$. Nextly, we shall examine effective χsb parameters; F_π^2 and vacuum energy ΔV_{eff}. Both quantities become finite after the Miransky Limit; $F_\pi^2 = 0.047 N_f B_0^2$ and $\Delta V_{eff} = -0.0069 N_f B_0^4$, where N_f means the related fermion number. These are good properties for a dynamical Higgs mechanism because they are observable.

In our main subjects (stability eigenvalue and bound states spectra), we shall continue to investigate the Miransky Limit. We have inverse propagators D^{-1}s for scalar, pseudo-scalar etc. in our effective tneory. The inverse propagators D^{-1}s are 2nd order partial differential operators and we need to evaluate their eigenvalues. According to following procedure, we go to find the mass pole,

(i) Calculate the complex variable S-D equation because relative momentum q_μ is spacelike and there exist ordinary mass poles in the time-like region.

(ii) Solve the 4-dimensional Schrödinger-like eigenvalue problem in the momentum space.

(iii) Search for zeros of inverse propagator eigenvalue E.

Then, we can obtain mass poles.

For scalar and pseudoscalar cases, the inverse propagator operators, D^{-1}s, is given by

$$D^{-1} = \text{"KINETIC TERM"} + \text{"POTENTIAL TERM"}. \qquad (5)$$

The common first term is made of such a differential operator, $(-\frac{2}{3C}\frac{\partial^2}{\partial q_\mu \partial q_\mu})$, and the second term is constructed by the fermion propagator, S_F. But one should not confuse our inverse propagators, D^{-1}s, in the momentum space with usual Schrödinger equation in the real space. For the scalar case (Fig.1), the potential term is given by

$$\text{"POTENTIAL TERM"} = \frac{1}{2}\text{tr } 1 S_F(q + \frac{iP}{2}) 1 S_F(q - \frac{iP}{2}), \qquad (6)$$

and for the pseudoscalar case (Fig.2) is by

$$\text{"POTENTIAL TERM"} = \frac{1}{2}\text{tr } i\gamma_5 S_F(q + \frac{iP}{2}) i\gamma_5 S_F(q - \frac{iP}{2}), \qquad (7)$$

where iP_μ means a time-like total momentum.

The following figures (Fig.1 and Fig.2) imply that so-called potential terms at $P_\mu = 1.0$ less than those at $P_\mu = 0.0$ for scalar 0^+ and pseudoscalar 0^-.

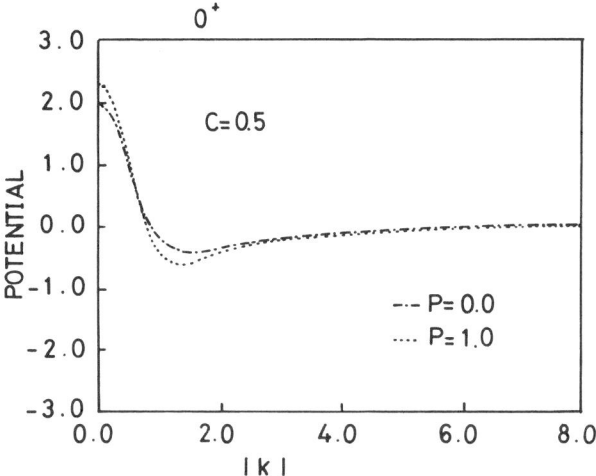

Figure 1. " Potential term" in a scalar bound state for the relative momentum.

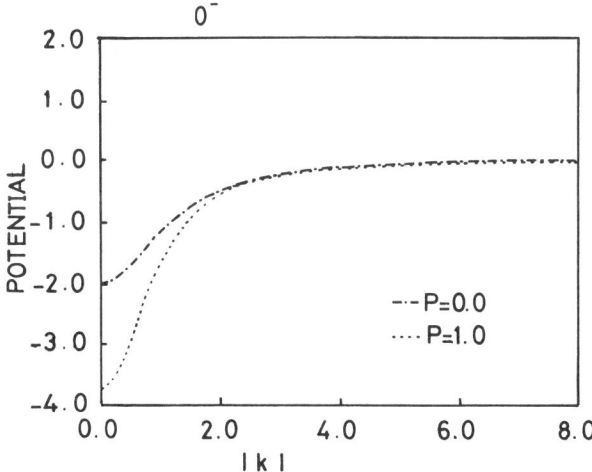

Figure 2. " Potential term" in a pseudo-scalar bound state for the relative momentum.

Let's sketch the eigenvalue of 0^+, 0^-, 1^-, and 1^+ inverse propagators, roughly (Fig.3(a)~(d)).

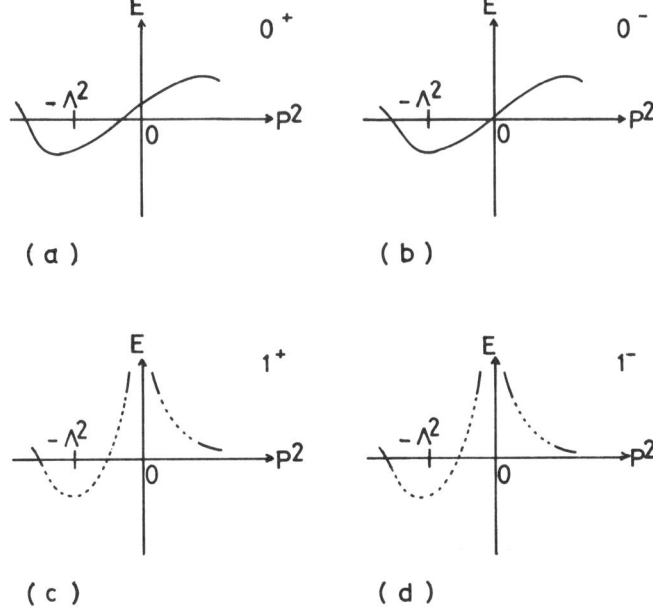

Figure 3.(a)~(d) Lowest eigenvalues of inverse propagator operators in various bound states.

Solid lines are drawn according to analytical and numerical calcurations. Broken lines are are drawn according to extraporations. One can see that there are two mass poles in any case. But left pole in the timelike region is larger than cutoff. So it does not affect the mass spectra after the continuum limit. In the connection with the Miransky Limit, for 0^- case, we can easily know $m = 0$ and it is trivial as for the limit. For 0^+ case, at $P = 0$, D^{-1} operator is stability one, $\delta^2 V_{eff}$, or curvature of V_{eff} and naively proportional to a square of inverse cutoff Λ. Even after the Miransky Limit, numerical calcurations lead us to the following Λ dependence,

$$\text{curvature of } V_{eff} \propto \left(\frac{B_0}{\Lambda}\right)^\delta \Lambda^{-2}, \tag{8}$$

where δ is a very small positive constant. Finally, we want to mention scalar mass pole. The mass pole is expected to be finite after the Miransky Limit. Our result is obtained by the method that operators, D^{-1}s, are diagonalized with finite base (five gauss functions which saticefy infrared and ultravioret boudary conditions) and we can only know the upper bound of it because its method is understood as a kind of variational one. Our preliminary result is written by the following table,

$C = \frac{\alpha}{\pi}$	0.5	0.45	0.4
$s = \frac{m_{0^+}}{B_0}$	≤ 2.5	≤ 3.3	≤ 9
$\frac{\Lambda}{B_0}$	17	39	210

Table 1. Coupling and cutoff dependence of the scalar bound state mass.

If you play a game, *i.e.*, to choose U(1) as Technicolor group, you can obtain a preliminary upper bound of Higgs scalar particle mass by the following argument. Using M_W = 81.8 Gev and M_Z = 92.6 Gev, we can obtain $F_\pi \sim 250$ Gev . Our SCQED scale (B_0) and the upper bound of scalar bound state (Higgs scalar particle) are determined as

$$B_0 = 1.2 \text{ Tev}/\sqrt{N_2}, \qquad (9)$$

where N_2 means the weak doublet number of technifermions and

$$m_{0^+} \leq 4.0 \text{ Tev}/\sqrt{N_2}, \qquad (10)$$

where s = 3.3 is used but it is not serious because of the approximations.

4. SUMMARY AND DISCUSSION

We shall summarize our talk. Firstly, we want to stress that our effective action has not only chiral symmetry but also partially hidden symmetry because of rich field components. The latter symmetry is not exact but is realized in a part of the field configuration space. We can also show our stability operator is positive and χsb restores very slowly and continuously in taking B_0 to zero. Thirdly, we find poles of bound states but can not check the Miransky Limit for them because of preliminaly results. Finally, there are remaining problems.

(α) What is the Miransky Limit? B_0 is not a real fermion mass but just a dynamical scale.

(β) All physical quantities must be finite, though there coexist finite values and infinite values of physical quantities after the Miransky Limit. Therefore, the simple Miransky Limit may be imperfect in the renormalization procedure.

These feature may teach us certain new method of renormalization procedure in SCQED or Ultraviolet Fixed Point Theory.

ACKNOWLEDGEMENT

We would like to thank to Prof. D. Atkinson for valuable discussions and to Prof. W. A. Bardeen for useful comments. We also thank to Prof. H. Kanada for reading our manuscript and TeX guide and to Dr. K. Yabana for TeX guide.

REFERENCES

1. For review article, see Formin, P. I., Gusynin, V. P., Miransky, V. A. and Sitenko, Yu. A., Riv. Nuovo Cim. 6, 1 (1983)

2. Fomin, P. I., Gusynin, V. P. and Miransky, V. A., Phys. Lett. 78B, 136 (1978)

   Miransky, V. A., Nuovo Cim. A90, 149 (1985)

3. Goldman, T. and Haymaker, R., Phys. Rev. D24, 724 (1981)

4. Roberts, C. D. and Cahill, R. T., Phys. Rev. D33, 1755 (1986)

5. Morozumi, T. and So, H., Prog. Theor. Phys. 77, 1434 (1987)

6. Kugo, T., Terao, H. and Uehara, S., Prog. Theor. Phys. Supplement 85, 122 (1985)

7. Fukuda, R. and Kugo, T., Nucl. Phys. B117, 250 (1976)

8. Atkinson, D. and Blatt, D. W. E., Nucl. Phys. B151, 342 (1979)

On the strong coupling phase of QED

T. Morozumi [†]

National Laboratory for High Energy Physics (KEK)
Oho, Tsukuba, Ibaraki 305, Japan

H. So

Department of Physics, Niigata University,
Niigata 960-21, Japan

[†] Reported by T.Morozumi

ABSTRACT

Chiral symmetry breaking of strong coupling QED is reexamined by means of effective action. The relation between the existence of Nambu Goldstone boson and the non-trivial solution of fermion mass generation is clarified in this scheme. The study of effective potential leads to the necessity of *Miranskys'* continuum limit. The main part of this talk is based on the collaboration with H.So.[1]

I. INTRODUCTION

Chiral symmetry breaking of QED has been studied with the ladder approximation by many authors.[2345] One of the reason why the ladder approximation has been used is it satisfies chiral Ward Takahashi Identity. In this talk the effective action which preserves the chiral WT identity is introduced. I like to emphasize the following points,

1) The relation between the existence of massless NG boson and the non-trivial solutions of fermion mass.

2) The stability of the ground state; the necessity of *Miranskys'* interpretation of continuum limit.

II. Effective Action

The effective action introduced here is constructed with the bilocal auxiliary fields[6,7,8,9] which are bound states of fermion and antifermion. Then the effective action corresponds to meson theory and these mesons are interacting via fermion loops.

I briefly explain how to construct effective action. For simplicity of notation I concentrate on Landau gauge for photon propagator.

We start with standard QED action.

$$\int [dA_\mu, d\psi, dB] e^{S[\psi, A_\mu, B]}, \tag{1}$$

$$S[\psi, A_\mu, B] = \int d^4x \bar{\psi}(\partial + ieA + m_0)\psi + \frac{1}{4}(\partial_\mu A_\nu - \partial_\nu A_\mu)^2 + \partial_\mu A_\mu B, \tag{2}$$

where

$$A = A_\mu \gamma_\mu,$$
$$\partial = \partial_\mu \gamma_\mu.$$

We also include breaking term of chiral symmetry in Eq.(2). First the photon field (A, B) is integrated,

$$\int [d\psi] e^{S[\psi]}, \tag{3}$$

$$S[\psi] = \int d^4x \bar{\psi}(\partial + m_0)\psi + \frac{e^2}{2}\int d^4x d^4y J_\mu(x) D_{\mu\nu}(x-y) J_\nu(y). \tag{4}$$

where J_μ is U(1) electromagnetic vector current and $D_{\mu\nu}$ is the photon propagator in the Landau gauge. Then the interaction is a non-local current-current interaction. Fierz transformation of the current-current interaction leads to the five kinds of non-local four fermi interactions,such as scalar,pseudo-scalar, vector,axial vector and tensor,

$$\int [d\psi] e^{S_4[\psi]}, \tag{5}$$

$$\begin{aligned} S_4 &= \int d^4x \bar{\psi}(\partial + m_0)\psi \\ &+ \frac{3}{8}e^2 \int d^4x d^4y (\bar{\psi}(x)\psi(y)\partial^{-2}(x-y)\bar{\psi}(y)\psi(x) + \\ &\quad \bar{\psi}(x)i\gamma_5\psi(y)\partial^{-2}(x-y)\bar{\psi}(y)i\gamma_5\psi(x)) \\ &- \frac{1}{8}e^2 \int d^4x d^4y \bar{\psi}(x)\gamma_\mu\psi(y)\partial^{-2}(x-y)\bar{\psi}(y)\gamma_\mu\psi(x) \\ &- \frac{1}{4}e^2 \int d^4x d^4y \bar{\psi}(x)\gamma_\mu\psi(y)\partial_{\mu\nu}^{-2}(x-y)\bar{\psi}(y)\gamma_\nu\psi(y) \\ &+ Axial\, vector\, and\, Tensor\, terms, \end{aligned} \tag{6}$$

where

$$\partial^{-2}(x) = -\frac{1}{4\pi^2 x^2},$$
$$\partial_{\mu\nu}^{-2}(x) = -\frac{1}{8\pi^2}(\frac{\delta_{\mu\nu}}{x^2} - \frac{2x_\mu x_\nu}{x^4}).$$

Since fermion terms are bilinear in the action , we integrate out the fermion fields.

$$\int [d\sigma d\pi dV dA dT] e^{S[\sigma,\pi,V,A,T]}, \tag{7}$$

$$\begin{aligned} & S[\sigma,\pi,V,A,T] \\ &= -\frac{2}{3e^2} \int d^4x d^4y [\sigma(x,y)\partial^{-2}(x-y)\sigma(y,x) + \pi(x,y)\partial^{-2}(x-y)\pi(y,x) + ...] \\ &- trlog[(\partial + m_0)\delta^4(x-y) + \sigma(x,y) + i\gamma_5\pi(x,y) + ...]. \end{aligned} \tag{8}$$

where ...means contribution from vector ,axial vector ,and tensor terms.
$Trlog$ term includes interaction terms via fermion loops. In the following we move to momemtum space where the bilocal fields are expanded by double fourier

transformation,

$$\sigma(x,y) = \int \frac{d^4P}{(2\pi)^4} \frac{d^4q}{(2\pi)^4} \sigma(q,P) exp[iP\frac{x+y}{2} + iq(x-y)], \qquad (9)$$

where P is total momemtum and q is relative momemtum repectively. The same transformation is also done for $\pi(x,y), V(x,y)$ etc.

III. Effective potential

The effective potential[1] is a functional of zero mode of total momemtum for the bilocal auxiliary fields.

$$\begin{aligned}\sigma(q,P) &= (2\pi)^4\delta^4(P)\sigma(q),\\ \pi(q,P) &= (2\pi)^4\delta^4(P)\pi(q),\end{aligned} \qquad (10)$$

$$\begin{aligned}V_{eff}[\sigma(q),\pi(q)] = \\ &-\frac{8\pi^2}{3e^2}\int \frac{d^4q}{(2\pi)^4}[\sigma(q)\partial_\mu\partial_\mu\sigma(q) + \pi(q)\partial_\mu\partial_\mu\pi(q)] \\ &-2\int \frac{d^4q}{(2\pi)^4}log(q^2 + (\sigma(q)+m_0)^2 + \pi(q)^2),\end{aligned} \qquad (11)$$

where

$$\partial_\mu = \frac{\partial}{\partial q_\mu}.$$

The effective potential obtained here satisfies the following important features,
1) $V_{eff}(m_0 = 0)$ has chiral symmetry.
2) The last term in Eq.(11) is negative ,because it is the vaccum energy of Dirac sea.
3) The stationary condition $\frac{\delta V}{\delta \sigma} = 0$ determines the vaccuum.
4) Expanding S_{eff} around the vacuum , we obtain the inverse propagators of bound states and interaction parts between them .

IV. Gap equation and the wave function of Nambu Goldstone boson

The stationary conditon of V_{eff} is,

$$\left.\frac{\delta V}{\delta \sigma}\right|_{\sigma(q)=B(q),\pi(q)=0} = 0,$$
$$\rightarrow \partial_\mu \partial_\mu B(q^2) = -\frac{3e^2}{4\pi^2}\frac{B(q^2)}{q^2+B^2(q^2)}, \quad (12)$$

with the boundary conditions,

$$\left.q^4 \frac{dB(q^2)}{dq^2}\right|_{q^2=0} = 0,$$
$$\left.\frac{d}{dq^2}(q^2 B(q^2))\right|_{q^2=\Lambda^2} = m_0. \quad (13)$$

Inverse propagator of Nambu Goldstone boson is obtained from,

$$S_{eff}^{free} = \int \frac{d^4 P}{(2\pi)^4}\frac{d^4 q}{(2\pi)^4}\pi(q,P)[-\frac{8\pi^2}{3e^2}\partial_\mu\partial_\mu + \frac{1}{2}tr[S_f(-\frac{P}{2}+q)i\gamma_5 S_f(\frac{P}{2}+q)i\gamma_5]]\pi(q,-P), \quad (14)$$

where $S_f(q) = iq_\mu \gamma_\mu + B(q^2)$.

Then the wave function of zero mass NG boson $\pi(q^2) \equiv \pi(q, P=0)$ must satisfy the following equation,

$$\partial_\mu \partial_\mu \pi(q^2) = -\frac{3e^2}{4\pi^2}\frac{\pi(q^2)}{q^2+B^2(q^2)}, \quad (15)$$

with boundary conditions,

$$\left.q^4 \frac{d\pi(q^2)}{dq^2}\right|_{q^2=0} = 0,$$
$$\left.\frac{d}{dq^2}(q^2 \pi(q^2))\right|_{q^2=\Lambda^2} = 0. \quad (16)$$

It should be noted that the gap equation Eq.(12) coincides with the differntial equation discussed by Fukuda and Kugo.[3] Further if the gap equation has nontrivial solution which satisfies the homogeneous boundary condition ($m_0 = 0$ in Eq(13)), massless Nambu Goldstone boson exists as expected. (In this case the wave function of NG boson $\pi(q^2)$ is proportional to the solution of the gap equation $B(q^2)$.)

As discussed in Ref.(3), in the weak coupling phase $e^2 < e_c^2$; $\frac{e_c^2}{4\pi^2} = \frac{1}{3}$ there is no nontrivial solution which satisfies the homogeneous boundary condition ($m_0 = 0$ in Eq.(13)) for finite Λ. Then chiral symmetry is explicitly broken. On the otherhand, in the strong coupling phase $e^2 > e_c^2$ the homogeneous boundary condition is satisfied for finite Λ. Therefore chiral symmetry is spontaneously broken and massless NG boson appears.

V. The stability of the vacuum.

In the strong coupling phase, the solutions which satisfy boundary condition are characterised by the nummber of nodes of $B(q^2)(0 \leq q^2 \leq \Lambda^2)$. A nodeless solution $B_0 \equiv B_0(q^2 = 0)$ satisfies following relation,

$$B_0 = \Lambda \kappa, \qquad (17)$$

where

$$\kappa = exp[-\frac{\pi}{\sqrt{\frac{e^2}{e_c^2} - 1}}]. \qquad (18)$$

The solution with n nodes $B_n(q^2)$ are related to the nodeless one through the oridginal scale invariance of gap equation[5] that is partially broken due to the introduction of cut off Λ.

$$B_n = B_0 \kappa^n, \qquad (19)$$

where

$$B_n \equiv B_n(q^2 = 0).$$

Then using the effective potential we determine which of the solutions is the true ground stste in the strong coupling phase.[159]

$$\Delta V_{eff} = V_{eff}(B(q^2)) - V_{eff}(0)$$
$$= \frac{1}{8\pi^2} \int dx x \left[\frac{B(x)^2}{x + B(x)^2} - \log(1 + \frac{B(x)^2}{x}) \right] < 0, \qquad (20)$$

$$V_{eff}(B_n) < V_{eff}(B_{n+1}). \qquad (21)$$

Therefore the nodeless solution is a true ground state. On the oherhand the trivial one $B(q^2) \equiv 0$ has the highest energy.

The same conclusion is derived by means of Cornwall Jackiw Tomboulis effective potential.[5,10,11]

The instability of the trivial solution is seen from the appearance of the tacyionic pole for the bound state as studied by Miransky et.al.[4] (See Ref.(13) for general discussions of the appearece of the tachyonic pole and the stability of the vacuum.) In this way the ground state energy is estimated,

$$\Delta V \simeq -B_0^4. \qquad (22)$$

From Eq.(17) and (22), the ground state energy is unbounded if we take the naiive continuum limit;($\Lambda \to \infty$ with e^2 fixed). However we can restore the stability of ground state if we take the $Miranskys'$ continuum limit.[45] $\Lambda \to \infty$, $e^2 \to e_c^2$ with B_0 fixed.) Furthermore if we use this procedure, oscillating wave functions for gap equation and NG boson $(B(q^2), \pi(q^2))$ which have nodes in the region $q^2 \subset [0, \Lambda^2]$ disappear and only the nodeless solution survives.

VI. Conclusions

1) Effective theory of strong coupling QED is constructed by means of effective action of bilocal fields.

2) The effective theory is mesons theory which interact via fermion loops.

3) The effective potential derived here has the chiral symmetry as the linear σ model has.

4) The relation between the existence of NG boson and non trivial solution of gap equation is clarified.

5) The ground state energy is finite if we take the Miranskys' continuum limit.

The following problems remain unsolved,

1) Are masses of the bound states also kept finite in the the Miranskys' continuum limit ? This problem is pursued by H.So and M.Suwa. (See the article by H.So in this proceedings.)

2) The physical coupling of QED is defined as usual renormalization procedure; $e_{physical}^2 = Z_3 e^2$. Is it possible to satisfy the both of the equations ? ; $\frac{de_{pyhsical}}{d\Lambda} = 0$, $\frac{dB0}{d\Lambda} = 0$.

ACKNOWLEDGMENTS

The author likes to thank Drs. H.So and K.Higashigima for discussions.

REFERENCES

1. T.Morozumi and H.So ,Prog.Theor.Phys.77(1987),1434

2. T.Maskawa and H.Nakajima ,Prog.Theor.Phys.52(1974),1326;
   Prog.Theor.Phys.54(1975),860

3. R.Fukuda and T.Kugo ,Nucl.Phys.B117(1976),250.

4. P.I.Fomin, V.P.Gusynin and V.A.Miransky,Phys.Lett.78B(1978),136
   P.I.Fomin, V.P.Gusynin V.A.Miransky and Yu.A.Sitenko,
   Riv.Nuovo Cim.,No.5(1983),1 V.A.Miransky,Nuovo Cim.A90 (1985),149

5. W.A.Bardeen, C.N.Leung and S.T.Love,
   Phys.Rev.Lett.56(1986),1230; Nucl.Phys.B273(1986)649

6. H.Kleinert,Phys.Lett.62B(1976),429;
   E.Shrauner,Phys.Rev.D16(1977),1877

7. T.Kugo,Phys.Lett.76B(1978),625

8. T.Goldman and R.Haymaker Phys.Rev.D24(1981),724

9. C.Roberts and H.Cahill,Phys.Rev.D33(1986),1775

10. J.M.Cornwall,R.Jackiw,and E.Tomboulis,Phys.Rev.D10(1974),2428

11. M.Inoue,H.Katata,T.Muta,and K.Shimizu, Prog.Theor.Phys.79(1988),519

12. R.Fukuda,Prog.Theor.Phys.78(1987),1487

A NEW PHASE OF QED AND NARROW PEAKS IN HEAVY-ION COLLISIONS

D. G. Caldi*

*Department of Physics, University of Connecticut, Storrs, CT 06269
and Department of Physics, Yale University, New Haven, CT 06511*

Dedicated to Heinz Pagels, 1939-1988, Mentor and Friend

ABSTRACT

After briefly reviewing the experimental evidence for narrow peaks in e^+e^- coincidence data from low-energy collisions of heavy ions and various theoretical explanations, we concentrate on the scenario that the narrow peaks are well explained by composite states of e^+e^- in a new phase of QED with chiral symmetry spontaneously broken. Evidence for a new phase from lattice and continuum studies is discussed, as well as the question of whether background fields can induce the phase transition. The spectrum obtained from a strong-coupling calculation in the new phase is given, which agrees well with the experimental peaks. A model for gamma-ray bursts based on the new phase is also discussed.

1. Introduction - Brief Review of the Experiments

A remarkable set of data has emerged in the past few years from a series of experiments at GSI involving heavy-ion collisions with the detection of positrons at first, and then in later experiments electrons and positrons in coincidence[1,2]. The unexpected and quite puzzling feature of these data is the appearance of narrow peaks whose position is stable to changes in the combined Z of the colliding nuclei. In this talk I will concentrate on the explanation[3] that these peaks are due to e^+e^- bound states in a new phase of QED.

The experiments which started this whole business were done at the GSI (Gesellschaft für Schwerionenforschung) in Darmstadt using the UNILAC accelerator. Two groups with rather different detectors were

---

* DOE Outstanding Junior Investigator. Work supported in part by U. S. Department of Energy, contract DE-AC02-79ER10336.

involved: the EPOS (Electron POsitron Solenoid) spectrometer group and the ORANGE spectrometer group. The original motivation for the experiments was to look for spontaneous positron emission, so-called sparking the vacuum. This effect, which had been predicted[4] for a hypothetical atom with a nucleus having a Z > 173 and a vacancy in the 1s orbital, entails the energetically favorable spontaneous e^+e^- pair production, with the electron going into the 1s state and the emission of a monoenergetic positron. Since such atoms are not known, the next best thing was to try to produce such a nucleus at least transiently in a collision of sufficiently large nuclei so as to achieve a supercritical Z. Due to the short lifetime of the compound nucleus, some broadening of the positron peak was expected (instead of a few keV, about 300 keV). However, a spectacular signature was expected to be the very strong dependence of the position of the peak on Z, i. e. Z^{20}.

So the initial experiments were done with U, Cm, and Th beams of about 6 MeV/nucleon on similar targets in all possible combinations. Initially only positrons were detected (besides the scattered ions). Indeed narrow positron peaks were found by both groups, but with essentially no Z-dependence and later below the critical Z (e. g. thorium on tantalum, Z = 163). For these and other reasons these positron peaks, whatever they were, did not seem to be the result of spontaneous positron emission. Because the data suggested a neutral particle of mass $\approx$ 1.7 MeV decaying to e^+e^- at rest in the center of mass frame, the EPOS group modified their spectrometer so that they could detect electrons in coincidence with the positrons. The result was that they found additional peaks. An important feature of their data is that the e^+ and e^- have equal energies (to within about 30 keV) and emerge back-to-back. There are now at least three peaks observed in the EPOS e^+e^- coincidence data with sum energies (including rest masses) of approximately 1.640 MeV, 1.770 MeV, and 1.830 MeV. The first and third also correspond to positron peaks seen by the ORANGE group, who claim to see an additional peak at positron kinetic energy around 250 keV. The EPOS group reports widths $\Gamma \leq 25 - 40$ keV (within experimental resolution). So the range of the lifetimes for these states is 10^{-19} s $< \tau < 10^{-9}$ s (the latter from the fiducial volume).

Subsequent to these experiments a Stanford-LBL group at the HILAC did a U-Th $\gamma\gamma$ coincidence experiment, reporting[5] a peak at 1.062 MeV.

However, on further analysis they feel[6] this can be explained by a known nuclear transition. There are now a considerable number of experiments[7] looking at low-energy e^+e^- scattering in hopes of seeing the same states observed in the heavy-ion collisions. Although there is a disputed positive report, the status appears to be that nothing has yet been definitively seen. If the scenario presented below is correct, this is perhaps not surprising.

2. What the Peaks Aren't

A large number of explanations have been proposed for these peaks, most of which have been found wanting even on phenomenological grounds[8]. With no claim of completeness, I will mention some of these in order to give a feeling for how mysterious these peaks are and how much they demand something really new to explain them.

First to go was the original motivation for the experiments, sparking the vacuum. As we have seen, there is no Z-dependence, the peaks are too narrow, and they appear below the critical Z. (There is actually some dispute[9] about what Z_{crit} is.) (Unfortunately, one of the unknowns in all this is how low in Z one can go and still see the peaks.) The final and crushing blow to this explanation was, of course, the appearance of electrons in coincidence.

Various attempts at explanations from nuclear physics effects have foundered on the following facts: the Z-independence, the narrowness, that the e^+e^- come out with equal energies and back-to-back, and finally the appearance of many peaks. Similar problems are encountered by attempts at atomic physics explanations.

Probably the most examined of the failed explanations was that the data were evidence for a new elementary particle - most notably the axion. This scenario had particular difficulties with the evidence that the state decays essentially at rest in the center of mass frame, that many other searches (especially beam dump experiments) although sensitive enough found nothing, and finally again that there are many peaks.

So after all these more or less conventional explanations failed, it seemed to some of us that, at the least, we had to assume some sort of composite system, especially because there are many states. The question then was, a composite system of what?

3. A "New" Phase of QED

Since the data appear to support the conclusion that some composite state decays to e^+e^-, the simplest and most conservative assumption is that the state is a composite of e^+ and e^-. So we are thinking of something like positronium, but not positronium itself since all its levels are known to be below the $2m_e$ threshold. Thus we were led to postulate[3] that these states are bound states of e^+ and e^- in a new non-perturbative phase of QED which has a new scale associated with it. The most familiar way to generate a new scale is to have chiral symmetry spontaneously broken in the new phase, so that the electron has an additional contribution to its mass and hence the bound states can lie in the neighborhood of 1.7 MeV.

This new-phase scenario is able to explain most (we hope eventually all) of the heretofore puzzling features of the data. It answers the question why the states are seen in the heavy-ion collisions and not elsewhere, by taking note of the strong and rapidly varying background electromagnetic fields as the high-Z nuclei collide. The idea is that these unusual background field environments induce a phase transition to the new (but already theoretically known, see below) QED vacuum. After the ions separate and the fields die off, the new vacuum is metastable but, since it is now false vacuum, it eventually decays, liberating the usual e^+e^- in the normal phase. Since there are no large fields now, the e^+ and e^- can come out with equal energies and back-to-back, as seen in the data. This picture also explains why there is some preference for decay into e^+e^- rather than photons. Of course, we also see why there are several states, since a composite system naturally exhibits several levels. Finally, a clear prediction of this scenario is the existence of an electro-pion which should be somewhat lighter than the other states. Although the $\gamma\gamma$ state at 1.062 MeV was a good candidate for such an electro-pion, its disappearance should not discourage the search for a relatively low-lying pseudoscalar state decaying to $\gamma\gamma$.

What is the theoretical evidence for a new, non-perturbative phase of QED? It comes from both lattice and continuum studies, and indeed much of this workshop is devoted to this topic. For quite some time it has been known[10] that as one increases the coupling to order one, lattice

QED in various guises undergoes a phase transition. For the pure gauge compact U(1) theory, there is rigorous proof that the new phase is a confining one. When fermions are added, both the compact and non-compact versions of lattice QED appear to have the new phase with chiral symmetry spontaneously broken, and this appears to persist even with dynamical fermions. (It should be noted that the compact case seems to have a first-order transition and so it is harder to relate to the continuum.)

In the continuum there have been a number of studies,[11] using the Schwinger-Dyson equation for the electron self-energy, which also show a spontaneously broken chiral phase when $\alpha \approx 1$, at least in the quenched, planar, ladder approximation. There are many papers in these proceedings on this topic. One of the significant developments has been the realization[12] that four-fermion operators are renormalizable in the new phase and hence relevant. Recent studies, both in the continuum Schwinger-Dyson equation approach[13] and in the Hamiltonian lattice formulation[14], of the two-coupling phase diagram indicate that the phase transition can occur for weak gauge coupling (including $\alpha = 1/137$) if the four-fermion coupling is strong enough.

In order for this theoretically known phase transition to be operative in the heavy-ion collision context, instead of varying the coupling (be it gauge or fermion), one must demonstrate that background field configurations present in the heavy-ion collisions can themselves induce the phase transition. There are a number of studies[15] beginning to explore this question so central to the new-phase explanation. However, so far they are inconclusive.

Pending definitive results from these ongoing efforts, one of the best pieces of indirect evidence for our scenario, besides its phenomenological successes, is the agreement of the spectrum calculated in the new phase and the observed states. My collaborators and I have done[16] a fourth-order ($1/g^8$) calculation of the spectrum using a strong-coupling expansion in Hamiltonian lattice QED with Kogut-Susskind fermions. We included a chirally invariant four-fermion interaction, whose coupling parameter we call A. The value A = 1.0 gives a reasonable fit to the heavy-ion data. Using [1,1] Padé approximants, I found[14] that the phase transition for A = 1.0 takes place at gauge coupling g = 0.585. Forming mass ratios for a slightly larger g, and fitting the lowest energy state to the peak observed at 1.64 MeV, we obtain the spectrum presented in Table I.

Table I. Spectrum of the low-lying electro-mesons

$m_\pi = 1.68$ MeV	$m_B = 1.82$ MeV
$m_\rho = 1.64$ MeV	$m_{A_1} = 1.64$ MeV
$m_\sigma = 1.69$ MeV	$m_f = 1.78$ MeV

We see quite good agreement with the experimental peaks, plus a prediction for additional states. The exact position of these states depends somewhat on the fitting state as well as on the accuracy of the fourth-order calculation. It remains to be seen whether the experiments can resolve these additional states. If one varies A from 0.5 to 1.5, the spectrum is fairly stable. We note that we are at an $\alpha_{lattice} = 0.0325$, i. e., quite weak gauge coupling. (What this corresponds to in the continuum has not yet been worked out.) We also note an obvious problem with the spectrum, namely that the electro-pion is too heavy. This is due to the well-known difficulty that Kogut-Susskind fermions do not have continuous chiral symmetry. The same problem is observed in the ρ-A_1 degeneracy. This may be cured in higher order, and is resolved in the continuous chiral symmetry of staggered fermions on the Euclidean lattice, as reported by Kogut[17].

4. γ-Ray Bursters

Quite recently my collaborators and I have found[18] another system, this time extraterrestial, in which these ideas about a new phase of QED again appear to explain a puzzling series of data. The observation of gamma-ray bursts with power of about 10^{34} erg/s has intrigued astrophysicists for almost two decades. Most believe that they originate in neutron stars. But what triggers these bursts and what powers them have been the subjects of much speculation. Because neutron stars have very large and rapidly fluctuating electromagnetic fields, a natural candidate for the power source is the new phase of QED. The model we have developed is that a sizable fraction of the neutron star or its magnetosphere is usually in this new phase. At some point in the star's life one of the many proposed triggering mechanisms (e. g., a star quake) produces a sudden, localized disturbance in the magnetosphere leading to a fluctuation in the electromagnetic field large enough to induce a phase transition back to the usual, perturbative vacuum of QED. The decay of the e^+e^- bound states over the relatively rapid time scale 10^{-19} to 10^{-9} seconds produces the gamma-ray bursts. That the emitting volumes come out to be only a very small fraction of the total volume of the star is an important advantageous feature of our model, since finding powering mechanisms for producing bursts with about 10^{34} ergs/s has proved difficult without exhausting most or all of the available energy in the star. Other interesting details of the gamma-ray bursts are also explained by our new-phase model.

5. Conclusions

It is an exciting prospect that what we had all thought to be our most understood quantum field theory may have a completely different side to it - a new, non-perturbative phase with chiral symmetry spontaneously broken and confinement, not just theoretically but actually realized both in the laboratory and in neutron stars. So far it is the only essentially complete explanation of the heavy-ion data, and now it also explains many of the aspects of gamma-ray bursts. The major problem is still to show that background electromagnetic fields can induce the phase transition. In addition, a tractable, realistic formalism for calculating things like lifetimes is needed. Finally we are continuing our search for yet other systems and experiments where new QED will put on its thought-provoking show.

Acknowledgements

Much of the work described here was done in collaboration mainly with A. Chodos, as well as with F. Accetta, K. Everding, D. Owen, and S. Vafaeisefat. It is a pleasure to thank T. Appelquist, W. Bardeen, J. Greenberg, R. Holdom, J. Kogut, Y. Nambu, and H. Minakata for stimulating and enlightening discussions.

References

1. J. Schweppe et al., *Phys. Rev. Lett.* **51**, 2261 (1983); M. Clemente et al., *Phys. Lett.* **137B**, 41 (1984); T. Cowan et al., *Phys. Rev. Lett.* **54**, 1761 (1985); T. Cowan et al., *Phys. Rev. Lett.* **56**, 444 (1986); H. Tsertos et al., *Phys. Lett.* **162B**, 273 (1985); H. Tsertos et al., *Z. Phys. A* **326**, 235 (1987).
2. For a review, see T. Cowan and J. Greenberg, in *Physics of Strong Fields*, ed. by W. Greiner (Plenum, New York, 1987).
3. D. G. Caldi and A. Chodos, *Phys. Rev.* **D36**, 2876 (1987); D. G. Caldi, A. Chodos, K. Everding, D. A. Owen, and S. Vafaeisefat, Yale preprint, YCTP-P21-88 (Sept. 1988); L. S. Celenza, V. K. Mishra, C. M. Shakin, and K. F. Liu, *Phys. Rev. Lett.* **57**, 55 (1986); L. S. Celenza, C. R. Ji, and C. M. Shakin, *Phys. Rev.* **D36**, 2144 (1987); Y. J. Ng and Y. Kikuchi, *Phys. Rev.* **D36**, 2880 (1987); Y. Kikuchi and Y. J. Ng, North Carolina preprint, IFP-314-UNC (April, 1988).
4. Ya. B. Zeldovich and V. S. Popov, *Usp. Fiz. Nauk* **105**, 403 (1971) [*Sov. Phys. Usp.* **14**, 673 (1972)]; B. Müller, J. Rafelski, and W. Greiner, *Z. Phys.* **257**, 62 (1972) and **257**, 183 (1972); W. Greiner, B. Müller, and J. Rafelski, *Quantum Electrodynamics of Strong Fields* (Springer-Verlag, Berlin, 1985) and references therein.
5. K. Danzmann et al., *Phys. Rev. Lett.* **59**, 1885 (1987).
6. W. E. Meyerhof, private communication.
7. A. P. Mills Jr. and J. Levy, *Phys. Rev.* **D36**, 707 (1987); U. von Wimmersperg et al., *Phys. Rev. Lett.* **59**, 266 (1987); K. Maier et al., *Z. Phys. A* **326**, 527 (1987) and **330**, 173 (1988); S. H. Connell et al., *Phys. Rev. Lett.* **60**, 2242 (1988); H. Tsertos et al., *Phys. Lett.* **207B**, 273 (1988); J. van Klinken et al., *Phys. Lett.* **205B**, 223 (1988); M. Minowa et al., Univ. of Tokyo preprint (1988); E. Lorenz et al., *Phys. Lett.* **B**, to be published; J. Greenberg et al., in progress at Brookhaven.
8. See A. Chodos, *Comments Nucl. Part. Phys.* **17**, 211 (1987) and references therein.
9. Y. Hirata and H. Minakata, *Phys. Rev.* **D34**, 2493 (1986), and private communication.
10. A. Guth, *Phys. Rev.* **D21**, 2291 (1980); J. Fröhlich and T. Spencer, *Commun. Math. Phys.* **83**, 411 (1982); C. B. Lang, *Nucl. Phys.* **B280 [FS18]**, 225 (1987), and references therein; E. Dagotto and J. Kogut, *Phys. Rev. Lett.* **59**, 617 (1987); *Nucl. Phys. B*, to be published; J. B. Kogut, E. Dagotto, and A. Kocic, *Phys. Rev. Lett.* **60**, 772 (1988).
11. T. Maskawa and H. Nakajima, *Prog. Theor. Phys.* **52**, 1326 (1974); **54**, 860 (1975); R. Fukuda and T. Kugo, *Nucl. Phys.* **B117**, 250 (1976); V. A. Miransky, *Il Nuovo Cim.* **90A**, 149 (1985); V. A. Miransky and P. I. Fomin, *Sov. J. Part. Nucl.* **16**, 203 (1985).
12. W. A. Bardeen, C. N. Leung, and S. T. Love, *Phys. Rev. Lett.* **56**, 1230 (1986); *Nucl. Phys.* **B273**, 649 (1986).
13. T. Appelquist, M. Soldate, T. Takeuchi, and L. C. R. Wijewardhana, Yale preprint, YCTP-P19-88 (August, 1988); K. Kondo, H. Mino, and K. Yamawaki, Nagoya preprint, DPNU-88-18 (June, 1988); W. A. Bardeen, C. N. Leung, and S. T. Love, these proceedings.
14. D. G. Caldi, *Phys. Lett.* **B**, to be published.
15. E. Dagotto and H. W. Wyld, *Phys. Lett.* **205B**, 73 (1988); A. Chodos, D. Owen, and C. Sommerfield, *Phys. Lett.* **B**, to be published, R. D. Peccei, J. Solà, and C. Wetterich, *Phys. Rev.* **D37**, 3206 (1988); Y. J. Ng, these proceedings; H. Minakata, these proceedings.
16. D. G. Caldi, A. Chodos, K. Everding, D. A. Owen, and S. Vafaeisefat, ref. 3.
17. J. B. Kogut, these proceedings.
18. F. S. Accetta, D. G. Caldi, and A. Chodos, Yale preprint, YCTP-P24-88 (November, 1988).

STRONG COUPLING QED AND ITS POSSIBLE RELATION
TO ANOMALOUS HEAVY-ION EVENTS

Y. Jack Ng[*]

Institute of Field Physics
Department of Physcis and Astronomy
University of North Carolina
Chapel Hill, NC 27599
USA

ABSTRACT

We suggest that the e^+e^- peaks observed in heavy-ion collisions at GSI are due to the decay of a bound e^+e^- system formed in a new QED confining phase which is induced by the electromagnetic fields of the large-Z ions. The mass spectrum for this new positronium system with a linear-potential is discussed. We also present a qualitative analysis of the Schwinger-Dyson equation in the presence of a semi-realistic background field to see if there possibly exist background electromagnetic fields (hopefully like those in heavy-ion collisions) that can induce a phase transition in QED at relatively weak couplings. Unfortunately, the (constant and weak) background field in our example does not seem to affect the QED critical coupling. Better approximations or some more realistic background fields may change the result.

1. INTRODUCTION

The recent observations of multiple narrow correlated e^+e^- peaks by both the EPOS[1] and the Orange[2] groups at GSI in Darmstadt and of a narrow two-photon peak by the Stanford-Berkeley-Livermore collaboration[3] at the Berkeley Super-HILAC in large-Z heavy-ion collisions have generated considerable theoretical work[4-14]. I will discuss the conjecture[4,5] that the peaks are due to the decay of a bound e^+e^- system formed in a new strong-coupling phase of QED. (See Ref. 6 for a similar idea involving the decay of a nontopological soliton formed in a new vacuum phase.) I will also speculate on the

---
[*]Supported in part by the U.S. Department of Energy under Grant No. DE-FG05-85ER-40219.

existence of states consisting of strongly bound photons only[4,14]. Then a qualitative study[11] of the Schwinger-Dyson equation in the presence of a weak photon condensate background field is presented. The result seems to suggest that the value of the QED ultraviolet fixed point (which separates the strong coupling regime from the weak coupling regime) is not modified by our semi-realistic background field. We do not know if a more accurate treatment of our example or some more realistic background fields (like those in the heavy-ion collisions) can cause the fixed point to move towards weaker coupling. Further investigations are warranted.

2. ANOMALOUS HEAVY-ION EVENTS AND A POSSIBLE THEORETICAL INTERPRETATION

2.1 Summary of Experimental Results

Both the EPOS[1] and the Orange[2] groups have reported evidence for correlated narrow-peak structures in electron and positron spectra. These multiple structures in the range of 1.5-1.8 MeV have been seen in U + Th, U + U and other heavy-ion collisions and are relatively Z independent. The energies of the e^+ and e^- are roughly equal (to within ~ 30 keV). They are emitted back to back (with the angle between them greater than 150°); the peak corresponding to the sum of the electron and positron energies is considerably narrower than that of the individual electron and positron peaks. (The sum peak is probably less than the instrumental resolution.) The experimental data is consistent with the two-body decay of a neutral object at rest in the center-of-mass frame of the heavy-ion system, with 10^{-19} sec $\leq \tau \leq 10^{-9}$ sec. (In comparison, the characteristic time in which the heavy ions interact strongly is only about ~ 10^{-21} sec.) Recently the Stanford-Berkeley-Livermore collaboration[3] at LBL has reported evidence of a correlated two-photon decay line in U + Th collisions with a sum energy of 1062 ± 1 keV and an intrinsic width less than 2.5 keV. The experimental data is summarized in Table 1.

Table 1. Heavy-ion Results

Mass (keV)	Width (keV)	Type of Peaks	Experimental Groups
1062±1	≤2.5	γγ	SBL collaboration
1498±20		e^+e^-	ORANGE
1646±10	≤25	e^+e^-	EPOS
1782±20		e^+e^-	EPOS, ORANGE
1837±10	≤40	e^+e^-	EPOS, ORANGE

2.2 Theoretical Interpretation

The multiple structure in the e^+e^- spectra observed at GSI is indicative of the formation of a bound e^+e^- state. But that bound state cannot be the familiar conventional positronium system with a Coulomb potential since they have very different energy spectra. A possible scenario is that the e^+e^- peaks are due to the decay of a bound e^+e^- system formed in a new phase of QED[4,5,6] and that this new phase is induced by the strong and rapidly varying electromagnetic fields of the large-Z ions. Conceivably[4,5], shortly after the scattering of the heavy ions, in the absence of the strong electromagnetic fields the new positronium system is left in a metastable vacuum; hence, it decays to the normal weak-coupling phase via the Coleman-Frampton tunneling mechanism[15] into the observed e^+e^- pairs. This idea fits nicely with several features of the experimental data enumerated above. It automatically explains why those e^+e^- peaks appear only in heavy-ion collisions and not elsewhere. Besides, the proposal[4,5] is a very economical one in that it dispenses with the need to postulate any new particles or fields and it makes use of a new phase of QED, the existence of which has strong support from earlier studies of the Schwinger-Dyson equations[16,17] and lattice calculations[18].

To consider the mass spectrum of the confined positronium system let us follow Ref. 4 and use

$$V(\vec{r}) = \lambda r, \qquad (1)$$

the linear potential between two charged particles obtained in the (static) strong-coupling regime of lattice QED. I will follow Cea[14] in using a modified Schrödinger-equation which takes into account relativistic kinematics in the form

$$\left[\sqrt{-\nabla^2+m_1^2} + \sqrt{-\nabla^2+m_2^2} + V(\vec{r})\right]\psi(\vec{r}) = M\psi(\vec{r}) \qquad (2)$$

where

$$m_1 = m_2 = m_e \simeq 511 \text{ keV} \qquad (3)$$

Cea fixed the string tension λ by fitting the available experimental e^+e^- spectrum (not including the $\gamma\gamma$ state found at Berkeley) and found $\sqrt{\lambda} \simeq 252$ keV. He also calculated the root mean square radius $\overline{r^2}$ for the different states. The result of his calculations is shown in Table II. The linear potential model[4,14] agrees well with the experimental spectrum.

TABLE II Low-lying e^+e^- bound states compared with the available experimental states. The potential is $V(r) = \lambda r$, $\sqrt{\lambda} = 252$ keV.

(n,l)	$M_{n,l}$(keV)	$\bar{r}$(Fermi)	M_{exp}(keV)
(1,0)	1471	983	1498±20
(1,1)	1660	1343	1646±10
(2,0)	1785	1651	1782±20
(1,2)	1818	1649	1837±10
(2,1)	1925	1934	–
(3,0)	2032	2167	–

Cea also argued that since the confining QED phase is triggered by the strong time-dependent nuclear Coulomb field[4,5,6] one expects that states with root mean radius greater than a certain critical value $\bar{r}_c$ are not produced. If the observed 1837 keV state is the most massive e^+e^- state then from Table II one has

$$\bar{r} < \bar{r}_c \simeq 1900 \text{ Fermi} \qquad (4)$$

In addition to the new positronium system one expects that there are states consisting of strongly bound photons only[4]. Since they are the QED counterparts of glueballs in QCD let us call them photoballs (or lightballs). The existence of photoballs was predicted[4] on the ground that in the lattice QED Lagrangian there is a term $\cos \theta_p$ where θ_p is proportional to $F_{\mu\nu}$, hence photons are coupled to themselves to form photoballs.

To calculate the mass spectrum of the photoballs let us simple-mindedly follow Cea[14] in assuming that photons are also trapped in the postulated new QED phase and that there are photon-photon bound states that obey Eqs. (1) and (2) with $m_1 = m_2 = 0$ and $\sqrt{\lambda} = 252$ keV (fixed by the e^+e^- spectrum). The result of Cea's calculations is shown in Table III.

Table III Low-lying photoball spectrum compared with the experimental available state.

(n,l)	$M_{n,l}$(keV)	$\bar{r}$(Fermi)	M_{exp}(keV)
(1,0)	796	1379	-
(1,1)	1062	1775	1062±1
(2,0)	1187	2108	-
(1,2)	1273	2088	-

We see that there are only two states with $\bar{r} < \bar{r}_c$, namely the states with mass 796 keV and 1062 keV. With regard to the 796 keV state one observes that this state is lighter than $2m_e$ so the corresponding γγ line is masked by the large background of nuclear cascades. This led Cea to argue that one can only observe the 1062 keV state. But this is exactly the state found by the Stanford-Berkeley-Livermore collaboration! Admittedly Cea's assumptions and procedures are open to questions, his results are nonetheless intriguing and amusing. We should add that an alternative and more plausible interpretation of the 1062 keV state is that it is the lightest positronium state, the counterpart of the pion in hadron physics.[4,5]

3. STRONG COUPLING QED IN THE PRESENCE OF A BACKGROUND FIELD

The attractive theoretical scenario described above hinges on the speculation that there is a connection between the strong (as well as rapidly-varying) electromagnetic fields in large-Z heavy-ion collisions and strong-coupling QED. But this connection has not been proven so far. In fact, it is known that constant strong background fields (in one-loop approximations) are not likely to lead to strong-coupling QED.[7,12] It is of interest, therefore, to investigate, using non-perturbative techniques, whether there exist electromagnetic background fields that can potentially induce a phase transition from the weak-coupling regime to the strong-coupling regime of QED at relatively weak couplings. I am now going to discuss a qualitative study[11] of the non-perturbative Schwinger-Dyson equation for the fermion self-energy $\Sigma(p)$ in the presence of a simple semi-realistic electromagnetic background field which takes the form of a weak photon-condensate.

Before we discuss the effects of an electromagnetic background field on the Schwinger-Dyson equation let us recall some of the analyses[16,17] of the SD equation in the absence of a background field. In the "ladder" approximation, after making a Wick rotation and adopting the Landau gauge for the photon propagator, the integral SD equation can be converted into the following differential equation[16] ($\Box$ is the four-dimensional Laplace operator)

$$\Box \Sigma(p) = -\frac{\alpha}{\alpha_c} \frac{\Sigma(p)}{p^2 + \Sigma^2(p)} \tag{5}$$

satisfying the boundary conditions (with m_o as the bare fermion mass)

$$\lim_{p^2 \to \infty} \left(\frac{d\Sigma(p)}{d(\ln p^2)} + \Sigma(p) \right) = m_o$$

$$\lim_{p^2 \to 0} \frac{d\Sigma(p)}{d(\ln p^2)} = 0 \tag{6}$$

In Eq. (5) we have used $\alpha = \frac{e^2}{4\pi}$ and $\alpha_c = \frac{1}{3}\pi$. The solutions to the Schwinger-Dyson equation (Eqs. (5) and (6)) were analyzed by Fukuda and Kugo [16] for both weak coupling ($\alpha<\alpha_c$) and strong coupling ($\alpha>\alpha_c$). In the ultra-violet limit they found

$$\Sigma(p) = \begin{cases} p^{-1 + \sqrt{1-\alpha/\alpha_c}} & \alpha \leq \alpha_c \\ p^{-1} \cos(\sqrt{\alpha/\alpha_c - 1} \ln p) & \alpha > \alpha_c \end{cases} \quad (7)$$

In order to have a sensible infrared limit for the theory it was pointed out by Miransky [16] that $\alpha = \alpha_c$ should be viewed as an ultra-violet fixed point.

An electromagnetic background field $A_\mu(x)$ can be easily incorporated into the Schwinger-Dyson equation. For ease of calculations, let us take the background field to have the form of a weak constant photon-condensate given by

$$\phi' = <0|2\pi\alpha \, F_{\mu\nu} F^{\mu\nu}|0> \quad (8)$$

Then Eq. (5) is replaced by

$$\Box \Sigma(p) = -\frac{\alpha}{\alpha_c} \frac{\Sigma(p) + B(p)}{p^2(1-A(p))^2 + (\Sigma(p)+B(p))^2} \quad (9)$$

where

$$A(q) = \phi \Sigma^2(q)/(q^2 + \Sigma^2(q))^3$$
$$B(q) = \phi \Sigma(q) q^2 /(q^2 + \Sigma^2(q))^3 \quad (10)$$

with

$$\phi = 2\pi\alpha <\vec{H}^2 + \vec{E}^2> \quad (11)$$

In the ultra-violet ($\phi/(p^2)^2 \ll 1$; $\frac{\Sigma(p)}{p} \ll 1$) limit we can approximate Eq. (9) as

$$\Box \Sigma(p) = -\frac{\alpha}{\alpha_c} (1 + \frac{\phi}{(p^2)^2}) \frac{\Sigma(p)}{p^2 + \Sigma^2(p)}$$

$$= -\frac{\alpha}{\tilde{\alpha}_c} \frac{\Sigma(p)}{p^2 + \Sigma^2(p)} \qquad (12)$$

with

$$\tilde{\alpha}_c \simeq \alpha_c (1 - \frac{\phi}{(p^2)^2}) < \alpha_c \qquad (13)$$

But Eq. (12) has the same form as the SD equation (5) for the case of vanishing external background fields. Naively the only difference is in the value of the critical point. For the case of a photon-condensate background field it is less than $\alpha_c = \frac{1}{3}\pi$. (We believe that the mixing of four-fermion interactions with the electrodynamic interactions as discussed by, e.g., Leung et al in Ref. 17 will not change this qualitative result.) This <u>naive</u> conclusion[19]) seems to be bolstered by the fact that Eq. (12) is solved, e.g., in the ultra-violet limit for the weak coupling regime, by

$$\Sigma(p) = p^{-1 + \sqrt{1-\alpha/\alpha_{c,eff}}} \qquad p \to \infty, \text{ weak coupling} \qquad (14)$$

with

$$\alpha_{c,eff} = \alpha_c \left[1 - \frac{1}{2} \frac{\sqrt{1-\alpha/\alpha_c}}{2-\sqrt{1-\alpha/\alpha_c}} \frac{\phi}{(\ln p^2)(p^2)^2} \right] < \alpha_c \qquad (15)$$

However, since $\alpha_{c,eff}$ depends on $\sqrt{1-\alpha/\alpha_c}$ it is not suitable to play the role of the (effective) critical coupling. The critical coupling can be properly identified by exhibiting the solutions to Eq. (12) in the ultra-violet limit as

$$\Sigma(p) = p^{-1 + \sqrt{1-\alpha/\alpha_c}} (1 + a \frac{\phi}{(p^2)^2}) \qquad \alpha \leq \alpha_c \qquad (16a)$$

$$\Sigma(p) = p^{-1} \left[(1+b \frac{\phi}{(p^2)^2}) \cos(\sqrt{\alpha/\alpha_c - 1} \ln p) \right.$$
$$\left. - \frac{1}{2} b \frac{\phi}{(p^2)^2} \sqrt{\alpha/\alpha_c - 1} \sin(\sqrt{\alpha/\alpha_c - 1} \ln p) \right] \qquad \alpha > \alpha_c \quad (16b)$$

where

$$a = -\frac{\alpha}{8\alpha_c} \frac{1}{2 - \sqrt{1-\alpha/\alpha_c}}$$
$$b = -\frac{1}{4} \frac{\alpha}{\alpha_c} \frac{1}{3 + \alpha/\alpha_c} \qquad (17)$$

We conclude that our semi-realistic background field modifies the solutions to the SD equation without changing the critical coupling.

The relevance of the result we have just obtained to the heavy-ion experiments is hard to assess. For one thing the background field in our example is too simplistic and not realistic enough; we have taken the background field to be constant and weak (instead of strong and rapidly-varying as in the heavy-ion experiments); we have kept only the leading order in the background-field; and we have used the ladder approximation in the SD equation. Perhaps in a proper treatment of the SD equation some realistic background fields[20] can be found to drive QED to a new phase. The problem is hard and much more work remains to be done. But it is exciting to contemplate the possibility that the anomalous heavy-ion events we have been discussing may turn out to be the first physical evidence of a phase transition in a gauge theory.

Some years ago Prof. J. Schwinger suggested to me a problem on the (conventional) positronium system as part of my thesis. That experience with the positronium inspired me to conjecture the theoretical scenario described in this talk to explain the anomalous heavy-ion events when I first heard of them (although it was T. Applequist who later convinced me to take seriously the multiple e^+e^- structure.) Furthermore the idea that four-dimensional QED can undergo a phase transition was already prefigured in Schwinger's work on the Schwinger model. Even our on-going program to study the effects of background fields on the Schwinger-Dyson equation will probably employ the strong-field techniques he taught me. This talk is appropriately dedicated to this great scientist and teacher on the occasion of his official (and fake) retirement in witness of my appreciation and affection for him.

NOTE ADDED

I just learned from W. E. Meyerhof that the ~6 MeV/N U + Th, U + U, Th + Th experiments at the Berkeley Super-HILAC this year have shown that the 1062 keV line is caused by a 543 + 519 keV nuclear cascade from a 32^+ state in U. The narrowness of the line is due to restricted kinematics needed to excite a high-spin state. Therefore, the 1062 keV line no longer represents an anomalous heavy-ion event.

REFERENCES

1. For a general review see Cowan, T. and Greenberg, J. S. in "Physics of Strong Fields" ed. by Greiner, W., Plenum Press, New York (1987), p. 111.
2. For a general review see Kienle, P. in "Physics of Strong Fields" ed. by Greiner, W., Plenum Press, New York (1987), p. 979.
3. Danzmann, K. et al, Phys. Rev. Lett. $\underline{59}$, 1885 (1987).
4. Ng, Y. J. and Kikuchi, Y., Phys. Rev. $\underline{D36}$, 2880 (1987).
5. Caldi, D. G. and Chodos, A., Phys. Rev. $\underline{D36}$, 2876 (1987).
6. Celenza, L. S., Mishra, V. K., Shakin, C. M. and Liu, K. F., Phys. Rev. Lett. $\underline{57}$, 55 (1986).
7. Peccei, R. D., Sola, J. and Wetterich, C., Phys. Rev. $\underline{D37}$, 2492 (1988).
8. See, e.g., Carrier, D., Chodos, A. and Wijewardhana, L. C. R., Phys. Rev. $\underline{D34}$, 1332 (1986); Müller, B. and Rafelski, J., ibid $\underline{D34}$, 2896 (1986); Krauss, L. M. and Zeller, M., ibid. $\underline{D34}$, 3385 (1986); Lane, K., Phys. Lett. $\underline{B169}$, 97 (1986); Wong, C. Y. and Becker, R. L., ibid $\underline{B182}$, 251 (1986); Müller, B., Reinhardt, J., Greiner, W. and Schäfer, A., J. Phys. $\underline{G12}$, L109 (1986); Celenza, L. S., Ji, C. R. and Shakin, C. M., Phys. Rev. $\underline{D36}$, 2144 (1987); Wong, C. W., ibid $\underline{D37}$, 3206 (1988).
9. Kogut, J. B., Dagotta, E., and Kocic, A., Phys. Rev. Lett. $\underline{60}$, 772 (1988).
10. Dagotto, E. and Wyld, H. W., Phys. Lett. $\underline{B205}$, 73 (1988).
11. Kikuchi, Y. and Ng, Y. J., Univ. of North Carolina preprint, IFP-314-UNC (1988). Part of this work is based upon Ishizuka, W. and Kikuchi, Y., Phys. Lett. $\underline{B127}$, 251 (1983).

12. Chodos, A., Owen, D. A. and Sommerfield, C. M., Yale University preprint, YCTP-P11-88 (1988).
13. Celenza, L. S., Pantziris, A., Shakin, C. M. and Wang, H. W., Brooklyn College preprint, B.C.C.N.T. 88/031/175 (1988).
14. Cea, P., Univ. of Bari preprint, BARI-TH/88-28 (1988).
15. Frampton, P. H., Phys. Rev. Lett. $\underline{37}$, 1378 (1976); Phys. Rev. $\underline{D15}$, 2922 (1977); Coleman, S. ibid. $\underline{15}$, 2929 (1977); Callan, C. G. and Coleman, S., ibid. $\underline{16}$, 1762 (1977).
16. Fukuda, R. and Kugo, T., Nucl. Phys. $\underline{B117}$, 250 (1976); Fomin, P., Gusynin, V., Miransky, V. and Sitenko, Y., Riv. Nuovo Cimento $\underline{6}$, 1 (1983); Miransky, V. A., Il. Nuovo Cim. $\underline{90A}$, 149 (1985); Miransky, V. A. and Fomin, P. I., Sov. J. Part. Nucl. $\underline{16}$, 203 (1985).
17. Johnson, K., Baker, M. and Willey, R., Phys. Rev. $\underline{136}$, B1111 (1964); Maskawa, T. and Nakajima, H., Prog. Theor. Phys. $\underline{52}$, 1326 (1974); Leung, C. N., Love, S. T. and Bardeen, W. A., Nucl. Phys. $\underline{B273}$, 649 (1986); Cornwall, J. M., Jackiw, R. and Tomboulis, E., Phys. Rev. $\underline{D10}$, 2428 (1974); Yamawaki, K., Bando, M. and Matumoto, K., Phys. Rev. Lett. $\underline{56}$, 1335 (1986).
18. Wilson, K. G., Phys. Rev. $\underline{D10}$, 2445 (1974); Creutz, M., Phys. Rev. Lett. $\underline{43}$, 553 (1979); Guth, A. H., Phys. Rev. $\underline{D21}$, 2291 (1980); Degrand, T. A. and Toussaint, D., ibid $\underline{D24}$, 466 (1981); Kogut, J. et al, Phys. Rev. Lett. $\underline{50}$, 393 (1983); Bartholomew, J. et al, Nucl. Phys. $\underline{B230}$, 222 (1984). See Kogut, J., Rev. Mod. Phys. $\underline{55}$, 775 (1983) for more references.
19. That the conclusion we have tentatively drawn so far is too naive was pointed out to me by T. Muta.
20. Not all interesting realistic background fields are necessarily like those in the heavy-ion collisions. We should look for any physical system or experiment that demonstrates the phase transition which we believe can occur in QED.

Does QED4 Exist?

J. B. Kogut

Department of Physics
University of Illinois at Urbana-Champaign
1110 West Green Street
Urbana, Illinois 61801
U.S.A.

ABSTRACT

We consider the analytic and numerical evidence that QED4 with N=0,2 or 4 species of Dirac fermions exist. The status of our understanding of chiral symmetry breaking fixed points at strong coupling is reviewed.

1. INTRODUCTION

Do interacting field theories which are not asymptotically free exist in four dimensions? The evidence, both analytic and numerical, is overwhelmingly negative for $\lambda\phi^4$. Even if the bare coupling λ_o is taken to infinity it appears that the renormalized charge is zero. There are famous arguments initiated by the Moscow School[1] which suggest that QED4 suffers the same trivial fate. In perturbation theory, improved perhaps by the renormalization group, the renormalized charge is found to vanish no matter how large the bare charge is chosen due to screening in the photon propagator. These analyses are not convincing at strong coupling of course, but only recently has a physically interesting loophole been suggested.[2] That loophole comes from studies, both analytic and numerical, in the quenched (N=0) version of QED4 which have uncovered a new source of coupling constant renormalization at strong coupling which leads to an ultra-violet stable/infra-red unstable fixed point that drives chiral symmetry breaking.[3] The dynamical mechanism, "collapse of the wavefunction", has been discussed in another contribution to this conference and will not be reviewed here. The point we are most concerned about is whether this new source of coupling constant renormalization and its associated fixed point survive the inclusion of fermions into the theory's dynamics. This is a difficult question that we only know how to study with computer simulation techniques. It is now possible to use those algorithms developed to simulate lattice QCD with light, dynamical fermions for a wide class of strongly coupled fermion problems.[4] Those algorithms which are based on stochastic differential equations have controlled errors and when combined with the staggered fermion lattice form of the Dirac equation allow one to study chiral symmetry breaking as a natural symmetry for any number of species N.

It is very important to develop analytic methods to discuss chiral symmetry breaking in QED4 beyond the quenched ladder approximation. This must be done in a systematic, self-consistent fashion which captures the essential physics. No doubt there will be many attempts in this direction which use the Schwinger-Dyson formalism. How should this system of equations be truncated and what ansatze should be used (or avoided!) for the various vertex parts, etc? Can we write down a tractable set of Schwinger-Dyson equations which captures the essence of the competition between the "collapse of the wave function" mechanism that drives chiral symmetry breaking in the N=0 theory and the screening that suppresses chiral symmetry breaking when N≠0? Recall that there are particularly compelling arguments that the Landau ghost and the triviality of QED4 is unavoidable when N→∞.[5] This result should come out of one's truncated set of Schwinger-Dyson equations.

Before delving into the body of this talk, I want to remind the Schwinger-Dyson enthusiast of some features of screening. There is a strong prejudice to assume that fermion vacuum polarization comes exclusively from insertions into the photon propagator. The massive Schwinger model gives a simple counter-example to this bit of conventional wisdom. Its abelian nature leaves the notion of confinement precise even in the case of screening. Place static charges $+\varepsilon$ and $-\varepsilon$ a distance r apart. Then there is a potential between them $V(r) = \sigma(\varepsilon)|r|$ where the string tension $\sigma(\varepsilon)$ is a periodic function of ε which vanishes at the integers $\varepsilon=0, \pm 1, \pm 2,...$ Single photon exchange with vacuum polarization corrections cannot give this behavior. Instead, multi-photon exchanges connected through a fermion loop are essential.[6] Recall also that the massive Schwinger model can be bosonized and its spectrum calculated. The theory consists of just massive mesons which interact through short range forces. The underlying photon (the $|r|$ potential in 1+1 dimension) does not show up in the spectrum or the Langrangian although it is responsible for the composite structure of the mesons. If we calculate the photon propagator, however, in the Landau gauge we would find a $1/q^2$ massless pole. Of course, all the physical states of the theory are exactly chargeless (this is a model of confinement) so the photon does not coupled at zero momentum transfer to physical states of the model. Of course, one "sees" the presence of the photon in the underlying dynamics by calculating the heavy quark potential $V(r) = \sigma(\varepsilon)|r|$ for fractional charges ε.

2. QUENCHED (N=0) DYNAMICS

The quenched limit of QED4 has some pathological features. The limit evades some of the consistency requirements of a real field theory. For example, as discussed extensively at this conference, the model develops an infinite correlation length at the critical point $\alpha = \alpha_c = \pi/3$ and has non-trivial chiral dynamics.[2] Nonetheless, it has no real temperature dependence[7] because the underlying dynamics (free photons) is free and scale invariant. Once fermion feedback is included in the dynamics one expects consistency here.

The chiral dynamics at zero temperature has the following interesting properties[2]:
1. Chiral symmetry breaking for strong coupling, $\alpha > \alpha_c = \pi/3$.
2. A dynamically generated correlation length with a non-trivial scaling law. The fermion dynamical mass reads,

$$\Sigma(0) = \Lambda \exp\left(-\pi/\sqrt{\alpha/\alpha_c - 1}\right) \qquad (2.1)$$

where Λ is the cutoff and $\alpha \gtrsim \alpha_c$.

3. An ultra-violet stable/infra-red unstable fixed point at $\alpha = \alpha_c$. Rewriting Eq. (2.1) we have the coupling constant flow,

$$\frac{\alpha}{\alpha_c} = 1 + \frac{\pi^2}{\ln^2(\Lambda/\Sigma(0))} \qquad (2.2)$$

4. The pion is a Goldstone boson. It is a $q\bar{q}$ composite with $f_\pi \neq 0$ in accord with low energy chiral theorems.

The scaling law Eq. (2.1) should come as quite a surprise to the reader knowledgeable in critical phenomena. One would have expected a mean field result $\Sigma(0) \sim \sqrt{\alpha - \alpha_{MF}}$ since we are in four dimensions and the system has a local order parameter. In fact one obtains Eq. (2.1) from a Schwinger-Dyson equation for $\alpha > \alpha_c$ where one assumes that $\Sigma(0) \ll \Lambda$, the ultra-violet cutoff. If instead one works at strong coupling $\alpha > \alpha_c$ and anticipates that the dynamical fermion mass should be large $\Sigma(0) \sim \Lambda$, the one finds from the same Schwinger-Dyson equation,[6]

$$\Sigma(0) = \Lambda \sqrt{\alpha - \alpha_{MF}} \qquad (2.3)$$

with $\alpha_{MF} = 4\alpha_c$. So, the equation illustrates a typical cross-over phenomenon between mean field theory in a region of parameter space where fluctuations are suppressed to a non-trivial scaling region where critical fluctuations dominate.

Extensive computer simulations have been done to test these N=0 predictions.[8] The mean field relation Eq. (2.3) has been confirmed well and some evidence has been found for the essential singularity Eq. (2.2). This is more difficult to accomplish numerically because of the accompanying fluctuations. Additional larger scale simulations will be done here to broaden the data in ref. 8. The reader is referred to that reference for tables of results and plots on $8^3 \cdot 16$ lattices.[8] The reader is also reminded that these and all simulations reported here were done in the <u>non-compact</u> version of lattice QED. This lattice method best approximates the continuum, non-compact theory and it avoids first order transitions that bedevil most forms of the compact theory on the lattice.[9]

Another crucial feature of the N=0 continuum model is the fact that four-fermi operators become renormalizable at $\alpha = \alpha_c$.[10] The scale dimension of $\bar{\psi}\psi$ is calculated in the quenched, planar approximation to be $d_{\bar{\psi}\psi} = 2 + \sqrt{1 - \alpha/\alpha_c}$ for $\alpha < \alpha_c$ while the scale dimension of $(\bar{\psi}\psi)^2$ is $2d_{\bar{\psi}\psi}$. These large derivatives from free field scaling have been related to the physical picture underlying chiral symmetry breaking in this model -- "collapse of the wavefunction".[3] It would be interesting to verify the formula for $d_{\bar{\psi}\psi}$ directly in a numerical simulation. Much larger lattices than $8^3 \cdot 16$ appear to be needed here.

3. FINITE SIZE EFFECTS IN THE N=2 AND 4 THEORIES

How can we distinguish between nontrivial theories which have divergent correlation lengths and those which do not? One way is to study the finite size effects in both cases and compare them. We made extensive measurements of $\langle \bar{\psi}\psi \rangle$ in the N=2 and 4 theories on 6^4, 8^4 and 10^4 lattices.[11] At a given coupling $\beta = 1/e^2$ and quark mass the condensate $\langle \bar{\psi}\psi \rangle$ was always larger on the larger lattices. By comparison we considered a free, quenched vector meson theory whose mass was chosen so that the theory's average plaquette was the same as the N=2 or 4 theory we were comparing with.[11] In this way the trivial model and the real theory had identical short distance behaviors. Typically the vector meson mass was O(1) in lattice units, reflecting the fact that the average plaquette is ultra-violet divergent. Measurements of $\langle \bar{\psi}\psi \rangle$ on 6^4, 8^4 and 10^4 lattices in the free theory gave the <u>opposite</u> systematics compared to the N=2 and 4 theories. Why? The free theory with a large screening mass has interactions only over distances comparable to the cutoff length, the lattice spacing. The model cannot develop a large correlation length. So, on larger lattices the interactions in such theories are less significant and the condensate falls. The fact that the N=2 and 4 theories have opposite systematics is our first, crude indication that they have interacting continuum limits.

Finite temperature studies show the same trend.[11] Field theories at temperature T are simulated by chosing an asymmetric lattice $N_\tau \times N^3$ with $N \gg N_\tau$. The temporal extent N_τ is related to the temperature T and the lattice spacing a, $N_\tau = 1/aT$. One sees in an interacting theory that chiral symmetry breaking disappears at large T by studying one lattice size $N_\tau \times N^3$ and varying the coupling β. As β approaches β_c the correlation length ξ grows and eventually reaches $N_\tau = 1/aT$. Then $\langle \bar\psi\psi \rangle$ decreases to zero rapidly because the finite size of the system becomes relevant and the broken symmetry must be restored. In this way one can measure the rate at which ξ grows as $\beta \to \beta_c$ by studying lattices of different N_τ. In an interacting theory the β value where chiral symmetry is restored shifts from strong coupling to β_c as N_τ increases. In a trivial theory there is little size dependence, of course, and it may in fact have the opposite systematic trend. Simulations on 2×8^3, 4×8^3, 6×12^3 and 8×16^3 showed these effects quite clearly. Roughly speaking the N=4 theory had finite temperature dependence which resembled QCD except shifted from zero coupling (asymptotic freedom) to a finite value. The trivial massive vector meson theory showed no temperature dependence for $N_\tau = 6$ or 8.

4. BULK CRITICAL BEHAVIOR IN THE N=0, 2 AND 4 THEORIES

In the N=0 theory the Schwinger-Dyson equations and the computer simulations show a clear cross-over from mean field behavior at relatively strong coupling to an essential singularity "tail" near the critical point. Similar measurements of $\langle \bar\psi\psi \rangle$ on 8^4 and 10^4 lattices for the N=2 and 4 theories suggest the same systematics.[11] In this case, however, we have much less theoretical guidance. Certainly mean field behavior is expected and is found at relatively strong coupling, but the scaling law near the critical point can only be guessed. Perhaps the N=2 and 4 data are consistent with essential singularities as in the N=0 case. Much better simulations are needed to confirm or refute this important point.

5. SPECTROSCOPY

Using methods familiar from QCD spectrum calculations[12] we measured the π, σ, ρ and A_1 masses on $8^3 \cdot 16$ lattices in the chiral limit $m \to 0$ of the quenched N=0 model. We were unable to chose β particularly close to the critical point β_c because of large finite size effects. However, at relatively strong coupling $\beta = .24$ we found,

$$m_\pi^2 = .069, \; m_\sigma^2 = .339, \; m_\rho^2 = .759, \; m_{A_1}^2 = 1.256 \tag{5.1}$$

The pion isn't quite massless due to finite size effects. The σ particle has a non-zero mass. If this theory had a massless dilation it would have appeared here.[10] If we could make the measurements of Eq. (5.1) closer to the critical point, the continuum limit, we could rule out the existence of such a massless state.

Next we did a spectroscopy study of the N=4 theory. A smaller lattice $6^3 \cdot 12$ was used because the inclusion of dynamical fermions in the dynamics slow down the algorithm considerably. Also a bare quark mass was fixed to m=.025 in lattice units and the coupling was chosen at β=.19, a relatively strong coupling. We would have preferred to choose a larger β closer to the critical point, but large finite size effects precluded this. We also would have preferred to take data at several fermion masses so an extrapolation to the chiral limit $m \to 0$ could have been considered. Nonetheless, the results for the spectrum were,

$$m_\pi^2 = .283, \; m_\sigma^2 = .566, \; m_\rho^2 = 2.122, \; m_{A_1}^2 = 2.304 \tag{5.2}$$

The pion mass is non-zero because of considerable finite size effects. Nonetheless, the σ, ρ and A_1 masses are considerably higher in all cases.

Perhaps the most important result of the N=4 spectrum calculation was that in the Landau gauge we found a very clear signal for a massless photon. There is no evidence for a Higgs effect in QED4 with light, dynamical fermions.

We would next like to measure the renormalized electric charge in the model to decide whether it is trivial or interacting in the most convincing, direct way. We are engaged in exploratory studies at this time. We are also considering methods to measure induced four point interactions that are predicted to be renormalizable in the quenched model and may also occur for $N \neq 0$.[3] It would be very interesting to find such Nambu-Jona Lasinio terms in a four dimensional theory. Clearly, we have more analytic and numerical work ahead of us than behind us in this quest!

ACKNOWLEDGEMENT

This work was done in collaboration with E. Dagotto and A. Kocic. JBK is supported in part by the National Science Foundation grant PHY87-01775.

REFERENCES

1. Landau, L.D. and Pomeranchuk, I. Ya., Dokl. Akad. Nank. 102, 489 (1955).
2. Miransky, V. N., Il. Nuovo Cimento 90A 149 (1985) and references therein.
3. Kogut, J.B., Dagotto, E. and Kocic, A., Phys. Rev. Lett. 60, 772 (1988).
4. Duane, S. and Kogut, J.B., Nucl. Phys. B275 [FS17], 398 (1986).
5. Kirzhnits, D. A. and Linde, A.D., Phys. Lett. 73B, 323 (1978).
6. Susskind, L., Coleman, S. and Jackiw, R., Annuals of Physics 93, 267 (1975).
7. Bartholomew, J., et. al., Nucl. Phys. B230, 222 (1984).
8. Kogut, J.B., Dagotto, E. and Kocic, A., "Strongly Coupled Quenched QED4", Ill-(TH)-88-#32, Aug. 1988.
9. Dagotto, E. and J. B. Kogut, Phys. Rev. Lett. 60, 772 (1988).
10. Bardeen, W., Leung, C. and Love, S., Nucl. Phys. B273, 649 (1986).
11. Kogut, J.B., Dagotto, E. and Kocic, A., "A Supercomputer Study of Strongly Coupled QED", Ill-(TH)-88-#31, Aug. 1988.
12. Grady, M., Kogut, J. B. and Sinclair, D. K., Phys. Lett. 200B, 149 (1988).

TRICRITICAL POINT IN COMPACT QED

M. Okawa

National Laboratory for High Energy

Physics (KEK),Tsukuba, Ibaraki 305, Japan

ABSTRUCT

I briefly review a present status in a search for new fixed points in compact $U(1)$ lattice gauge theory with dynamical fermion.

1. INTRODUCTION

The subject of this paper is the search for tricritical points in compact QED. I am going to discuss the phase diagram of this model determined by the Monte Carlo simulations and comment on the remaining problems to be studied in a further work.

2. ACTION AND PHASE DIAGRAM OF COMPACT QED

I studied the $U(1)$ gauge theory with N_f dynamical fermions defined by the partition function

$$Z = \int [dU] \, e^{S_W(U)} \, \det\{M^\dagger(U)M(U)\}^{N_f/4} . \tag{1}$$

Here S_W is the pure gauge action parameterized by two couplings β and γ;

$$S_W = \sum_P (\beta \cos\theta_P + \gamma \cos 2\theta_P) \tag{2}$$

with θ_P the plaquette angle. I used staggered fermions, so the matrix $M(U)$ is given by

$$M(U)_{i,j} = \frac{1}{2} \sum_{\mu} \eta_{i,\mu} [U_{i,\mu} \delta_{j,i+\mu} - U^*_{j,\mu} \delta_{j,i-\mu}] + m\delta_{i,j} , \qquad (3)$$

where the notation is standard. The simulations have been done on a 8^4 lattice by making use of the hybrid-molecular-dynamics algorithm.[1]

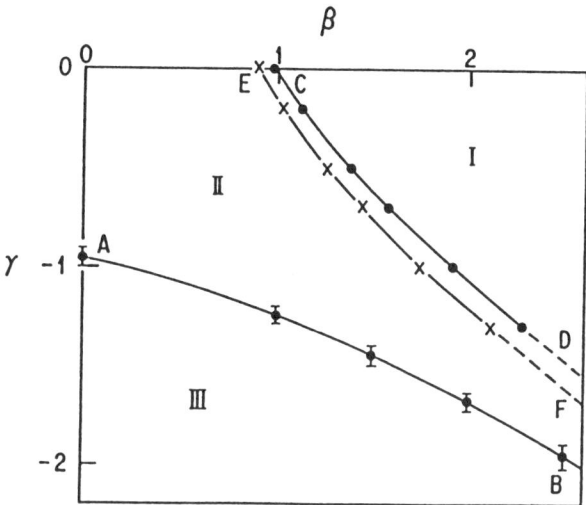

Fig. 1. Phase diagram of the theory in the β, γ plane.

Fig. 1 shows the phase diagram of this model with fixed mass $m = 0.1$ for two values of flavors $N_f = 1$ and 4. The diagram is divided into three regions. Region I corresponds to the chiral symmetric phase. In the chiral limit $m = 0$, $< \bar{\psi}\psi >$ should become identically zero. Region II corresponds to the phase where chiral symmetry is spontaneously broken. Region III is the antiferromagnetic phase. On the line AB which separates regions II and III, the transitions are always first order. Two transition lines for $N_f = 1$ and 4 are overlapping within large error bars. The line CD separates regions I and II for $N_f = 1$ and the line EF for

$N_f=4$. The upper half plane corresponding to $1/g_0^2 \equiv \beta + 4\gamma \geq 0$ has been studied by Dagotto and Kogut.[2] In this region, they found that the lines CD and EF are always first order and thus useless.

To study the nature of the transitions along the lines CD and EF, I measured discontinuities appearing in $E = \beta <\cos\theta_P> + \gamma <\cos 2\theta_P>$ and $<\bar{\psi}\psi>$.[3] I found that the sizes of discontinuities decrease as γ decreases and that at $\gamma = -1.3$ they disappear. This fact strongly suggests that there are tricritical points near $\gamma = -1.3$, where the transitions become second order from first order. As one approaches to the second order critical lines, correlation functions diverge and continuum field theories can be constructed along these lines.

Now the remaining problem is whether these continuum theories are unitary and non-trivial. I am currently attacking this problem by the Monte Carlo renormalization group method. Unitary theory should have correlation functions satisfying reflection positivity and non-trivial theory should have non-gaussian critical indices.

REFERENCES

1. S. Gottlieb, W. Liu, D. Toussaint, R. L. Renken, and R. L. Sugar, **Phys. Rev. D35**, 2531 (1987).

2. E. Dagotto and J. Kogut, **Nucl. Phys.** B295[FS21], 123 (1988).

3. M. Okawa, KEK preprint KEK-TH-204.

Phase Structure of Compact Lattice QED with Dynamical Fermions

H.C. Hege and A. Nakamura
FB Physik, Freie Universität Berlin
D-1000 Berlin 33, FRG

Abstract

The phase structure of compact lattice QED with dynamical Wilson fermions is studied by Monte Carlo simulation. The fermion contribution is calculated using an exact algorithm, capable of large Markov steps in configuration space.

A first order phase transition line in the (β,κ) plane, ending at the quenched point $(\beta=1.0, \kappa=0)$ has been found. The phases are explored by measuring gauge dependent quantities besides the more conventional gauge invariant observables. For this we applied a non-iterative gauge fixing algorithm.

From the phenomenological point of view weak coupling QED is the most successful physical theory due to the excellent agreement of experimental results and those obtained within the perturbative approach. However, to understand QED as a complete field theory it is necessary to get insight into the theory at strong coupling, too. Recently it has been suggested, based upon Schwinger-Dyson calculations in the quenched ladder approximation [1], that in the strong coupling regime a nontrivial zero of the beta function exists. If the fixed point really exists and survives in the presence of dynamical fermions, it may also have some phenomenological consequences. It has been pointed out, that such an unexpected behaviour of strong coupling QED may revive certain technicolor models by suppresssion of undesired flavour-changing neutral currents [2]. Another speculation, that the sharp lines observed in positron and electron spectra in heavy ion collisions may be caused by a new phase of QED [3] (but see i.e. [4]), raised the interest in non-perturbative investigations of QED, too. The lattice formulation of QED allows non-perturbative calculations which may be more reliable than the ladder approximation.

For the pure gauge part of lattice QED there are two formulations, the compact (periodic) and the non-compact one. While the non-compact version models conventional weak coupling continuum theory directly, the compact theory contains additional topological excitations (monopoles, whose density vanishes exponentially in the weak coupling region), but fits into the general lattice gauge theory scheme. Since the gauge invariant lattice Dirac operators are customarily defined in terms of compact gauge field variables $U_{m,\mu}=\exp(i\theta_{m,\mu})$ which are elements of the U(1) group, the compact version looks more natural from a conceptual point of view.

Up to now, only staggered fermions have been employed in Monte Carlo simulations of lattice QED. Recently Dagotto and Kogut found phase transitions of first order in compact QED [5] and of second order in noncompact QED [6] with staggered fermions coupled to compact gauge field variables.

Here we report the first results of a compact lattice QED Monte Carlo simulation with dynamical Wilson fermions, that is with action

$$S = \beta \sum_{m,\mu\nu} \text{Re } U_{m,\mu} U_{m+\mu,\nu} U^\dagger_{m+\mu+\nu,\mu} U^\dagger_{m+\nu,\nu} + \overline{\Psi}_m (1 - \kappa Q)_{mn'} \Psi_{n'} \quad (1)$$

$$Q_{mn'} = \sum_\mu \left\{ (1 - \gamma^\mu) U_{m,\mu} \delta_{m+\mu,n'} + (1 + \gamma^\mu) U^\dagger_{m,\mu} \delta_{m-\mu,n'} \right\} \quad (2)$$

where n denotes sites and μ,ν directions in the four dimensional hypercubic lattice. We always used one flavour and antiperiodic boundary conditions for fermions.

The four dimensional <u>pure</u> U(1) lattice gauge theory, i.e. compact lattice QED, already exhibits an interesting phase structure: a strong-coupling confining phase and a weak-coupling deconfined Coulombic phase. The dynamics of the separating phase transition is well understood: in the confinement phase the presence of a gas of monopoles and antimonopoles causes the narrow electric flux tube producing a linear potential. In the weak coupling phase the density of monopoles vanishes exponentially, the static potential is Coulombic and the model has a line of fixed points, each assigned to a trivial model with another renormalized fine structure constant. The order of the phase transition point (which extends to a phase transition line in the fundamental-adjoint parameter space) is still controversial. Most recent MCRG calculations [7] indicate a discontinuos phase transition.

On finite lattices monopole loops which are closed only due to the periodic boundary conditions, and Dirac sheets (world lines of Dirac strings) without monopole loops as boundary and closed due to boundary conditions occur. These divide the configuration space in different sectors, which local Monte Carlo algorithms have problems to cross [8]. Simulating compact QED in the disordered phase or near the deconfining phase transition it is therefore important to use an algorithm capable of large Markov steps in configuration space.

We use a Metropolis-type algorithm without any systematic bias and allowing large Markov steps with customary acceptance rates. It updates the links on a hypercube several times (here 10 times) before proceeding to the next one, and thereby decorrelates successive configurations. The columns of $(1-\chi Q)^{-1}$, required when stepping from one hypercube to the next, are calculated by a preconditioned CR/CG solver, starting from a 4th order hopping parameter expansion. Details of this algorithm are discussed in Ref.[9]. Configurations produced by this algorithm, proved to be weakly correlated, even near first order phase transitions, as was observed in a QCD simulation [10].

In fig.1 we show the value of the plaquette energy as function of Monte-Carlo sweeps at $\beta=0.9$ and $\chi=0.175$ on 4^4 and 4^3*8 lattices each with ordered and disordered start. These histories display two co-existing phases and therefore suggest strongly the occurence of a first order phase transition. We found jumps in many observables for instance in $<\bar{\Psi}\Psi>$, all of them strictly correlated with the jumps in the plaquette action. The eigenvalue distribution of the fermion matrix is drastically different for each phase, see fig.2. The "positronium" masses also jump: from $m_{para}=1.37\pm0.08$ and $m_{orth.}=1.67\pm0.07$ in the disordered phase, to $m_{para}=1.84\pm0.07$ and $m_{orth.}=2.01\pm0.06$ in the ordered one (in units of the inverse lattice spacing).

In fig.3a,b and c we plot the plaquette energy as function of κ for $\beta=0.90$, 0.95 and 1.0 on 4^4 and 4^3*8 lattices. The discontinous phase transition observed at $\beta=0.90$ persists and moves to larger fermion mass with increasing β. At $\beta=1.0$ the observables display for all κ the ordered phase as expected, since the quenched phase transition point for our lattice sizes lies at $\beta \simeq 0.995$. The phase transition continues to small κ (fig.3d), and reaches the quenched (probably weakly first order) phase transition point. At $\chi=0.1$ the fluctuation of the fermion determinant ratio around 1 is within a few percent, nevertheless the phase transition is most probably of first order. Dynamical fermions seem to strengthen this phase transition.

We repeated this kind of analysis and observed several phase
transition points, shown in fig.4. One is tempting to relate the
transition with the zero mode due to the small fermion mass; we
investigated the eigenvalue distribution but found no eigenvalue
very close to zero. In Ref.[4] the phase diagram of lattice QED
with staggered fermions is pictured as follows: the phase transition line, also running with decreasing fermion mass to lower β,
is broken at intermediate fermion mass and reappears for small
fermion mass as a chiral symmetry breaking transition. The phase
structure in fig.4 is similar to that in SU(3) with Wilson
fermions on finite lattices [11]. This seems to be the case also
for staggered fermions (compare Ref.[5] and e.g. Ref.[12]).

In order to clarify the physical meaning of each phase, we fix
now the gauge to the covariant Landau gauge. This opens the possibility to calculate gauge dependent, but fundamental quantities, such as the photon and electron propagator, familiar in
perturbation theory. In case of U(1), unlike non-Abelian lattice
gauge fields, the gauge transformation $A'_\mu(x) = A_\mu(x) + \partial_\mu \chi(x)$
which fixes the whole configuration exactly to Landau gauge is
easily found. Introducing forward/backward lattice derivatives,
$\partial_\mu^{(+)} f(x) = f(x+\mu) - f(x)$ and $\partial_\mu^{(-)} f(x) = f(x) - f(x-\mu)$, we set

$$\chi(x) = \sum_y \alpha(x-y) \partial_\nu^{(-)} A_\nu(y) \tag{3a}$$

where the lattice Green function is defined as

$$\alpha(x) = \frac{1}{\text{volume}} \sum_p e^{ipx} / 4 \sum_\mu \sin^2 p_\mu/2 . \tag{3b}$$

Then A'_μ satisfies the Landau gauge condition $\partial_\mu^{(-)} A'(x) \equiv 0$
exactly, unlike the iteration method in Ref.[13].

In Tab.I we show the photon and fermion masses at various
(β, κ) pairs on a $4^3 * 8$ lattice. Photon propagotors $<\sum_{\vec{t}} \sin A_i(\tau)$
$\sin A_i(0)>$ (i=1,2 and 3) are fitted with the usual cosh form,
while fermion propagators $G(\tau) = <\overline{\psi}_\alpha(\tau) \psi_\alpha(0)>$ are fitted with

$|G(\tau)| = |C*\sinh(N_t/2-\tau)|$. In the disordered phase the photon mass is $\simeq 1$. In the quenched case the photon becomes immediately massless when moving through the phase transition point into the deconfined phase (this was previously observed in [13]). With dynamical fermions, however, the photon remains massive even in the ordered phase. This is probably because the photon mass renormalization due to fermion loops produces a "plasmon" mass on a finite lattice. It would be interesting to see the photon mass with staggered fermions.

The fermion mass measured by propagators decreases as κ increases until $\kappa = 0.18$. At $\kappa = 0.175$, where we observe the phase transition, the fermion mass is far from zero. This is consistent with the behavior of the fermion matrix eigenvalues. Therefore this transition cannot originate from the chirality of fermions. Above $\kappa = 0.18$ the fermion mass suddenly takes large values, probably we are beyond κ_c.

In conclusion, using an exact fermion algorithm, we have found a phase transition line in compact lattice QED with Wilson fermions, which is most probably first order. The line seems to continue up to the quenched phase transition point where the order of the transition becomes weaker. In order to characterize the two phases also gauge dependent quantities have been measured using a method which allows to fix a configuration to Landau gauge in one step. These quantities behave as expected within the disordered phase. In the quenched case the photon becomes suddenly massless when entering the ordered phase, however, with dynamical fermions even at small κ it acquires a heavy mass. The phase structure becomes much more complex with dynamical fermions and requires further extensive analyses. Currently we are investigating the behaviour of the eigenvalues and monopoles in the (β, κ) plane [14].

We would like to acknowledge valuable discussions with G.Feuer, K.Kanaya, V.Linke, I.O.Stamatescu, and W.Theis. The calculations have been performed at ZIB Berlin (CRAY X-MP/24) and the computer center of university Kiel (CRAY X-MP/18).

[1] V.P. Gusynin and V.A. Miransky, Phys.Lett.B 191 (1987) 141;
V.A. Miransky, Nouvo Cimento 90A (1985) 149;
C.N. Leung, S.T. Love, and W.A. Bardeen, Nucl.Phys. B 273 (1986) 649.

[2] M. Bando, T. Morozumi, H. So, and K. Yamayaki, Phys.Rev.Lett. 59 (1987) 384, and references therein.

[3] L.S. Celenza, V.K. Mishra, C.M. Shakin, and K.F. Liu, Phys.Rev.Lett. 57 (1986) 55, and references in [4]

[4] R.D. Peccei, J. Sola, and C. Wetterich, Phys.Rev. D 37 (1988) 2492.

[5] J.B. Kogut and E. Dagatto, Phys. Rev. Lett. 59 (1987) 617;
E. Dagatto and J.B. Kogut, Nucl. Phys, B 295 (1988) 123.

[6] J.B. Kogut, E. Dagatto, and A. Kocic, Phys.Rev.Lett. 60 (1988) 772.

[7] A. Hasenfratz, Phys. Lett. B 201 (1988) 492.

[8] V. Grösch, K. Jansen, J. Jersak, C.B. Lang, T. Neuhaus, and C. Rebbi, Phys.Lett.B 162 (1985) 171.

[9] A. Nakamura, G. Feuer, H.C. Hege, V. Linke, and M. Haraguchi, to appear in Comp.Phys.Comm.

[10] Ph. de Forcrand, M. Haraguchi, H.C. Hege, V. Linke, A. Nakamura, and I.O. Stamatescu, Phys.Rev.Lett. 58 (1987) 2011.

[11] Ph. de Forcrand and I.O. Stamatescu, Nucl.Phys.B 304 (1988) 628

[12] J.B. Kogut, E.V.E. Kovacs, and D.K. Sinclair, Nucl.Phys. B 290 (1987) 431.

[13] P. Coddington, A. Hey, J. Mandula, and M. Ogilvie, Phys.Lett.B 197 (1987) 197.

[14] H.C. Hege, V. Linke and A. Nakamura (in preparation).

Table Caption
=============

Tab.I: Photon and fermion mass (in units of the inverse lattice spacing) extracted from propagator fits of about 100-200 equilibrated configurations (1000 in the quenched case) on a 4^3*8 lattice.

Figure Captions
===============

Fig.1 : Plaquette energy as function of sweeps (each involving 10 hits per link and 10 updates per hypercube) starting with an ordered/disordered configuration on a 4^3*8 lattice (4^4 lattice in the small figure).

Fig.2 : Typical eigenvalue distribution of the matrix $1-\kappa Q$ in the disordered phase (above) and ordered phase (below).

Fig.3 : Plaquette energy as function of κ for $\beta=0.90$ (a), 0.95 (b) and 1.0 (c) and as function of β for $\kappa=0.10$ (d) on 4^4 ($\diamond$) and 4^3*8 ($\circ$) lattices (empty: disordered start, filled: ordered start, half-filled: ordered and disordered start).

Fig.4 : Phase diagram of compact lattice QED with Wilson fermions in the (β,κ) plane.

Table 1
=======

β	κ	phase	m_{photon}	$m_{fermion}$
0.97	0	dis.	0.86 (5)	
1.00	0	ord.	0.25 (3)	
1.05	0	ord.	0.16 (2)	
1.01	0.100	ord.	1.08 (13)	
0.90	0.160	dis.	1.05 (19)	0.87 (5)
0.90	0.175	dis.	0.90 (45)	0.78 (10)
0.90	0.175	ord.	0.99 (03)	0.69 (6)
0.90	0.180	ord.	0.56 (08)	0.60 (12)
0.90	0.190	ord.		1.23 (3)

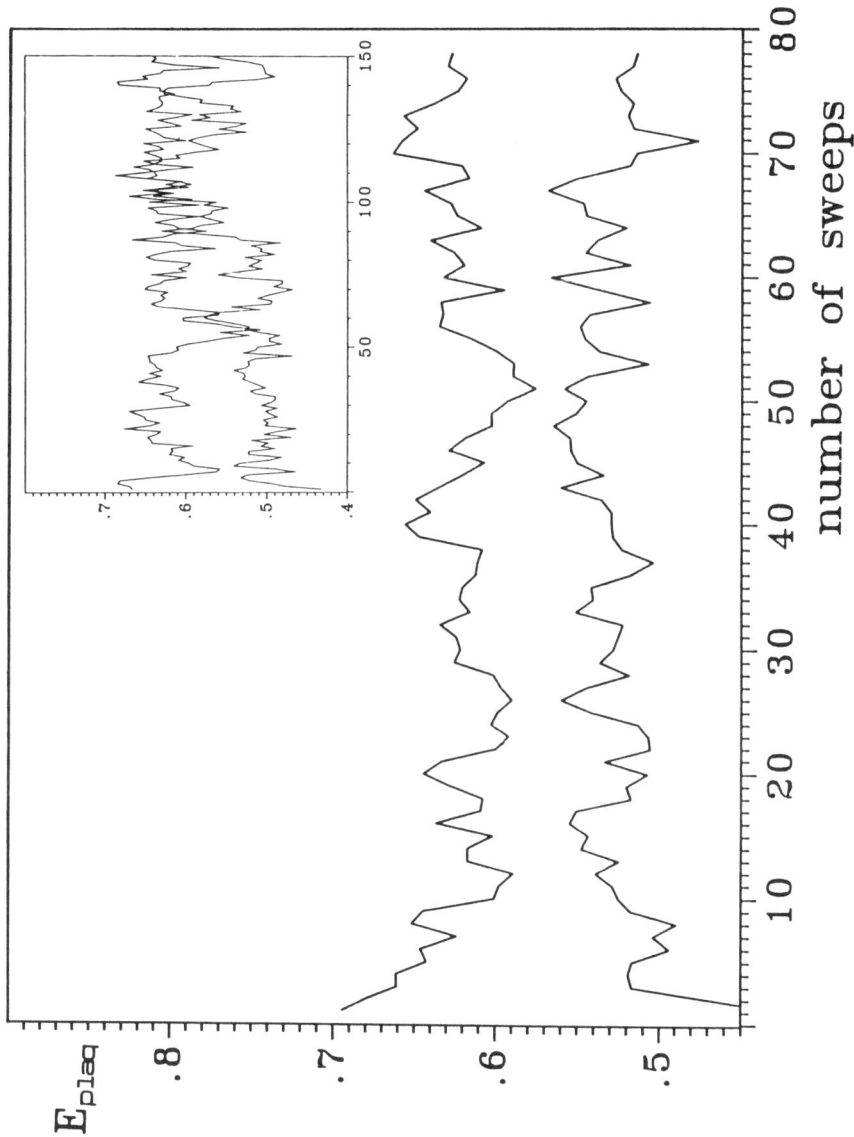

Fig. 1

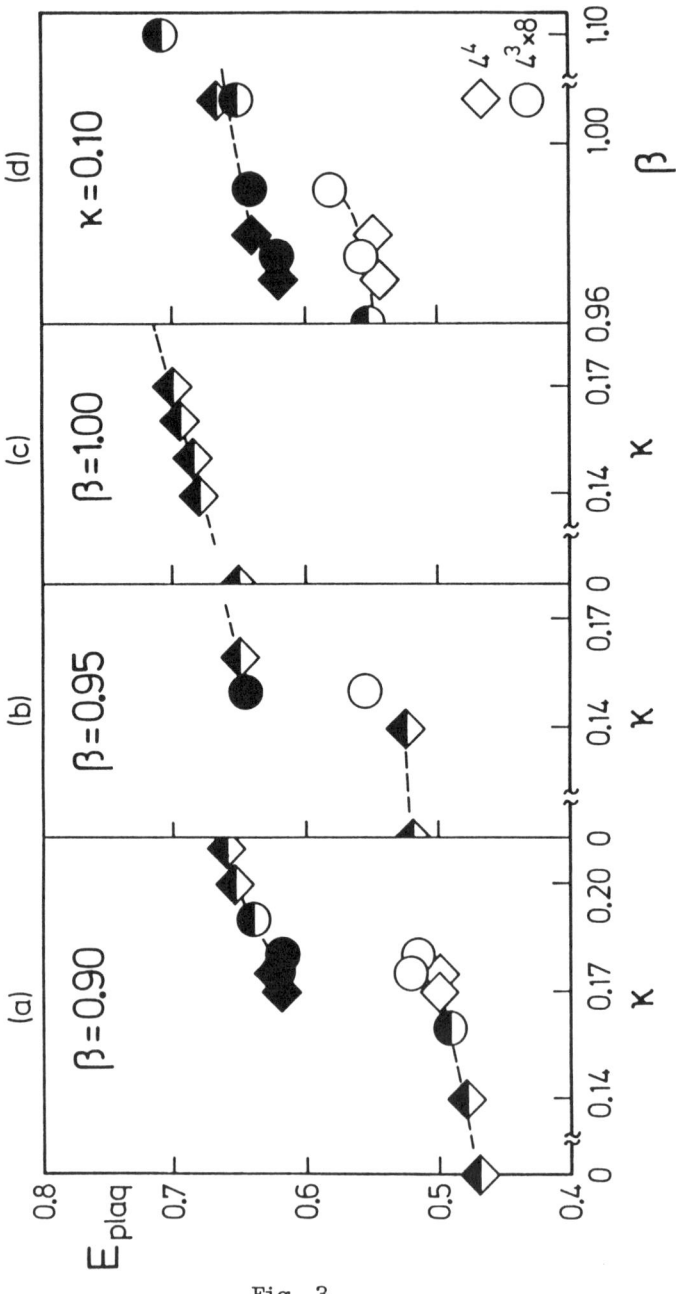

Fig. 3

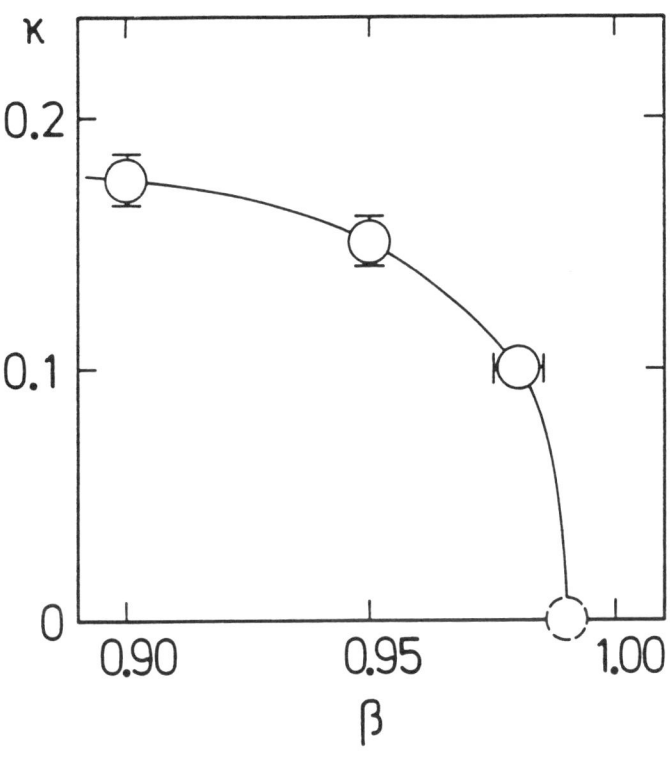

Fig. 4

CHIRAL SYMMETRY BREAKING IN (2+1)-DIMENSIONAL QED

Elbio Dagotto

Institute for Theoretical Physics
University of California at Santa Barbara
Santa Barbara, CA 93106

ABSTRACT

A short review is presented of a recent numerical simulation of QED in 3 dimensions with N four component massless fermions. In the continuum limit we found chiral symmetry breaking only for $N \leq 3$.

Quantum Electrodynamics in 2+1 dimensions (QED_3) is a very interesting field theory mainly because of its similarities with four dimensional models with dynamical symmetry breaking [1]. The Lagrangian of QED_3 in the continuum is

$$L = -\frac{1}{4}F_{\mu\nu}F^{\mu\nu} + \sum_{j=1}^{N} \bar{\chi}^j \gamma^\mu (i\partial_\mu - eA_\mu)\chi^j \qquad (1)$$

where we have included N species of four component massless fermions. We use this four component formulation because in this case we can define a γ^5 matrix and a chiral symmetry in the model [2]. The mass term does not break parity (as in the two component formulation) but only chiral symmetry as usual.

We report here a numerical study of QED_3 using the lattice techniques recently applied to the more difficult problem of the existence of multi-flavor

QED in 3+1 dimensions [3]. The present work has been done in collaboration with J. Kogut and A. Kocić[4]. Our simulations on relatively small lattices ($6^3, 8^3$ and limited data from 10^3 systems) indicate that QED_3 with N Dirac fermions breaks chiral symmetry only for a sufficiently small number of dynamical fermions. For $N \leq 3$, measurements of the chiral condensate $< \bar{\chi}\chi >$ indicate symmetry breaking in the continuum limit of the lattice theory. However, for $N \geq 4$ no symmetry breaking is found. Apparently in this case the inclusion of massless dynamical fermions into the pure photon theory leads to screening which overwhelms the attractive forces needed to form a chiral condensate.

There are some analytic studies of this model in the framework of the Schwinger-Dyson equations combined with the $1/N$ expansion. Previous results suggested chiral symmetry breaking for a large number of flavors although with a dynamical mass exponentially small with N [5]. We found no evidence of such behavior.

In fact our new numerical results are quantitatively similar to those of a recently revised [6] study of the continuum model in a $1/N$ expansion, where no chiral symmetry breaking was found for large N.

The lattice (Euclidean) version of this model using staggered fermions is given by the action

$$S = -\frac{\beta}{2} \sum_p \theta_p^2 + \sum_{j=1}^{N} \sum_{x,y} \bar{\psi}_x^j M_{x,y} \psi_y^j, \qquad (2)$$

where M is the Dirac operator on the lattice defined as

$$M_{x,y} = \frac{1}{2} \sum_\mu \eta_{x,\mu}[e^{i\theta_{x,\mu}}\delta_{y,x+\mu} - e^{-i\theta_{y,\mu}}\delta_{y,x-\mu}] + m\delta_{x,y} \qquad (3)$$

x, μ and p denote sites, directions and plaquettes of a three dimensional hypercubic lattice. $\theta_{x,\mu}$ are dimensionless fields on the links of the lattice that are proportional to the gauge fields through the relation $\theta_{x,\mu} = eaA_\mu(x)$, where a is the lattice spacing which acts as an ultraviolet cutoff. $\bar{\psi}_x, \psi_x$ are (dimensionless) one component Grassmann variables on sites (i.e. to reduce species doubling in the continuum we use staggered fermions). The lattice fermionic fields are related with their continuum counterparts through $\psi_x \to a\chi_x$. The rest of the notation is standard. A careful analysis of the naive continuum limit has been done in ref.[7] showing that the continuum fermions coming from eq.(2) correspond to the 4-component theory. The dimensionless coupling constant β of the lattice action is related with the charge by $\beta = 1/(e^2 a)$. Then, the continuum limit of the model is recovered at $\beta \to \infty$ if we assume that e^2, that sets the scale of the theory, is finite (for a different point of view see ref.[8]).

Note that in eq.(2) we use the noncompact formulation of lattice QED_3 which has proven to be very successful in the study of QED_4 [3]. As numerical method we use the hybrid technique with noisy fermions. Details of the method applied to QED has been presented in ref.[9].

We measured local observables such as the plaquette and the chiral condensate on $6^3, 8^3$ and 10^3 lattices. Our most accurate data was taken on the 8^3 lattices and will be discussed here. We simulated the theory with $N = 0$ (quenched),$1, 2, 3, 4$ and 5 Dirac flavors with bare fermion masses of $m = 0.050$ and 0.025 at β values ranging from zero to one. The chiral limit of the condensate $<\bar{\psi}\psi>$ was taken at each β by linearly extrapolating the finite mass results to $m = 0$. The hybrid algorithm was run with a time step $dt = 0.025$ for 10^5 to 2.5×10^5 sweeps at each β and m value. Statistical errors estimates included the sweep-to-sweep correlation effects inherent in our "small change" algorithm. Runs of 500,000 sweeps were made at several β and m values to check on statistical analyses.

Our results are summarized in fig.1 where we show the chiral condensate $<\bar{\psi}\psi>$ (extrapolated to massless quarks) vs. β, for $N = 0, 1, 2, 3, 4$ and 5. Note the "tails" in the $N = 0, 1, 2, 3$ curves which extend to weak coupling. These curves are smooth from strong coupling where chiral symmetry is broken (analytic methods such as the strong coupling expansions or mean field theory on the lattice prove this point at least for the compact formulation) to the weakest coupling we can probe. So, in these cases there appears to be no phase transitions in $<\bar{\psi}\psi>$ as we pass from strong to weak coupling. Therefore, the theory is predicted to reside in only one phase where chiral symmetry is broken.

From dimensional analysis it follows that the chiral condensate should behave like $<\bar{\psi}\psi> = A\beta^{-2}$ near the continuum limit. If the constant A is nonzero we have chiral symmetry breaking. We have search for this behavior in the quenched theory. There the scaling window is easily reached and is quite broad. Although we cannot take β literally to infinity to extract the continuum limit, the plateau in the $\beta^2 < \bar{\psi}\psi >$ vs. β curve, which extends from $\beta > 0.75$ to 2.00, is quite clear. For β larger than 2.0, the signal becomes very small and difficult to measure accurately by a statistical method. For $N = 1$ the chiral condensate is suppressed (due to screening) and more difficult to measure so we could not check in this case the existence of nontrivial scaling.

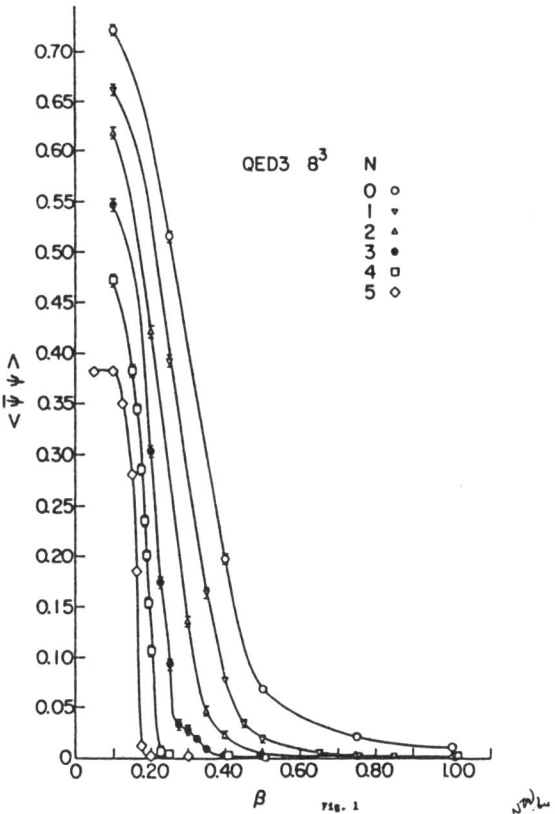

Fig. 1. $< \bar{\psi}\psi >$ vs. $\beta = 1/(e^2 a)$ for QED_3 on a 8^3 lattice for $N = 0, 1, 2, 3, 4$ and 5.

The smoothness of the $N = 2 <\bar{\psi}\psi>$ vs. β curve was checked carefully. By contrast, the $N = 4$ data showed a rather clear non-analytic behavior. Inspecting Fig.1 we see that $<\bar{\psi}\psi>$ is essentially zero for $\beta > 0.25$ but is non-zero for stronger coupling. In fact we found that the data for $<\bar{\psi}\psi>$ near the transition is compatible with a mean field behavior. This is a theoretically satisfying result. In fact, we expected either a first order or a mean field chiral transition in those cases where these models have a transition at finite β. Such phase transitions would not lead to an interacting relativistic field theory in the strongly cutoff lattice model. Finding an interacting continuum model at such a point would be very puzzling indeed.

How severely are our conclusions subject to finite size effects? We are finishing a simulation on a 10^3 lattice to answer this question but from our results for 6^3 and 8^3 lattices we believe that the value of the "critical" N (beyond which chiral symmetry is recovered) will not change much. We are studying carefully the special case $\beta = 0$ since it may occur that even in very strong coupling the chiral phase dissappears after enough number of fermions are introduced [10]

Can we develop a physical picture and a quantitative calculational technique to assimilate these numerical results? Models of chiral symmetry breaking in four dimensional Quantum Chromodynamics show how the long distance attraction due to flux tube formation leads to a negative self-mass for constituents quark which leads to a chiral condensate and the existence of a multiplet of Goldstone pions. Can such an idea be extended to QED_3 in the quenched $N = 0$ limit in such a way that the inclusion of N species of dynamical fermion can be seen to generate a finite N_c? Physical pictures of chiral symmetry breaking in QED_3 have been made starting at $N = \infty$ and developing a $1/N$ expansion. Judging from fig.1 a more compelling physical picture should be developed from just the opposite $(N = 0)$ extreme, in fact.

Note also that our results may be important for some attempts to theoretically understand the recently discovered high temperature superconductors. It has been found that the Heisenberg model (strong coupling limit of the Hubbard model) can be written as a lattice gauge theory with staggered fermions and gauge group SU(2) in 2+1 dimensions [11]. It would be very surprising if our results for a U(1) theory can not be qualitatively extended to a SU(2) theory. If this is correct then we conclude that the Hubbard model can not be studied from the large N limit of a lattice gauge theory [12] because there is a singularity in between (chiral symmetry in gauge theories is related with the global spin rotation symmetry of the Heisenberg model). These and related problems in statistical mechanics and field theory are under study.

This work was supported in part by the NSF grants PHY87-01775 and PHY82-17853 and supplemented by funds from NASA. The computer simulations were done on the CRAY X-MP/48 of the National Center for Supercomputing Applications at Urbana, IL.

References

1) J. Cornwall, Phys. Rev. **D22**, 1452 (1980); T. Appelquist and R. Pisarski, Phys. Rev. **D23**, 2305 (1981); R. Jackiw and S. Templeton, Phys. Rev. **D23**, 2291 (1981); T. Appelquist and U. Heinz, Phys. Rev. **D24**, 2169 (1981).

2) T. Appelquist, M. Bowick, D. Karabali and L. Wijewardhana, Phys. Rev. **D 33**, 3704 (1986).

3) J.Kogut, E. Dagotto and A. Kocić, Phys. Rev. Lett.**60**, 772 (1988).

4) E. Dagotto, J. Kogut and A. Kocić, "A computer simulation of chiral symmetry breaking in (2+1)-dimensional QED with N flavors", ILL-(TH)-88-30, submitted to Phys. Rev. Letters.

5) R. Pisarski, Phys. Rev. **D29**, 2423 (1984); K. Stam, Phys. Rev. **D34**, 2517 (1986).

6) T. Appelquist, D. Nash and L. Wijewardhana, Phys. Rev. Lett. **60**, 2575 (1988); T. Matsuki, L. Miao and K. Viswanathan, Simon Fraser Univ. preprint, June 1987 (revised version: May 1988).

7) C. Burden and N. Burkitt, Europhys. Lett. **3**, 545 (1987).

8) For attempts to transform a superrenormalizable theory to a nontrivial renormalizable one, see K. Wilson, Phys. Rev. **D7**, 2911 (1973) and S. Shei and T. Yan, Phys. Rev. **D8**, 2457 (1973).

9) E. Dagotto and J. Kogut, Nucl. Phys. **B295**, 123 (1988).

10) One may claim that if the exponential behavior predicted in ref. 5 begins at $N = 0$, then the value of the chiral condensate at large N is so small that can hardly be distinguish from zero (D. Atkinson, P. Johnson and M. Pennington, "Dynamical mass generation in three-dimensional QED", BNL preprint and P. Johnson, private communication). This is a potential problem for a computer simulation. But if we numerically show that even in strong coupling there is no phase that breaks chiral symmetry then those claims are invalidated.

11) G. Baskaran and P.W. Anderson, Phys. Rev. **B37**, 580 (1988); E. Dagotto, E. Fradkin and A. Moreo, Phys. Rev. **B38**, 2926 (1988).

12) I. Affleck and J. Marston, Phys. Rev. **B37**, 3774 (1988).

Migdal-Kadanoff Renormalization Group Analysis of Lattice QED

MASAHIRO IMACHI

Department of Physics, Kyushu University, 812   Japan

ABSTRACT

Lattice $(QED)_4$ is studied according to Migdal-Kadanoff renormalization group method. Kogut-Susskind fermion is used. It is shown that a four fermi coupling is induced through renormalization group transformations. Renormarization group flows and chiral order parameter are investigated.

The Migdal-Kadanoff renormalization group (MKRG)[1] method gives us, although qualitative, useful information about the phase structure of lattice gauge theory (LGT). We are able to see how various parameters contained in the action change under RG transformations. The set of parameters defines a point in the multidimensional space of parameters and it moves in parameter space by RG transformations defining a trajectory of this space, which is called "renormalized trajectory (RT)"[2]. Paying attention to the RT[3,4], we can obtain the Gell-Mann-Low function[3] through the analysis of correlation length for various bare theories (bare couplings). By this method we found the structure corresponding to the cross over phenomena. Moreover, when we start RG from mixed action (namely, the bare action containing not only fundamental representation but also adjoint representation), we found a very sharp stepwise transition[3] (at these couplings, MC analysis found "1 st order" phase transition).

In this paper we attempt to apply the MKRG method to the gauge system including fermions. Actually we consider here the lattice quantum electrodynamics $(QED)_4$. The action of lattice QED contains both gauge part and fermion part. The parameters obtained through RG transformations are called new and those before the transformations will be called old. The effects of (1) old parameters of gauge system to new parameters of gauge system abbreviated as "gauge to gauge ", (2) fermion to fermion (3) fermion to gauge and (4) gauge to fermion should be considered. (1) was extensively studied in previous analyses [3][4]. A method to take into account the effect of (2) and (3) was given in ref.(5) and this gives coupling constant renormalization due to fermion loops ="vacuum polarization" .

We propose a method to include the effect of (4), *i.e.*, "fermion self energy" contribution into the MKRG approach. (4) is crucial to consider the chiral symmetry breaking.

As a lattice fermion we adopt the staggered fermion, since it is chiral symmetric in massless limit. The staggered fermion bare action is given by

$$S_f = -B_0 \sum_x \bar{\psi}(x)\psi(x) + A_0 \sum_{x,\mu}(-\bar{\psi}(x)U_\mu(x)\psi(x+\mu)$$
$$+ \bar{\psi}(x+\mu)U_\mu^\dagger(x)\psi(x))\eta_\mu(x) + C_0 \sum_{x,\mu} \bar{\psi}(x)U_\mu\psi(x+\mu)\,\bar{\psi}(x+\mu)U_\mu^\dagger\psi(x), \quad (1)$$

where $\bar{\psi}$ and ψ denote staggered fermions and $\eta_\mu(x)$ is the Kogut-Susskind analogue of γ-matrix. A, B and C are parameters corresponding to hopping term, mass term and four fermion term respectively.

The gauge action is

$$S_G = -\sum_{q,\mu\nu}(1 - \mathrm{Re}\chi_q(\theta)\,)\beta_g^q \quad (2)$$

where β_g^q and $\chi_q(\theta) = \mathrm{Tr}UUU^\dagger U^\dagger = \exp(iq\theta)$ is inverse coupling constant and the real part of the character of q-irreducible representation of $U(1)$ gauge group, respectively. The partition function is defined as

$$Z = \int [d\psi d\bar{\psi}][dU]e^{S_G+S_f}. \quad (3)$$

The integrand of partition function, $\exp[S_G(\theta)]$ is expanded by irreducible characters as

$$e^{S_G(\theta)} = \sum_{q=0}^{\infty} \tilde{f}_q(L)\chi_q(\theta), \quad (4)$$

where L denotes the length scale at some stage of RG transformations. Consider the fermion contributions $N_0 \exp(L_{01})$ and $N_0 \exp(L_{12})$ related to sites 0,1 and 1,2 to Z,

$$\begin{cases} L_{01} = & -B_0(0)\bar{\psi}(0)\psi(0) - B_0(1)\bar{\psi}(1)\psi(1) \\ & + A_0(0)(\varepsilon_{0+}\bar{\psi}(0)U_{\mu=1}(0)\psi(1) + \varepsilon_{0-}\bar{\psi}(1)U_{\mu=1}^\dagger(0)\psi(0)) \\ & + C_0(0)\bar{\psi}(0)U_1(0)\psi(1)\bar{\psi}(1)U_1^\dagger(0)\psi(0), \\ L_{12} = & -B_0(1)\bar{\psi}(1)\psi(1) - B_0(2)\bar{\psi}(2)\psi(2) \\ & + A_0(1)(\varepsilon_{0+}\bar{\psi}(1)U_{\mu=1}(1)\psi(2) + \varepsilon_{0-}\bar{\psi}(2)U_{\mu=1}^\dagger(1)\psi(1)) \\ & + C_0(1)\bar{\psi}(1)U_1(1)\psi(2)\bar{\psi}(2)U_1^\dagger(1)\psi(1), \end{cases} \quad (5)$$

and perform the integration over Grassmann numbers $\psi(1)$ and $\bar{\psi}(1)$, which we call fermion decimation, with the use of

$$\int d\psi(1)d\bar{\psi}(1)\exp(-p\bar{\psi}(1)\psi(1) + \bar{\psi}(1)\xi + \bar{\zeta}\psi(1)) = p\exp((1/p)\,\bar{\zeta}\xi). \quad (6)$$

After exponentiating the result we obtain new parameters in terms of A_0, B_0 and C_0;

$$\begin{cases} N'(0) = N_0 N_0 (B_0(1) + B_0(1)) \\ A'(0) = A_0(0)A_0(1)/(B_0(1) + B_0(1)) \\ B'(0) = B_0(0) + \Delta B_1(0) \\ C'(0) = \Delta B_1(0)\Delta B_1(2) - A'(0)^2 \varepsilon_{1+}\varepsilon_{1-} \end{cases} \quad (7)$$

where $\Delta B_1(0) = -[A_0^2(0)\varepsilon_{0+}\varepsilon_{0-} + C_0(0)]/(B_0(1) + B_0(1))$, $\eta'(0) = \eta(0)\eta(1)$ and $\varepsilon_{1\pm} = \varepsilon_{0\pm}\varepsilon_{0\pm}$. This process defines a "fermion decimation". By repeated use of this fermion decimation, we have new fermion action at scale λL. Effect (2) is taken into account through this process.

The vacuum polarization is calculated through similar process; after the decimation at sites 1,2 and 3, we equate 0 and 4 sites and integrate over $d\psi(0)d\bar{\psi}(0)$, we have fermion loop. Convoluting it with the original gauge plaquette and integrate the gauge variables of inner links we obtain new plaquette action at scale λL. The number of flavor N_{f_1} is taken into account by this process. The effects (1) and (3) are included. In order to take into account the effect (4), we should consider "fermion self energy correction". Consider a fermion path shown in Fig 2. Fermion integrations and gauge integrations are successively performed. We adopt those paths whose defining plane is perpendicular to the plane of gauge decimation in order to avoid complexity. Through this process we are able to include the effect (4);

$$\begin{cases} A_G(\lambda L) = A(\lambda L)\,(\dfrac{\tilde{f}_1(L)}{\tilde{f}_0(L)})^\lambda \\ B_G(\lambda L) = B(\lambda L) \\ C_G(\lambda L) = C(\lambda L) + (1 - (\dfrac{\tilde{f}_1(L)}{\tilde{f}_0(L)})^{2\lambda})A^2(\lambda L) \end{cases} \quad (8)$$

where the suffix G means that the effect due to gauge action is taken into account and $\tilde{f}_q$ is defined in eq.(4). $A(\lambda L), B(\lambda L)$ and $C(\lambda L)$ are parameters obtained when fermion self energy correction is neglected. The number of fermion flavor contribution to Fig.2 is denoted as N_{f_2}. When fermion self energy correction is neglected, the renormalized parameter $C(\lambda L)$ of new scale λL is proportional to that of old scale L. In this case, C(renormalized)=0 when we set C(bare)=0. But when fermion self energy correction is taken into account, new parameter C_G becomes non zero even if C(renormalized)=0, since $\tilde{f}_1/\tilde{f}_0 \neq 1$ ($\tilde{f}_1/\tilde{f}_0$ ($\propto \beta_{1g}$) is very small compared with unity in strong coupling regime and near to unity in weak coupling regime).

The chiral order parameter, $<\bar{\psi}\psi>_{B_0=0}$, where $<\bar{\psi}\psi>_{B_0}$ is given by

$$\langle\bar{\psi}\psi\rangle_{B_0} = \frac{-1}{N_S}\left[\frac{\partial Z}{\partial B_0}/Z\right]_{B_0}, \qquad (9)$$

where N_s is the total number of lattice sites λ^{dM} (when M iterations in d-spacetime dimension are performed) and B_0 denotes the fermion bare mass parameter.

Now we will present the numerical results. We choose $\lambda = 3$ hereafter.

Fig.3 shows the flow diagrams for cases (i)pure gauge system ($N_{f_1} = N_{f_2} = 0$) (ii) vacuum polarization included($N_{f_1} \neq 0, N_{f_2} = 0$) (iii) both vacuumpolarization and fermion self energy correction are included($N_{f_1} \neq 0, N_{f_2} \neq 0$). The strong coupling region is controlled by IR fixed point $\beta_g = 0$ (confinement phase)and in weak coupling region the coupling constant is almost fixed,i.e.,the beta function is quite near to zero for $N_{f_1} = 0$ (Coulomb phase). The introduction of fermion loops ($N_{f_1} \neq 0$) makes the coupling constant moving to weaker region ("shielding" effect due to vacuum polarization). Introduction of fermion self energy($N_{f_2} \neq 0$) leads to a slight shift to strong coupling region (Fig.3(c)).

The next task is to obtain $<\bar{\psi}\psi>$ (=chiral order parameter). The value of $<\bar{\psi}\psi>$ obtained in MKRG shows slight but clear indication of chiral symmetry breaking. $<\bar{\psi}\psi>$ at $B_0 = 0$ is not directly calculable since the recursion formula contains B_0 as a denominator. $<\bar{\psi}\psi>_0$ is defined as the linearly extrapolated value of $<\bar{\psi}\psi>_{B_0}$ and $<\bar{\psi}\psi>_{2B_0}$ ($B_0 \neq 0$). The magnitude of $<\bar{\psi}\psi>_0$ obtained in this way is, although very small compared with $<\bar{\psi}\psi>_{B_0}$ or $<\bar{\psi}\psi>_{2B_0}$, shows rather clearly the feature of chiral transition, namely, $<\bar{\psi}\psi>$ in strong coupling regime is much larger than that of $<\bar{\psi}\psi>$ in weak coupling regime. The chiral order parameter $<\bar{\psi}\psi>_0$ as shown on Fig.4 for C(bare)=0. Transition occurs at $\beta_g \sim 1$. The value $<\bar{\psi}\psi>_0$ is almost independent of N_{f_1}, but strongly depends on N_{f_2}. In this figure, β_2 dependence of $<\bar{\psi}\psi>$ is also shown; for $\beta_2 > 0$, chiral transition is very strong and for $\beta_2 < 0$, it becomes moderate. $\beta_2 = -\beta_1/4$ (< 0) is related to the gaussian action $S_G \propto \theta^2$. When the fermion self energy correction is switched off ($N_{f_2} = 0$), we obtain very small value of $<\bar{\psi}\psi>$, which is also shown in Fig.4.

Fig.5 shows the flow diagrams projected on $\beta_1 - G$ plane, where $G \equiv C_G/A_G^2$ is the normalized four fermi coupling (A_G corresponds to the fermion wave function renormalization). For $C^b = 0$ ($A^b = 1$ is taken for all cases shown), phase transition separating strong and weak four fermi coupling regions is seen around $\beta_1^b \sim 2$. For $C^b > 0$, weak coupling region occupies wider region. For $C^b < 0$, strong coupling four fermi region becomes dominant. Especially, for $C^b < -2$, all the regions of β_1^b belong to strong four fermi coupling phase even for weak gauge coupling ($\beta_1^b = 10.1$),which might be related to Nambu-Jona-Lasinio model[7) 8)].

Finally we show the flows projected onto $G - K$ plane, where K is defined by $K \equiv B/A_G$. In strong coupling regions $G - K$ flows seem to be given by a universal curve independent of the bare couplings. The trajectories starting from various bare β_1 and C are absorbed by a trajectory (Fig.6), on which $\log G \propto \log K$ namely $K \propto G^p$ ($p \sim 0.8$).

In conclusion, the phase structure of lattice (QED)$_4$ is studied with MKRG method. The value of $< \bar{\psi}\psi >$ shows slight but clear indication of chiral transition. Fermion self energy correction is crucial to the chiral symmetry breaking. Even in the weak coupling region of gauge coupling β_g, the system shows strong coupling behavior about four fermi coupling when the bare four fermi coupling exceeds a critical value.

ACKNOWLEGDMENTS

The author would like to thank H.Yoneyama, K.Ghoroku,T.Kashiwa, K.Funakubo, H.So and K.Matumoto for valuable discussions. He is also indebted to T.Otofuji and A.Kakuto for their support in uses of TEX .

REFERENCES

(1) Migdal,A.A.,*Soviet Phys.* JETP**42**,413;743(1976);
    Kadanoff,L.P.,*Ann.Phys.*(N.Y.)**100**,359(1976).
(2) Wilson,K.G.and Kogut,J.,*Phys.Reports*,**12c**,75(1974).
(3) Imachi,M., Kawabe,S.and Yoneyama,H.,*Prog.Theor.Phys.***69**, 221;1005(1983);
    Imachi,M. and Yoneyama,H.,*Prog.Theor.***78**,623(1987) and references therein.
(4) Bitar,K.M., Gottlieb,S. and Zachos,C.K.,*Phys.Rev.***D26**,2853(1982); *Phys.Lett.***121B**, 163(1983).
(5) Caracciolo,S.and Menotti,P.,*Ann.Phys.*(N.Y.)**122**, 74(1979);
    See also Matsui,T., *Nucl.Phys.***B136**, 277(1978).
(6) Kogut,J. and Susskind,L., *Phys. Rev.***D11**, 395(1975);
    Susskind,L.,*Phys.Rev.***D16**, 3031(1977).
(7) Nambu,Y. and Jona-Lasinio,G., *Phy.Rev.***122**, 345(1961).
(8) Leung,C.N., Love,S.T. and Bardeen,W.A., *Nucl. Phys.***B273**, 649(1986).
    Kondo,K-I.,Mino,H. and Yamawaki,K., preprint DPNU-88-18.
    Aoki,K-I., preprint RIFP-758.

FIGURE CAPTIONS

Fig. 1. Gauge plaquette decimation; vacuum polarization contained. Crosses denote the fermion decimations.

Fig. 2. Fermion self energy correction.

Fig. 3. Flow diagrams of gauge coupling constants a)pure gauge b)vacuum polarization contained c)vacuum polarization and fermion self energy contained.

Fig. 4. Chiral order parameter $< -\bar{\psi}\psi >^{1/3}$.
$N_{f_1} = N_{f_2} = 1; \beta_2 =$ a) -1.0 b) 0 c)1.0 d)2.0.
e)$N_{f_1} = 0, N_{f_2} = 1, \beta_2 = 0$("quenched"). f)$N_{f_1} = 1, N_{f_2} = 0, \beta_2 = 0$.

Fig. 5. Flow digrams projected onto β_1-G plane.

Fig. 6. Flow digrams projected onto G-K plane.

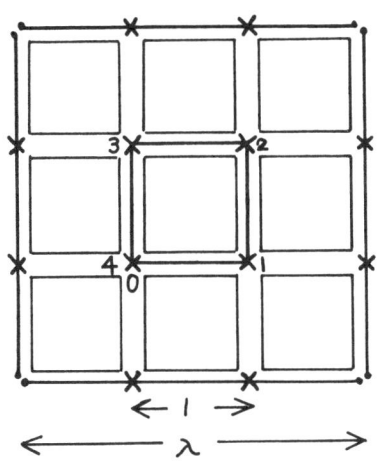

Fig.1

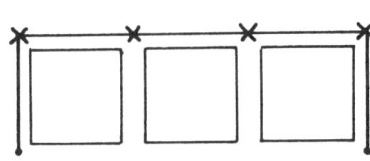

Fig.2

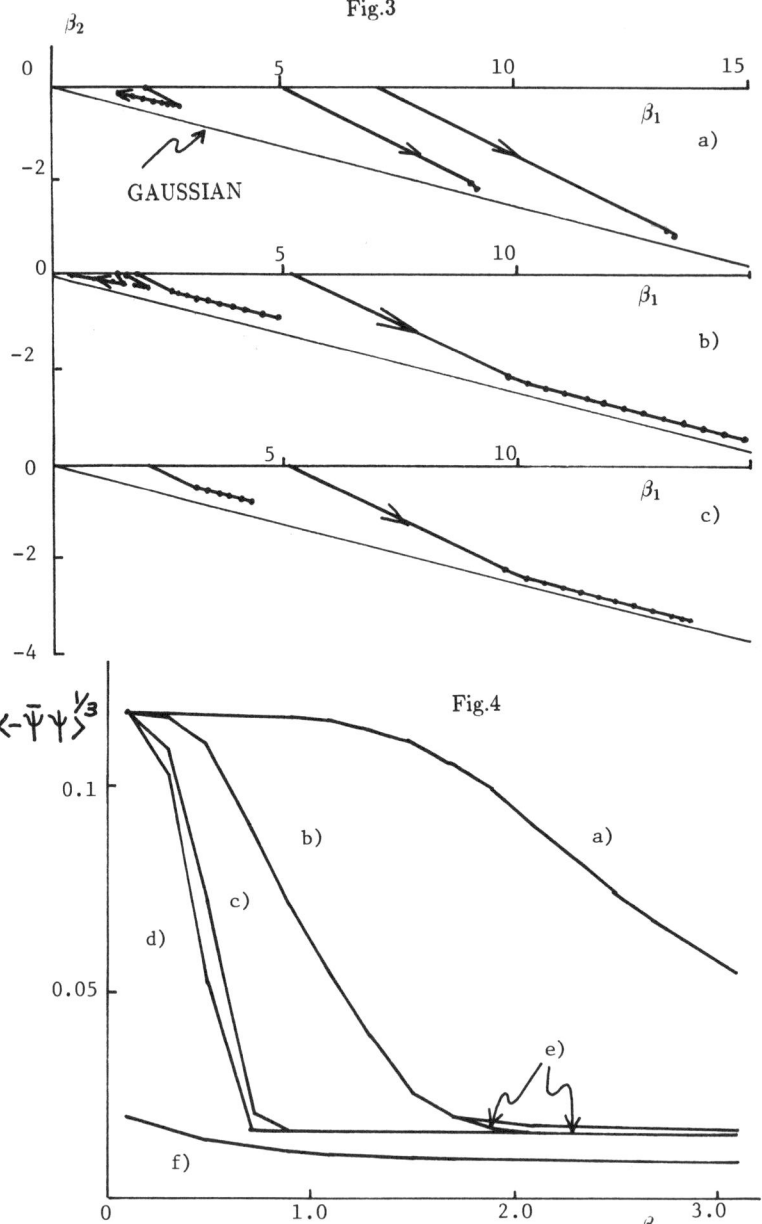

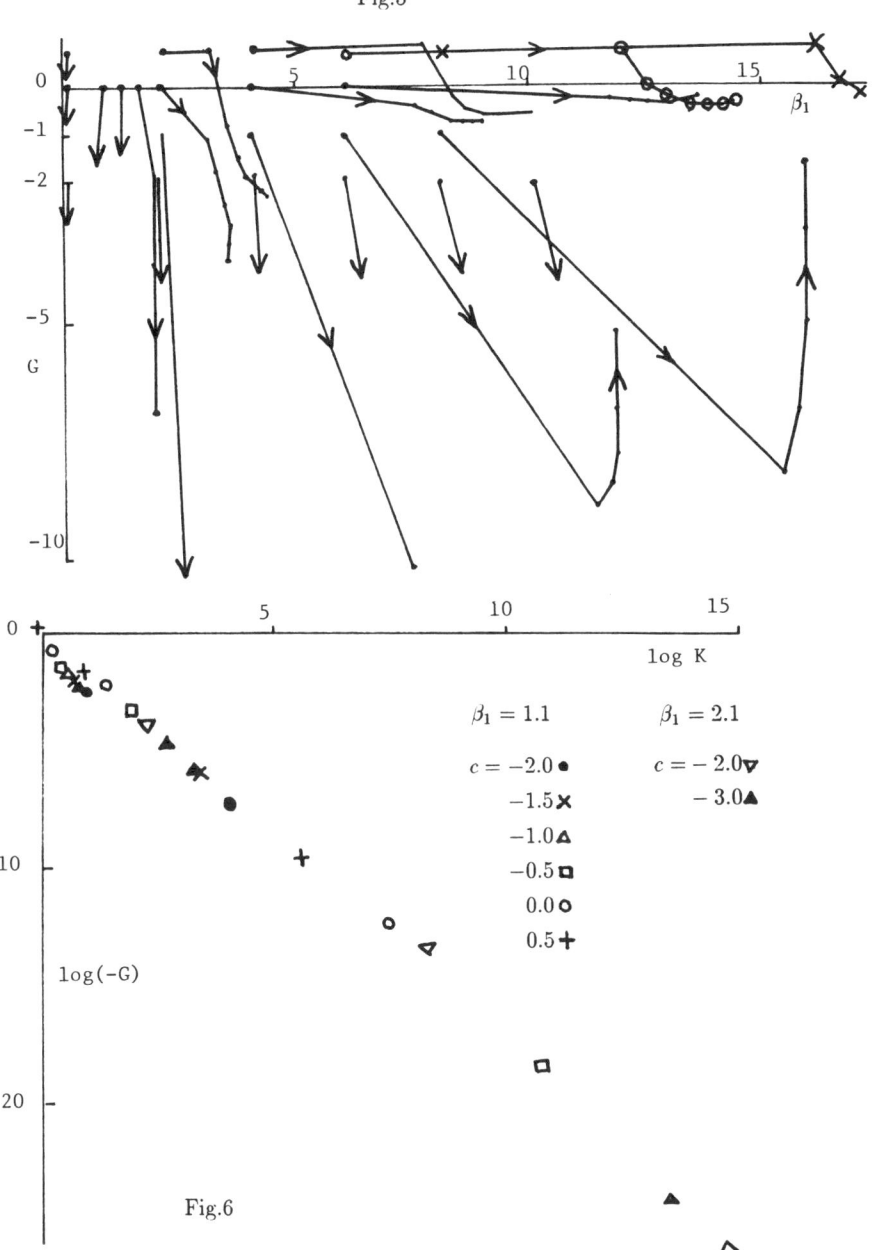

Fig.5

Fig.6

Effective Potential for Bilocal Composite Fields and its Ambiguity[*]

T. Muta[†]

Department of Physics, Hiroshima University
Naka-ku, Hiroshima 730

ABSTRACT

It is pointed out that an ambiguity exists in the definition of the effective potential for bilocal composite fields which is an indispensable tool to discuss dynamical symmetry breaking. The ambiguity gives warning to arguments on the stability of ground states based on the curvature of the effective potential.

---

[*] The talk based on the work in collaboration with M. Inoue, H. Katata and K. Shimizu: Prog. Theor. Phys. **79** (1988) 519.

[†] Supported in part by the Grant-in-Aid for Scientific Research from the Ministry of Education, Science and Culture (No. 63540221).

1. Introduction

One of the major achievements of modern quantum field theory is the discovery of spontaneous symmetry breaking.[1] Spontaneous symmetry breaking takes place either by elementary scalar fields $\langle\phi\rangle$ or through dynamical effects without such elementary scalars, i.e. through the formation of bound states, e.g. fermion-antifermion pairs $\bar{\psi}\psi$. The latter mechanism for spontaneous symmetry breaking is referred to as dynamical symmetry breaking.

To describe spontaneous symmetry breaking it is the most convenient to rely on effective potentials which may be used to determine a ground state (vacuum) in the presence of spontaneous symmetry breaking. In the case of spontaneous symmetry breaking through elementary scalar fields with nonvanishing vacuum expectation value, we use the effective potential for the vacuum expectation value of the scalar field under the influence of an external source. On the other hand we deal with the effective potential for the vacuum expectation value of composite fields $\bar{\psi}\psi$ in the case of dynamical symmetry breaking. The composite field may be either local $(\bar{\psi}(x)\psi(x))$ or bilocal $(\bar{\psi}(x)\psi(y))$.

If the local composite field $\bar{\psi}(x)\psi(x)$ is adopted for the definition of the effective potential, we are naturally led to the renormalization of the composite field for removing the divergence due to the singularity of the product of fields at the same space-time point. This renormalization procedure introduces counter terms made of polynomials of a source function for the composite field and results in an ambiguity in the finite part of the effective potential.[2,3,4,5] This ambiguity was the main reason for the fact that the effective potential for local composite fields was abandoned in the description of dynamical symmetry breaking.

In order to circumvent the above difficulty, the bilocal composite field $\bar{\psi}(x)\psi(y)$ is ordinarily employed for the description of dynamical symmetry breaking.[6] Since the bilocal composite field $\bar{\psi}(x)\psi(y)$ is free from the singularity arising from the product of fields at the same space-time point, the effective potential for the bilocal composite field does not suffer from the above-mentioned ambiguity. Unfortunately, however, it will be shown in the present talk that a new source of ambiguity exists in the definition of the effective potential for bilocal composite fields.[7] The ambiguity emerges from the fact that different effective potentials for bilocal composite fields result depending on different ways of coupling the bilocal composite field to the source function. In the following sections this ambiguity existing in the definition of the effective potential for bilocal composite fields will be explained and a possible choice of the most reasonable effective potential will be suggested.

2. Effective potential for bilocal composite fields

Consider, for simplicity of our argument, the theory of fermion field $\psi(x)$ interacting through scalar field $A(x)$. The Schwinger functional W generating connected Green function with insertion of bilocal composite field $\psi(x)\bar{\psi}(y)$, $\langle \psi(x)\bar{\psi}(y)\psi(x_1)\cdots A(x_j)\cdots \rangle$, is given by the relation

$$e^{iW} = \int [d\psi d\bar{\psi} dA]\, e^{iS}, \tag{2.1}$$

where $[d\psi d\bar{\psi} dA]$ represents path integrals over variables ψ, $\bar{\psi}$ and A, and S is the action,

$$S = \int dx\,(\mathcal{L} + \bar{\eta}\psi + \bar{\psi}\eta + JA) + \int dxdy\, \bar{\psi}(x) K(x,y)\psi(y), \tag{2.2}$$

with $\bar{\eta}$, η, J and K the source functions. In the following we are only interested in the source term of the composite field $\bar{\psi}_\alpha(x)\psi_\beta(y)$ and so we set $\bar{\eta} = \eta = J = 0$.

The effective action Γ is a Legendre transform of the Schwinger functional $W[K]$,

$$\Gamma[G] = W[K] - \int dxdy\, \frac{\delta W}{\delta K_{ab}(x,y)} K_{ab}(x,y), \tag{2.3}$$

where G is the fermion propagator in the presence of source K,

$$G_{ba}(y,x) = \frac{\delta W}{\delta K_{ab}(x,y)} = \langle \psi_b(y)\bar{\psi}_a(x)\rangle_K. \tag{2.4}$$

If the translational invariance of the theory is imposed, we have $K(x,y) = K(x-y)$

and hence $G(x,y) = G(x-y)$ by Eq.(2.4). Thus we have

$$\Gamma[G] = -V[G]\int dX, \qquad (2.5)$$

with $X = (x+y)/2$ and $V[G]$ the effective potential. Note that V is not an ordinary function but a functional of $G(x-y)$.

3. Ambiguity

In this section we shall show that an ambiguity exists in the definition of the effective potential for bilocal composite fields. This ambiguity will be shown to be ascribed to the arbitrariness of the choice of source terms for bilocal composite fields.

Let us start with an elementary example which consists of a system of noninteracting fermions under the influence of the bilocal external source $K(x,y)$. The action S for this system reads

$$S = \int dx dy\, \bar{\psi}(x)\,[iG_0^{-1}(x,y) - K(x,y)]\psi(y), \qquad (3.1)$$

where $G_0^{-1}(x,y)$ is the inverse of the free fermion propagator given by

$$G_0^{-1}(x,y) = \delta(x-y)(\not{\partial}_y + im). \qquad (3.2)$$

The Schwinger functional W corresponding to this action is readily calculated resulting in

$$W[K] = -iTr\ln(iG_0^{-1} - K), \qquad (3.3)$$

from which we have

$$\frac{\delta W}{\delta K} = i\,(iG_0^{-1} - K)^{-1} \equiv G. \qquad (3.4)$$

Making the Legendre transformation to Eq.(3.3) with the new variable G in Eq.(3.4) we obtain the effective action $\Gamma[G]$ in a closed form,

$$\Gamma[G] = -iTr[\ln(iG^{-1}) + GG_0^{-1} - 1]. \tag{3.5}$$

By imposing the translational invariance we define the effective potential as in Eq.(2.5). The resulting effective potential derived from Eq.(3.5) is known to be unbounded from below as a functional of $G(x - y)$.[3,8]

We shall now show that this embarrassing property of the effective potential can be attributed to our improper choice of the source term, i.e. $\bar{\psi}(x)K(x,y)\psi(y)$. For this purpose we first introduce the action $\hat{S}$ which differs from S in Eq.(3.1) by the term quadratic in K:

$$\hat{S} = S + \frac{1}{2}Tr[KCK], \tag{3.6}$$

where Tr refers to the trace in the coordinate and spin degrees of freedom and $C_{ab,cd}(x,y)$ is a function independent of fields that will be specified later. The Schwinger functional $\hat{W}$ corresponding to the action $\hat{S}$ of Eq.(3.6) is readily given by

$$\hat{W} = W + \frac{1}{2}Tr[KCK]. \tag{3.7}$$

Making the Legendre transformation and using the variable G as before we find

$$\hat{\Gamma} = -Tr[\ln(iG^{-1}) + GG_0^{-1} - 1] + \frac{1}{2}Tr[(G_0^{-1} - G^{-1})C(G_0^{-1} - G^{-1})]. \tag{3.8}$$

In Eq.(3.8) it is now possible to choose the function C so that the effective potential $\hat{V}$ defined through $\hat{\Gamma}$ is made bounded from below as a functional of G.

In summary we have two different effective potentials V and $\hat{V}$ which coincide with each other in the physical limit $K = 0$: $\hat{V}$ is bounded from below while V is unbounded ; the difference comes from the additional term depending only on the source K.

The above conclusion is derived from the example of noninteracting fermions. It may, however, be generalized to more general cases. In fact we have freedom of adding to the Schwinger functional W terms depending only on the source K without changing our physical predictions, i.e.,

$$\hat{W}[K] = W[K] + F[K], \qquad (3.9)$$

where $F[K]$ is an arbitrary functional of K satisfying

$$F[0] = 0. \qquad (3.10)$$

After the Legendre transformation we have the following effective action,

$$\hat{\Gamma}[\hat{G}] = \Gamma[G] + (F[K] - Tr[\frac{\delta F}{\delta K}K])_{K = iG_0^{-1} - iG^{-1}}. \qquad (3.11)$$

Obviously we see that the value of $\hat{\Gamma}$ is equal to that of Γ at the stationary point $K = 0$. To ensure the uniquencess of the stationary condition, we further require that

$$\frac{\delta F}{\delta K} = 0 \quad \text{for} \quad K = 0. \qquad (3.12)$$

4. An example of interacting fermions – QED

Here we choose quantum electrodynamics (QED) of massless fermions as an example of interacting fermions and examine the ambiguity in the effective potential which is pointed out in the last section.

The action for massless QED under the influence of bilocal external field $K(x,y)$ reads[6]

$$S = \int dx\,dy\,[\,\bar{\psi}(x)\,(iG_0^{-1}(x,y) - K(x,y))\,\psi(y) + \frac{1}{2}A^\mu(x)\,i\,D_{0\mu\nu}^{-1}(x,y)\,A^\nu(y)\,]$$
$$- e \int dx\,\bar{\psi}(x)\gamma_\mu\psi(x)A^\mu(x), \qquad (4.1)$$

where $G_0^{-1}(x,y)$ is the inverse of the free fermion propagator as given in Eq.(3.2) and $D_{0\mu\nu}^{-1}(x,y)$ is the inverse of the free photon propagator defined by

$$D_{0\mu\nu}^{-1}(x,y) = \delta(x-y)i(-g_{\mu\nu}\Box + (1-1/\alpha)\partial_\mu\partial_\nu). \tag{4.2}$$

The Schwinger functional $W[K]$ derived from the action S is found to be given by the relation,

$$e^{iW[K]} = \int [d\psi d\bar{\psi}]e^{iI}, \tag{4.3}$$

where we have performed the path integration on photon field A_μ after shifting the variable and

$$I = \int dx dy [\bar{\psi}(x)(iG_0^{-1}(x,y) - K(x,y))\psi(y) \\ + i\frac{e^2}{2}\bar{\psi}_a(x)\psi_b(x)D(x,y)_{ab,cd}\bar{\psi}_c(y)\psi_d(y)], \tag{4.4}$$

with

$$D(x,y)_{ab,cd} = \gamma^\mu_{ab}D_{0\mu\nu}(x,y)\gamma^\nu_{cd}. \tag{4.5}$$

The effective action $\Gamma[G]$ as a functional of fermion propagator $G = \delta W/\delta K$ may be obtained from $W[K]$ through the standard procedure. Denoting by $\Gamma_0[G]$ the effective action for noninteracting fermions as given in Eq.(3.5) we write

$$\Gamma[G] = \Gamma_0[G] + \Gamma_2[G], \tag{4.6}$$

where $\Gamma_2[G]$ is a two-particle-irreducible vacuum amplitude representing the effect of interactions.[6] Up to two-loop order neglecting the vertex part we have

$$\Gamma_2 = -\frac{i}{2}e^2 Tr[GDG]. \tag{4.7}$$

Hence $\Gamma[G]$ is given by[6]

$$\Gamma[G] = -iTr[\ln(iG^{-1}) + GG_0^{-1} - 1 - \frac{e^2}{2}GDG]. \tag{4.8}$$

It is easy to show that a stationary condition for the effective action (4.8) provides us with the Dyson-Schwinger equation in the quenched planar approximation.[6] The effective potential V derived from the effective action (4.8) is known to be unbounded from below in the chiral-symmetry-breaking phase.[9]

Let us consider an effective action $\hat{\Gamma}$ defined through the Schwinger functional

$$\hat{W}[K] = W[K] + \frac{1}{2}Tr[KCK], \tag{4.9}$$

as in the way explained in the previous section. In the present case we choose C such that

$$C = (i/e^2)D^{-1}, \tag{4.10}$$

where D^{-1} is defined by

$$D(x,y)_{ab,cd} D^{-1}(x,y)_{dc,fe} = \delta_{ae}\delta_{bf}. \tag{4.11}$$

The reason why the choice (4.10) is made will become clear soon. Solving Eq.(4.11) we obtain D^{-1} in an explicit form,

$$\begin{aligned}D^{-1}_{ab,cd}(x,0) =& \frac{\pi^2}{4} \frac{1}{\alpha(\alpha-1)(\alpha+3)}[(\alpha+1)(\alpha^2+6\alpha-3)x^2 g_{\mu\nu} \\&+ 2(\alpha+3)(\alpha^2-6\alpha+1)x_\mu x_\nu]\gamma^\mu_{ad}\gamma^\nu_{cb} \\&+ \frac{\pi^2}{4}\frac{(\alpha+1)^2}{\alpha(\alpha-1)(\alpha+3)}[(\alpha-3)x^2 g_{\mu\nu} + 2(\alpha+3)x_\mu x_\nu] \\&\times (\gamma_5\gamma^\mu)_{ad}(\gamma_5\gamma^\nu)_{cb}.\end{aligned} \tag{4.12}$$

Note that in Eq.(4.12) we set $y = 0$ since we know that $D^{-1}(x,y)$ is a function only of $x - y$. We realize in Eq.(4.12) that D^{-1} is singular for frequently-used gauges, i.e. $\alpha = 0, 1, -3$, and so C is ill-defined for these gauges.

The effective action $\hat{\Gamma}$ defined through $\hat{W}$ reads

$$\hat{\Gamma}[\hat{G}] = \Gamma[G] - \frac{i}{2e^2}Tr[KD^{-1}K], \tag{4.13}$$

where we used Eq.(4.10). The source function K is related to G by

$$K = -\frac{\delta\Gamma}{\delta G} = \Sigma + ie^2 DG, \tag{4.14}$$

where Σ is a self-energy part defined by

$$\Sigma = iG_0^{-1} - iG^{-1}. \tag{4.15}$$

For $\Gamma[G]$ given by Eq.(4.8) in two-loop approximation we obtain

$$\hat{\Gamma} = -iTr\ln(iG^{-1}) - \frac{i}{2e^2}Tr[\Sigma D^{-1}\Sigma]. \tag{4.16}$$

We assume the following form for Σ in conformity with the case of the Landau gauge,

$$\Sigma_{ab} = \delta_{ab}B. \tag{4.17}$$

Inserting Eq.(4.17) in Eq.(4.16) we obtain

$$\hat{\Gamma} = -iTr\ln(iG^{-1}) - \frac{i}{2e^2}\int dxdy B(x,y)D^{-1}_{aa,cc}(x,y)B(y,x), \tag{4.18}$$

with

$$D^{-1}_{aa,cc}(x,y) = \frac{16\pi^2}{\alpha+3}(x-y)^2. \tag{4.19}$$

It should be noted here that $D^{-1}_{aa,cc}$ given by Eq.(4.19) is singular only when $\alpha = -3$ although D^{-1} in Eq.(4.12) is singular for $\alpha = 0, 1, -3$.

The effective potential $\hat{V}$ derived from $\hat{\Gamma}$ is given in momentum space for Landau gauge by

$$\hat{V}[B] = -2\int \frac{dp}{(2\pi)^4}[\ln(1+\frac{B(p^2)^2}{p^2})+\frac{4\pi}{3e^2}B(p^2)\Box_p B(p^2)], \qquad (4.20)$$

where we regarded $\hat{V}$ as a functional of $B(p^2)$ which is a Fourier transform of $B(x,y)$ in Eq.(4.17) and the momentum integration in Eq.(4.20) is performed in Euclidean space with $\Box_p$ the d'Alembertian in Euclidean momentum. The effective potential $\hat{V}$ of Eq.(4.20) is apparently bounded from below. The stationary condition for $\hat{V}[B]$ is easily derived from Eq.(4.20) and reads

$$\frac{B}{p^2+B^2}+\frac{4\pi^2}{3e^2}\Box_p B = 0. \qquad (4.21)$$

Equation (4.21) is the Dyson-Schwinger equation in the differential form discussed by Fukuda and Kugo.[10] While for effective potential V corresponding to Γ the Dyson-Schwinger equation is in the form of an integral equation, the Dyson-Schwinger equation for $\hat{V}$ corresponding to $\hat{\Gamma}$ takes the form of a differential equation in momentum space. As shown in Ref.10 these two equations are equivalent if the boundary condition is properly chosen. Hence we conclude that, though effective potentials V and $\hat{V}$ are different in their shape, they make the same physical predictions. Since $\hat{V}$ is bounded from below, we realize that $\hat{V}$ is more convenient than V in the practical use.

It is important to note that $\hat{V}$ agrees with the potential obtained in the auxiliary field method.[11] Also noted is that the relation (4.9) between W and $\hat{W}$ with C given by Eq.(4.10) is essentially the same as the one obtained by Haymaker, Matsuki and Cooper[12] when they argued the relation between the auxiliary field method[13] and the Cornwall-Jackiw-Tomboulis method.[6] In the auxiliary field method the Schwinger functional W_A is given by

$$e^{iW_A} = \int [d\psi d\bar{\psi}d\phi]e^{iI_A}, \qquad (4.22)$$

where the path integration over A_μ is performed after the shift of variables, ϕ is the bilocal auxiliary field and I_A is given by

$$\begin{aligned}I_A = \int dxdy[&\bar{\psi}(x)(iG_0^{-1}(x,y) + ie^2 D(x,y)\phi(x,y))\psi(y) \\ &+ i\frac{e^2}{2}\phi(x,y)D(x,y)\phi(x,y) + \phi(x,y)K(x,y)].\end{aligned} \qquad (4.23)$$

Note that the external source K is coupled directly with the auxiliary field ϕ. This feature is an essential difference from the case of the Cornwall-Jackiw-Tomboulis method. In the Cornwall-Jackiw-Tomboulis method we start with Eq.(4.4) and introduce the bilocal auxiliary field ϕ. We then rewrite Eq.(4.4) in the form similar to Eq.(4.23). In this case, however, we have a source term different from the one in Eq.(4.23). By comparison

$$W_A = W + \frac{1}{2}Tr[K\frac{i}{e^2}DK]. \qquad (4.24)$$

According to Eqs.(4.9) and (4.24) we thus find

$$\hat{W} = W_A. \qquad (4.25)$$

Summing up, we have shown that, according to the arbitrariness of adding to the action a function only of the source function K, the effective potential for bilocal composite fields has an ambiguity. This ambiguity may be employed to convert an effective potential unbounded from below to the one bounded from below without changing its physical predictions. Examples of such modified effective potentials are found in the auxiliary field method[11] and in Ref.14. There is a recent attempt to find out the meaningful stability condition for the ground state.[15]

ACKNOWLEDGEMENTS

I would like to thank Bill Bardeen, Reijiro Fukuda, Masato Inoue and Hoi-Lai Yu for useful conversations.

REFERENCES

1. Y. Nambu and G. Jona-Lasinio, Phys. Rev. **22**, 345 (1961); **124**, 246 (1961).
2. D.J. Gross and A. Neveu, Phys. Rev. **D10**, 3235 (1974).
3. T. Banks and S. Raby, Phys. Rev. **D14**, 2182 (1976).
4. U. Ellwanger, Nucl. Phys. **B207**, 447 (1982).
5. M. Inoue, H. Katata, T. Muta and K. Shimizu, Perspectives on Particle Physics, ed. S. Matsuda, T. Muta and R. Sasaki (World Scientific Pub. Co., Singapore, 1989).
6. J. Cornwall, R. Jackiw and E. Tomboulis, Phys. Rev. **D10**, 2428 (1974).
7. M. Inoue, H. Katata, T. Muta and K. Shimuzu, Prog. Theor. Phys. 79, 519 (1988).
8. M.E. Peskin, Les Houches 1982, ed. J.B. Zuber and R. Stora (North-Holland Pub. Co., Amsterdam, 1984).
9. R.W. Haymaker and J. Perez-Mercader, Phys. Rev. **D27**, 1353 (1983); R.W. Haymaker and T. Matsuki, Phys. Rev. **D33**, 1137 (1986).
10. R. Fukuda and T. Kugo, Nucl. Phys. **B117**, 250 (1976).
11. T. Morozumi and H. So, Preprint RIFP-671 (1987); Prog. Theor. Phys. **77**, 1434 (1987).
12. R.W. Haymaker, T. Matsuki and F. Cooper, Phys. Rev. **D35**, 2567 (1987).
13. H. Kleinert, Phys. Lett. **62B**, 429 (1976). E. Schrauner, Phys. Rev. **D16**, 1887 (1977). T. Kugo, Phys. Lett. **76B** 625 (1978).
14. R. Casalbuoni, S. de Curtis, D. Dominici and R. Gatto, Phys. Lett. **140B**, 357 (1984); 150B, 295 (1985).
15. R. Fukuda, M. Komachiya and M. Ukita, Keio University preprint (1988)

Gauge Dependence of the Effective Potential for Bilocal Composite Operators in Abelian Gauge Theories

H.L. Yu[*]

Institute of Physics, Academia Sinica, Nankang, Taipei
TAIWAN

Abstract

We generalized the Nielsen-Fukuda-Kugo identity to the case of the effective potential for bilocal composite fields in quantum electro-dynamics and show that the ground-state energy in the presence of dynamical chiral symmetry breaking is indeed gauge independent. The derived identity is then employed to explore the gauge properties of various quantities, i.e., the fermion propagator and the dynamical mass of massless QED in the presence of spontaneous chiral symmetry breaking.

---

[*] Talk based on the work with N. Mitani and T. Muta, Hiroshima University preprint HUPD-8811.

1. Introduction

The effective potential (or effective action) has provided with us a very powerful tool in studying the dynamical problems in quantum field theory. In particular, when one wants to search for non-perturbative solution corresponding to dynamical symmetry breaking.

Being a powerful tool the effective potential suffers from diseases, i.e., the arbitrariness for introducing the bilocal external sources which is also the subject of Muta's talk in this Workshop. Yet, another ambiguity arises in the study of gauge theories[1,2,3] and hence may limit its predictive powers. In the case of elementary fields, the way how the effective potentials depends on the chosen gauge has been clarified in a general framework through the identity derived by Nielsen[4], Fukuda and Kugo[5] in quantum electro-dynamics(QED). We shall call this identity the Nielsen-Fukuda-Kugo(NFK) identity in the present paper. With this identity, one can obtain useful information on the gauge-dependence of the effective potential in QED and show that it is gauge-independent at its stationary point.

In the study of the spontaneous symmetry breaking of dynamical origin, i.e. the symmetry breaking whose order parameter is a vacuum expectation value of composite fields rather than elementary scalar fields one is then required to look at effective potentials that depend on composite fields. Cornwall, Jackiw and Tomboulis[6] had constructed an explicit example of effective potential for bilocal composite fields. In order to explore the gauge dependence of effective potentials for bilocal composite fields, we shall, in this present paper, generalize the NFK identity to accommodate bilocal composite fields[7].

We first generalize the NFK identity to incorporate bilocal composite fields and then discuss some applications of the resulting identity. We shall show that the effective potential for bilocal composite fields is in fact gauge-independent at its stationary point, i.e. on the ground state. Hinted by the derived identity we were able to write down an explicit relation that represents the gauge variance of fermion propagators.

2. NIELSEN-FUKUDA-KUGO IDENTITY FOR BILOCAL COMPOSITE FIELDS

Following the method described in Ref. 4 and Ref. 9 we can obtain a straight forward generalization of the NFK identity to accommodate bilocal composite fields. To be specific, we shall work in massless QED in covariant gauge, the Lagrangian of which is given by

$$\mathcal{L} = -\frac{1}{4}F_{\mu\nu}F^{\mu\nu} - \frac{1}{2\alpha}(\partial \cdot A)^2 + i\partial^\mu \chi_1 \partial_\mu \chi_2 + \overline{\psi}\gamma \cdot (i\partial + eA)\psi, \qquad (2.1)$$

where A_μ and ψ are photon and electron fields respectively, $F_{\mu\nu}$ is the photon field strength, α is the gauge parameter, χ_1 and χ_2 are Faddeev-Popov ghost fields which are introduced to maintain manifest BRS invariant[8].

The main trick to derive the identity is to observe that

$$\alpha \frac{\partial W}{\partial \alpha} = \frac{1}{Z} \int D[fields] \int d^4 x \frac{(\partial \cdot A)^2}{2\alpha} e^{iS},$$

$$= \left\langle \frac{(\partial \cdot A)^2}{2\alpha} \right\rangle, \qquad (2.2)$$

where $W = -i \ln Z$ is the generating functional for the connected Green functions in the standard notation, i.e. the explicit gauge dependence of the generating functional Z, W and the effective action Γ is entirely coming from the gauge-fixing term in the action S. In order to get $\left\langle \frac{(\partial \cdot A)^2}{2\alpha} \right\rangle$, we introduce the local composite field $\sigma(x)$[4] and the corresponding function $h(x)$:

$$\sigma(x) = -\frac{i}{2}\chi_1(x)\partial \cdot A(x), \qquad (2.3)$$

and under BRS transformation, we have

$$\delta\sigma(x) = \delta\hat{\sigma}; \hat{\sigma} = \frac{1}{2\alpha}(\partial \cdot A)^2 - \frac{i}{2}\chi_1 \Box \chi_2 \qquad (2.4)$$

With the introduction of the composite field $\sigma(x)$, we turn the problem of calculating $\left\langle \frac{(\partial \cdot A)^2}{2\alpha} \right\rangle$ into a calculation of $\langle \delta\sigma(x) \rangle$ i.e. $\langle \hat{\sigma} \rangle$ which can be done easily.

Let us start with the generating functionals Z for Green functions including bilocal composite fields,

$$Z[J, \xi_1, \xi_2, \eta, \overline{\eta}, \zeta\overline{\zeta}, K, L, h] = \int [dF]e^{iS}, \qquad (2.5)$$

where

$$[dF] = [dAd\chi_1 d\chi_2 d\psi d\overline{\psi}], \qquad (2.6)$$

and S is given by

$$S = \int d^4x[\mathcal{L} + ie\overline{\zeta}\chi_2\psi + ie\overline{\psi}\chi_2\zeta + J\cdot A + \xi_1\chi_1 + \xi_2\chi_2 + \overline{\eta}\psi + \overline{\psi}\eta + h\sigma]$$
$$+ \int d^4x d^4y \overline{\psi}(x)[K(x,y) + ie[\chi_2(x) - \chi_2(y)]L(x,y)]\psi(y) \qquad (2.7)$$

Here in Eq.(2.7), other then the source term $h\sigma$, J_μ, ξ_1, ξ_2, $\overline{\eta}$ and η are source functions of fields A_μ, χ_1, χ_2 ψ and $\overline{\psi}$ respectively and $K_{\alpha\beta}(x,y)$ is a source function of the bilocal composite field $\overline{\psi}_\alpha(x)\psi_\beta(y)$. We have also introduced source function $\overline{\zeta}$, ζ and L for the purpose of linearizing the resulting Ward-Takahashi identity[10]. Choose the BRS transformed variables as the new variables, we arrive at the following identity

$$\int [dF]e^{iS}\left[\int d^4x[\delta A^\mu J_\mu + \xi_1\delta\chi_1 + \xi_2\delta\chi_2 + \overline{\eta}\delta\psi + \delta\overline{\psi}\eta + h\delta\sigma]\right.$$
$$\left. + \int d^4xd^4y[\delta\overline{\psi}(x)K(x,y)\psi(y) + \overline{\psi}(x)K(x,y)\delta\psi(y)]\right] = 0 \qquad (2.8)$$

Substituting the BRS transformation (2.2) into Eq.(2.8), define W through,

$$Z = exp(iW) \qquad (2.9)$$

perform the Legendre transformation to introduce the effective action Γ,

$$\Gamma[A_\mu^c, \chi_1^c, \chi_2^c, \overline{\psi}^c, \psi^c, G, \zeta, \overline{\zeta}, L, h]$$
$$= W[J_\mu, \xi_1, \xi_2, \eta, \overline{\eta}, \zeta, \overline{\zeta}, K, L, h] - \int d^4x(J_\mu\frac{\delta W}{\delta J_\mu} + \cdots) \qquad (2.10)$$

where the classical fields are defined in the standard notation, finally, differentiate by $h(z)$ and then set $h = 0$, we then get an expression for $\langle \hat{\sigma} \rangle$, i.e.,

$$\langle \hat{\sigma}(z) \rangle = \int d^4x d^4y \left[\frac{\delta \Gamma(z)}{\delta L} \frac{\delta \Gamma}{\delta G} + \frac{\delta \Gamma}{\delta L} \frac{\delta \Gamma(z)}{\delta G} \right]$$

$$+ \int d^4x \left[\frac{\delta \Gamma(z)}{\delta A^c_\mu} \partial_\mu \chi^c_2 - (\frac{i}{\alpha}) \frac{\delta \Gamma(z)}{\delta \chi^c_1} \partial^\mu A^c_\mu \right] \quad (2.11)$$

$$- \frac{\delta \Gamma(z)}{\delta \psi^c} \frac{\delta \Gamma}{\delta \overline{\zeta}} - \frac{\delta \Gamma}{\delta \psi^c} \frac{\delta \Gamma(z)}{\delta \overline{\zeta}} + \frac{\delta \Gamma(z)}{\delta \zeta} \frac{\delta \Gamma}{\delta \overline{\psi}^c} - \frac{\delta \Gamma}{\delta \zeta} \frac{\delta \Gamma(z)}{\delta \overline{\psi}^c} \right]$$

where $\Gamma(z)$ is defined as

$$\Gamma(z) = \frac{\delta \Gamma}{\delta h(z)} \bigg|_{h=0} \quad (2.12)$$

On the other hand, we have the following expansion for $\langle \hat{\sigma} \rangle^{[11]}$:

$$\int d^4z \langle \hat{\sigma}(z) \rangle = \alpha \frac{\partial W}{\partial \alpha} - \frac{1}{2} \int d^4x \langle \chi_1(x) \rangle \xi_1(x) = \alpha \frac{\partial \Gamma}{\partial \alpha} - \frac{1}{2} \int d^4x \chi^c_1 \frac{\delta \Gamma}{\delta \chi^c_1} \quad (2.13)$$

where we have used the equation of motion for the ghost field χ_2: $i \Box \chi_2 = \xi_1$. Put Eq.(2.11) into Eq.(2.13) after integrating over z, set $\eta = \overline{\eta} = \zeta = \overline{\zeta} = \xi_1 = \overline{\xi}_2 = L = 0$, and apply conservation of ghost number, we finally arrive at

$$\alpha \frac{\partial \Gamma}{\partial \alpha} = - \int d^4x d^4y d^4z \frac{\delta \Gamma(z)}{\delta L} \frac{\delta \Gamma}{\delta G} + \frac{i}{\alpha} \int d^4x d^4z \frac{\delta \Gamma(z)}{\delta \chi^c_1} \partial \cdot A^c \quad (2.14)$$

For most cases that we are interested in, $\partial \cdot A^c$ is considered to be zero and therefore the last term on the right hand side of Eq.(2.14) disappears.

If we assume translation invariance, the Green function $G(x,y)$ becomes a function only of $x-y$, then the identity (2.14) when applied to the effective potential $V[G]$ will result in

$$\alpha \frac{\partial V[G]}{\partial \alpha} = \int d^4\xi c(G(\xi), \alpha) \frac{\delta v}{\delta G(\delta)} \quad (2.15)$$

where ξ is the relative coordinate $x - y$ and

$$C(G(\xi), \alpha) = - \int d^4z \frac{\delta \Gamma(z)}{\delta L(\xi)}. \quad (2.16)$$

the identity (2.15) is the NFK identity generalized to the case of the bilocal composite operator $\overline{\psi}_\alpha(x)\psi_\beta(y)$. One remark are in order before ending this session. The change of $V[G,\alpha]$ w.r.t the gauge parameter α is linearly proportional to the corresponding change of $V[G,\alpha]$ w.r.t. the propagator G. This linearity structure of the identity is crucial in proving the gauge invariance of the effective potential at the stationary point as we shall see in next session.

3. APPLICATIONS OF THE GENERALIZED NIELSEN-FUKUDA-KUGO IDENTITY

3.1 THE EFFECTIVE POTENTIAL AT THE STATIONARY POINT IS GAUGE INVARIANT.

Intuitively, it is expected that the effective potential at the stationary point where $K = 0$ be gauge-independent since it has a physical meaning of the vacuum energy. The stationary condition at $G = \overline{G}$ is

$$\left.\frac{\delta V[G,\alpha]}{\delta G}\right|_{G=\overline{G}} = 0. \tag{3.1}$$

Hence, at the stationary points, the linear structure of identity (2.15) will immediately give

$$\left.\frac{\partial V[G,\alpha]}{\partial \alpha}\right|_{G=\overline{G}} = 0 \tag{3.2}$$

Therefore, the effective potential at the stationary point $V[\overline{G}(\alpha),\alpha]$ is in fact gauge independent because

$$\frac{dV[\overline{G}(\alpha),\alpha]}{d\alpha} = \frac{\partial V[\overline{G}(\alpha),\alpha]}{\partial \alpha} + \frac{\delta V[\overline{G}(\alpha),\alpha]}{\delta \overline{G}}\frac{\delta \overline{G}}{\delta \alpha} = 0. \tag{3.3}$$

3.2 Deriving the gauge variance of fermion propagators.[12]

Integrating Eq.(2.16) over Z, we obtain

$$\int d^4z \langle \hat{\sigma}(z) \rangle + \int d^4x d^4y d^4z \frac{\delta^2 W}{\delta h(z)\delta L_{\alpha\beta}(x,y)} K_{\beta\alpha}(y,x) = 0. \qquad (3.4)$$

It is not difficult to show that

$$\int d^4x d^4y d^4z \frac{\delta^2 W}{\delta h \delta L} K$$

$$= exp(-iW) \int [dAd\psi d\overline{\psi}] \int d^4x d^4y d^4z \frac{e}{2}(det\Box)\overline{\psi}(x)K(x,y)\psi(y) \qquad (3.5)$$

$$\cdot (\partial_\mu A^\mu(z))[(\Box^{-1})_{xz} - (\Box^{-1})_{yz}] exp(iS),$$

where we have integrated out the ghost fields χ_1 and χ_2. $(\Box^{-1})_{xz}$ is defined by

$$(\Box^{-1})_{xz} = \int \frac{d^4p}{(2\pi)^4} \frac{1}{p^2} exp(ip \cdot (x-z)). \qquad (3.6)$$

To eliminate the $\partial \cdot A(z)$ factor in Eq.(3.5), we perform a change of variables that corresponds to a gauge transformation with the following gauge function,

$$\Lambda(z) = \frac{\Delta \alpha e}{2} \int d^4x d^4y \overline{\psi}(x) K(x,y) \psi(y) [(\Box^{-2})_{xz} - (\Box^{-2})_{yz}], \qquad (3.7)$$

after putting Eq.(3.5) into Eq.(3.4). Differentiating the resulting expression w.r.t. bilocal source $K(x,y)$, we obtain the relation

$$G(x,y)_{\alpha+\Delta\alpha} = G(x,y)_\alpha - e^2 \Delta\alpha[(\Box^{-2})_{x,y}(\Box^{-2})_{x,x}]G(x,y)_\alpha \qquad (3.8)$$

for the fermion propagators under the change of the gauge parameter α. It should be noted that:

(1) Eq.(3.8) indicates that under gauge transformation, the fermion propagator G will only rescale.

(2) $G(x,y)$ in question is the bare fermion propagator and hence the renormalized version of Eq.(3.9) will take a different form.

ACKNOWLEDGEMENTS

The author wants to thank the Physics Department of Hiroshima University for its support and hospitality during part of this research.

REFERENCES

1. R. Jackiw, Phys. Rev. $\underline{D9}$, 1686 (1974)

2. L. Dolan and R. Jackiw. Phys. Rev. $\underline{D9}$, 2904 (1974)

3. J.S. Kang, Phys. Rev. $\underline{D10}$, 3455 (1974)

4. N.K. Nielsen, Nucl. Phys. $\underline{B101}$, 173 (1974)

5. R. Fukuda and T. Kugo, Phys. Rev. $\underline{D13}$, 3469 (1976)

6. J.M. Cornwall, R. Jackiw and E. Tomboulis, Phys. Rev. $\underline{D10}$, 2428 (1974)

7. This problem has once been investigated by D.A. Jahnston(Nucl. Phys. $\underline{B283}$, 317 (1987)) in scalar QCD. Here we reconsider the problem in spinor QED on account of recent interests in the strong coupling phase QED.

8. C. Becchi, A. Rouet and R. Stora, Ann. Phys. $\underline{98}$, 287 (1976)

9. I.J.R. Aitchison and C.M. Fraser, Ann. Phys. $\underline{156}$, 1 (1984)

10. H. Kluberg-Stern and J.B. Zuber, Phys. Rev. $\underline{D12}$, 467 (1975)

11. The problem of "explicit" and "implicit" dependence of generating functionals of Γ on a source which is not Legendre transformed in passing from W to Γ had been discussed in the Appendix of Ref 9.

12. G. Thompson and H.L. Yu, Phys. Rev. $\underline{D31}$, 2141 (1985)

ON-SHELL EXPANSION OF THE EFFECTIVE ACTION

Reijiro Fukuda

Department of Physics, Faculty of Science and
Technology, Keio University, Yokohama 223, JAPAN

ABSTRACT

On shell expansion of the effective action is obtained by analogy with the classical analytical dynamics. The lowest order determines the particle spectrum excited above the chosen solution and the higher orders are related to the S-matrix elements of these particles.

In this talk, the on shell natures of the effective action are discussed. They are hitherto unobserved important properties that can be used in various fields of the quantum systems. My talk is based on the two papers listed in ref. 1) and 2).

Let us begin with the classical analytical dynamics and let $L(q,\dot{q})$ be the Lagrangian of the mechanical system of one particle with the coordinate q and $\dot{q} \equiv dq/dt$. The equation of motion is derived by the use of the action I[q]:

$$I[q] = \int dt\, L(q,\dot{q}) \qquad (1)$$

which is the functional of q. Now the functional derivative gives us the Euler-Lagrange equation;

$$0 = \frac{\delta I[q]}{\delta q(t)} - \frac{\partial L}{\partial q} - \frac{d}{dt}\frac{\partial L}{\partial \dot{q}}. \qquad (2)$$

One of the solutions to this equation is denoted by $q^{(0)}(t)$ and let us look for near this solution another solution $q(t)$ by writing

$$q(t) = q^{(0)}(t) + \Delta q(t) \qquad (3)$$

with small $\Delta q(t)$. To the first order, $\Delta q(t)$ satisfies

$$\int dt' \left(\frac{\delta^2 I[q]}{\delta q(t)\,\delta q(t')}\right)_0 \Delta q(t') = 0, \qquad (4)$$

where $(\cdots)_0$ implies the value of $(\cdots)$ evaluated at $q^{(0)}(t)$. Equation (4) determines $\Delta q(t)$ once the suitable boundary conditions are given. An example is given here as the simplest case. Suppose

$$L = \frac{m}{2}\dot{q}^2 - V(q),$$

where m is the mass of the particle and V(q) is the potential. Then

$$0 = -\frac{\delta I[q]}{\delta q(t)} = m\ddot{q}(t) + \frac{\partial V(q(t))}{\partial q(t)} \qquad (5)$$

and since

$$-\frac{\delta^2 I[q]}{\delta q(t)\,\delta q(t')} = m\,\delta''(t-t') + V''(q)\delta(t-t'), \qquad (6)$$

the eq.(4) takes a simple form when $q^{(0)}(t)$ is a constant solution $q^{(0)}$. We find in Fourier space

$$\{m\omega^2 - (V'')_0\}\Delta q(\omega) = 0. \qquad (7)$$

This is diagonal in ω so that the non-zero solution to (7) is easily found to be

$$\Delta q(\omega) \propto \delta(\omega^2 - \frac{(V'')_0}{m}). \qquad (8)$$

The mode or the particle spectrum around $q^{(o)}$ is determined in this way. For the general case where $q^{(o)}(t)$ depends on the time, the equation we have to solve is

$$det\left(\frac{\delta^2 I[q]}{\delta q(t) \delta q(t')}\right)_0 = 0, \qquad (9)$$

where the determinant is taken by regarding the indices t or t' as the matrix indices.

The higher order terms are expected to be related to the scattering among these modes. This is in fact the case as is shown below.

For the quantum system where q is the operator denoted by $\hat{q}$, the question is; what is the quantity that plays the same role as the classical action? The answer is the usual effective action $\Gamma[q]$. There is the complete parallelism between the classical and the quantum system.

Let us introduce $W[j]$ and $\Gamma[q]$ by

$$exp\, i W[j] = \int [dq]\, exp\, i \int dt \{L(q,\dot{q}) + j(t) q(t)\},$$
$$\Gamma[q] = W[j] - \int dt\, j(t) q(t),$$
$$q(t) = \frac{\delta W[j]}{\delta j(t)}.$$

The equation of motion is, as in (5),

$$\frac{\delta \Gamma[q]}{\delta q(t)} = 0. \qquad (10)$$

The mode of the small oscillation around one of the solution to (10) can be obtained by writing $q(t) = q^{(o)}(t) + \Delta q(t)$

and by solving

$$\int dt' \left(\frac{\delta^2 \Gamma[\bar{q}]}{\delta \bar{q}(t) \delta \bar{q}(t')} \right)_0 \Delta \bar{q}(t') = 0. \quad (11)$$

The following identity elucidates the meaning of eq. (11); Note the relation

$$\int dt' \frac{\delta^2 \Gamma[\bar{q}]}{\delta \bar{q}(t) \delta \bar{q}(t')} \cdot \frac{\delta^2 W[j]}{\delta j(t') \delta j(t'')} = -\delta(t-t'') \quad (12)$$

and the fact that the second derivative of W is the propagator in the presence of j(t)

$$\frac{\delta^2 W[j]}{\delta j(t) \delta j(t')} = \langle T \hat{q}(t) \hat{q}(t') \rangle_j . \quad (13)$$

Consider again the time translation invariant case. Then in Fourier space

$$\left(\frac{\delta^2 \Gamma[\bar{q}]}{\delta \bar{q}(t) \delta \bar{q}(t')} \right)_{0,\omega} = -\left(\frac{\delta^2 W[j]}{\delta j(t) \delta j(t')} \right)_{0,\omega}^{-1} , \quad (14)$$

so that in eq.(11) we are looking for the eigen-value crresponding to the pole of the propagator. The frequency of the small oscillation is determined by the Green's function — a reasonable result.

Generalization 1. The case of the relativistic field theory is discussed in the same way as above by the replacement $\hat{q} \to \hat{\phi}(x)$ where $x = (t, \mathbf{x})$. We get

$$\frac{\delta \Gamma[\phi]}{\delta \phi(x)} = 0, \quad (15)$$

$$\int d^4 y \left(\frac{\delta^2 \Gamma[\phi]}{\delta \phi(x) \delta \phi(y)} \right)_0 \Delta \phi(y) = 0 \quad (16)$$

for (10) and (11) respectively. For the space-time translation invariant case, eq. (16) is the eigen-value equation for the four momentum p^2 which is the same as the position of the pole of the propagator of $\hat{\phi}(x)$.

Generalization 2. $\hat{\phi}(x)$ is generalized to any local composite operator $\hat{O}(x)$. The equations (15) and (16) hold with the replacement $\phi(x) \to O(x)$. The pole is now the one for the propagator of $\hat{O}(x)$.

Generalization 3. $\hat{\phi}(x)$ is further generalized to the bilocal composite operator $\hat{\phi}(x)\hat{\phi}(y) \equiv \hat{D}(x,y)$. The effective action is now a function of $D(x,y)$. We have two equations given by, with the obvious notations,

$$\frac{\delta \Gamma[D]}{\delta D(x,y)} = 0, \tag{17}$$

$$\iint d^4x' d^4y' \left(\frac{\delta^2 \Gamma[D]}{\delta D(x,y) \delta D(x',y')} \right)_0 \Delta D(x',y') = 0 \tag{18}$$

The equation (17) is known to be the Schwinger-Dyson equation, while (18) turns out to be just the Nambu-Bethe-Salpeter equation. The reason is simple; by introducing the source $j(x,y)$ coupled to $D(x,y)$, we have

$$\iint d^4x' d^4y' \frac{\delta^2 \Gamma[D]}{\delta D(x,y)\delta D(x',y')} \frac{\delta^2 W[j]}{\delta j(x',y')\delta j(x'',y'')} = -\delta^4(x-x'')\delta^4(y-y''), \tag{19}$$

$$\frac{\delta^2 W[j]}{\delta j(x,y)\delta j(x',y')} = \langle T\phi(x)\phi(y)\phi(x')\phi(y')\rangle.$$

At the pole of the four point Green's function, eq.(18) has the non-zero $\Delta D(x,y)$. Another explicit proof the above fact is given by the formula for the effective action of the composite operator[3];

$$\Gamma[D] = -\frac{1}{2i} Tr \ln D - \frac{i}{2} Tr D_0^{-1} D + \gamma^{(2)}[D], \tag{20}$$

where D_0 is the free propagator and $\gamma^{(2)}[D]$ is the two particle irreducible vacuum graphs with the propagator D. In fact eq.(17) is the Schwinger-Dyson equation

$$(D^{-1})_{x,y} - (D_0^{-1})_{x,y} = \Sigma(x,y), \quad (21)$$

$$\Sigma(x,y) = 2i \frac{\delta \gamma^{(2)}[D]}{\delta D(x,y)}, \quad (22)$$

where Σ is the self energy and eq.(18) is the well known form of the two body bound state equation,

$$\iint d^4x' d^4y' \left[-D^{-1}(x,x') D^{-1}(y',y) + \left(T^{(4)}(x,y;x',y')\right)_0 \right] \Delta D(x',y') = 0, \quad (23)$$

$$\left(T^{(4)}(x,y;x',y')\right)_0 = -\left(\frac{\delta \Sigma(x,y)}{\delta D(x',y')}\right)_0, \quad (24)$$

where $T^{(4)}$ is just the two paricle irreducible four point function (the kernel) evaluated at $D=D^{(0)}$ which is one of the solution to eq.(21).

Generalization 4. We can generalize the above studies to the case of the finite temperature. For the local composite operator $O(t)$, we introduce the double paths effective action [4] $\Gamma[O_1, O_2]$. Then $O(t)$ is determined by

$$\left. \frac{\delta \Gamma[O_1, O_2]}{\delta O_1(t)} \right|_{O_1(t)=O_2(t)=O(t)} = 0 \quad (25)$$

and let one of the solution be $O^{(0)}(t)$. The mode determining equation is

$$\int dt' \left(\frac{\delta^2 \Gamma[O_1, O_2]}{\delta O_1(t) \delta O_1(t')} - \frac{\delta^2 \Gamma[O_1, O_2]}{\delta O_1(t) \delta O_2(t')} \right)_{O_1=O_2=O^{(0)}} \Delta O(t') = 0, \quad (26)$$

or

$$\int dt' \left(G^R\right)^{-1}(t,t') \Delta O(t') = 0. \tag{27}$$

Here G^R is the retarded Green's function. So, the mode is given by the pole of the retarded Green's function — a well known fact.

If we choose the bilocal operator $\hat{D}(x,y)$ we get the bound state equation at finite temperature. We do not reproduce it here, see ref.1) for details.

Now we come to the problem of the higher order. Consider, as a simplest example, the effective action of the elementary field ϕ of the relativistic field theory. Writing one of the solution to (15) by $\phi^{(c)}(x)$, and $\phi(x)$ by $\phi^{(c)}(x) + \delta\phi(x)$, we expand $\Gamma[\phi]$ as

$$\Gamma[\phi] = \Gamma[\phi^{(c)}] + \sum_{n=2}^{\infty} \frac{1}{n!} \Gamma^{(n)}(x_1,\cdots,x_n) \delta\phi(x_1) \cdots \delta\phi(x_n) \tag{28}$$

where $\Gamma^{(n)}$ is the one paricle irreducible Green's function. This is the off-shell expansion. The on-shell expansion is obtained by staying always at $j(x)=0$, where $j(x)$ is the source coupled to $\hat{\phi}(x)$. In order to do that we write

$$\phi(x) = \phi^{(c)}(x) + \Delta\phi(x) \tag{29}$$

and solve

$$0 = \frac{\delta \Gamma[\phi]}{\delta \phi(x)} = \frac{\delta \Gamma[\phi^{(c)} + \Delta\phi]}{\delta \phi(x)} \tag{30}$$

in a successive way by expanding

$$\Delta\phi(x) = \Delta\phi^{(1)}(x) + \Delta\phi^{(2)}(x) + \cdots. \tag{31}$$

We assume that $\Delta\phi^{(n)}(x)$ is of the order $(\Delta\phi^{(1)}(x))^n$. To order $\Delta\phi^{(1)}$, the equation obtained is of cource the same as eq.(15);

$$\int d^4y \left(\Gamma^{(2)}_{x,y}\right)_0 \Delta\phi^{(1)}(y) = 0, \qquad (32)$$

where we have used the notation

$$\Gamma^{(n)}_{x_1,x_2,\cdots,x_n} \equiv \frac{\delta^n \Gamma[\phi]}{\delta\phi(x_1)\cdots\delta\phi(x_n)}. \qquad (33)$$

With $\phi^{(0)}(x)$, the translationally invariant solution $\phi^{(0)}$, $\Delta\phi^{(1)}(p)$ in Fourier space has the form

$$\Delta\phi^{(1)}(p) = C(p)\,\delta(p_\mu^2 - m^2) \qquad (34)$$

Here m^2 is the assumed zero of $\Gamma^{(2)}(p_\mu^2)$ which is the pole of $W^{(2)}(p_\mu^2)$, and because of the δ-function the coefficients $C(p)$ can be written by

$$C(p) = C^+(p)\,\theta(p^0) + C^-(p)\,\theta(-p^0) \qquad (35)$$
$$\equiv \sum_{\varepsilon=\pm} C^\varepsilon(p)\,\theta(\varepsilon p^0)$$

In the course of getting the solution for $\Delta\phi^{(n)}$, the one particle reducible graphs are recovered and we find for $n \geq 2$,

$$\Delta\phi^{(n)}_{(x)} = \frac{1}{n!}\int\cdots\int d^4x_1\cdots d^4x_n\,d^4x_1'\cdots d^4x_n'\,W^{(n+1)}_{x,x_1,\cdots,x_n} \times \qquad (36)$$
$$\times W^{(2)-1}_{x_1,x_1'}\cdots W^{(2)-1}_{x_n,x_n'}\,\Delta\phi^{(1)}(x_1')\cdots\Delta\phi^{(1)}(x_n').$$

Here we have introduced

$$W^{(n)}_{x_1,\cdots,x_n} = \left(\frac{\delta^n W[j]}{\delta j(x_1)\cdots\delta j(x_n)}\right)_{j=0}. \qquad (37)$$

Equation (36) is the result of the straightforward application of the mathematical induction. Note the following structure; the connected Green's function $W^{(n+1)}$ is amputated by $W^{(2)-1}$ in its n-legs and then the on-shell projection is taken by $\Delta\phi^{(1)}$. So $\Delta\phi^{(n)}(x)$ is given by the n+1-point S-matrix element with one of its external legs is off the mass-shell and contains the pole.

Next we insert $\Delta\phi^{(n)}$ of eq.(36) into Γ itself and get the formula for Γ which is the on-shell expansion of the effective action;

$$\begin{aligned}\Gamma[\phi] &= \Gamma[\phi^{(0)} + \Delta\phi] \\ &= \Gamma[\phi^{(0)}] + \sum_{n=3}^{\infty}\frac{1}{n!}\Gamma^{(n)}_{x_1,\cdots,x_n}\Delta\phi(x_1)\cdots\Delta\phi(x_n) \qquad (38) \\ &= \Gamma[\phi^{(0)}]\end{aligned}$$

$$+ \sum_{n=3}^{\infty}\frac{1}{n!}\frac{1}{i^n}\sum_{\varepsilon_\alpha}\prod_{\alpha=1}^{n}\int\frac{d^3P_\alpha}{(2\pi)^4 2\omega(\vec{P}_\alpha^2)}\tilde{C}^{r\varepsilon_\alpha}(\vec{P}_\alpha)S^{(n)}_{p_1,\cdots,p_n}. \qquad (39)$$

Here $S^{(n)}$ is the usual S-matrix element with all the external legs on the mass-shell. $\tilde{C}^r(p)$ is the scaled version of $C(p)$ which gives us the correct normalization of $S^{(n)}$.

Our formula ss summerized in eqs.(32), (36) and (39). By these we can say that the effective action is also a generating functional of the on shell quantity. Compare the difference between eqs.(28) and (38).

One thing looking strange is the fact that since we are staying always on the solution satisfying eq.(30), the natural question is why we get $\Gamma[\phi] \neq \Gamma[\phi^{(0)}]$. The answer lies in the fact that the stationarity condition (32) does not determine $\phi^{(\omega)}(x)$ completely and $C^{\pm}(\mathbb{p})$ is the undetermined function. In fact we have used this arbitrariness in obtaining eq.(36). This corresponds to the freedom at the boundary in the classical analytical dynamics where the value $I[q]$ changes as we shift the boundary value of q, i.e. the initial conditions. In other words $I[q]$ is not stationary under the variation of the initial conditions.

Recently there are papers of the alternative approach to the on-shell expansion of the effective action.[5),6)] In particular, Jevicki and Lee[6)] derived the full order relation which expresses the S-matrix in terms of the effective action evaluated at the stationary solution. They rely on the back ground field which is a classical solution of motion. The difference of our approach from the back ground field approach is

1) Our scheme can be applied to any operator including the composite field. Thereby we find the S-matrix elements between the bound states.

2) Our expansion scheme naturally includes the spectrum determining conditions.

There are expected several applications of our formula especially to the case where we want to determine the on-shell coupling between various modes or particles.

REFERENCES

1. R. Fukuda, Prog. Theor. Phys. $\underline{78}$, 1487(1987).

2. R. Fukuda, M. Komachiya, M. Ukita, "On-shell Expantion of the Effective Action — S-matrix and the Ambiguity Free Stability Critrrion — " Keio Univ. Preprint(1988)

3. C. De Dominicis and P. C. Martin, J. Math. Phys. $\underline{5}$, 14(1964), ibid $\underline{5}$, 31(1964).

4. A. J. Niemi and G. W. Semenoff, Ann. of Phys. $\underline{152}$, 105(1984).
    K. Chou, Z. Su, B. Hao and L. Yu, Phys. Rep. $\underline{118}$, 1(1985).
    R. Fukuda, ref. 1.

5. L. D. Faddeev, in Methods in Field Theory, eds. R. Balian and J. Zinn-Justin (North Holland, 1976)

6. A. J. Niemi and C. K. Lee, Preprint Brown-HET-634(1987)

EFFECTIVE ACTION AND
THE AMBIGUITY-FREE STABILITY CRITERION

R. Fukuda, M. Komachiya and M. Ukita

Department of Physics, Faculty of Science and Technology,
Keio University, Yokohama 223, Japan

ABSTRACT

The correct stability criterion of the given solution is discussed in terms of the effective action. It is free from any ambiguities which are the subjects of recent controversy.

1. INTRODUCTION

For describing the non-perturbative phenomena, the effective potential formalism has been used by many people. But it has been also discussed that the effective potential possesses ambiguity especially when we choose a composite operator as the argument of it.[1,2,3,4,5] In fact the second derivative of the effective potential evaluated at the stationary solution is not unique and it depends on how one introduces the source term in calculating the effective potential.[1,2] Moreover, the effective potential is not bounded from below in some cases, even for the free field theory.[3,4,5] So, it is doubtful that the stability of the given solution can be discussed in terms of the effective potential. The purpose of this paper is to solve this problem.

Let us review a classical mechanical system with the Lagrangian,

$$L(q) = \frac{m}{2}\dot{q}^2 - V(q) . \tag{1}$$

We take one of the static solution of the equation of motion and we denote it $q^{(0)}$. The stability of $q^{(0)}$ can be discussed as usual. Inserting $q(t) = q^{(0)} + \Delta q(t)$ into the equation of motion and retaining the linear term in $\Delta q(t)$, we get

$$m\Delta\ddot{q}(t) = -V''(q^{(0)})\Delta q(t) \quad (V'' \equiv d^2V/dq^2) \ . \tag{2}$$

The requirement that $\Delta q(t)$ does not contain blowing up solutions is equivalent to $V''(q^{(0)}) > 0$ and, though the stability problem is essentially a time dependent phenomena, we can discuss it by knowing the curvature of V. But this is because the kinetic energy term is known to be $(m/2)\dot{q}^2$ and it is positive definite.

However, for the quantum field theory, the classical action is replaced by the effective action defined as

$$\Gamma[\phi_c] \equiv W[J] - \int d^4x \, J(x)\phi_c(x) \ , \tag{3}$$

where $W[J]$ is the generating functional of the Green's function and $\phi_c(x) \equiv \delta W[J]/\delta J(x)$. The local expansion of the effective action can be written as,

$$\Gamma[\phi_c] = \int d^4x \left\{- V_{eff}(\phi_c) + \frac{1}{2}(\partial_\mu \phi_c)^2 Z(\phi_c) + \cdots \right\} \ . \tag{4}$$

The first term is the effective potential and the rest involves the space-time derivatives. In this case, we must calculate $Z(\phi_c)$ and the other remaining terms. So it is clear that the stability criterion given by using only the effective potential is not correct and that we must use the effective action instead of the effective potential.

2. AMBIGUITY FREE STABILITY CRITERION

Using the effective action, we do the same discussion as in the classical case. The equation of motion is $\delta\Gamma[\phi_c]/\delta\phi_c = 0$ and let us take the space-time independent solution ϕ_0. In order to study the stability of it, we take another solution $\phi_c(x) = \phi_0 + \Delta\phi_c(x)$ (where $\Delta\phi_c$ is a small deviation). Then $\Delta\phi_c$ satisfies,

$$\int d^4y \left[\frac{\delta^2 \Gamma[\phi_c]}{\delta\phi_c(x)\delta\phi_c(y)}\right]_{\phi_c=\phi_0} \Delta\phi_c(y) \equiv \int d^4y \, \Gamma^{(2)}(x,y)|_{\phi_c=\phi_0} \Delta\phi_c(y) = 0 \ . \tag{5}$$

This is the fundamental equation we use below. In the momentum space, this equation becomes,

$$G(p^2)^{-1}\Delta\phi_c(p) = 0 \ , \tag{6}$$

where G is the propagator of ϕ and we use the following relations,

$$\int d^4 y \, \Gamma^{(2)}(x,y) W^{(2)}(y,z) = -\delta^4(x-z) , \qquad (7.1)$$

$$W^{(2)}(x,y)|_{J=0} = \left.\frac{\delta^2 W[J]}{\delta J(x)\delta J(y)}\right|_{J=0} = G(x,y) . \qquad (7.2)$$

If we assume the pole of G is given by $p^2 = m^2$, $\Delta\phi_c$ is now obtained in the form,

$$\Delta\phi_c \sim \exp(\pm i \sqrt{m^2 + \vec{p}^{\,2}}\, t) . \qquad (8)$$

Then we conclude, if $m^2 > 0$ the solution is stable and if $m^2 < 0$ it is not. The important point is the fact that the pole of the Green's function determines the stability. On the other hand, the curvature of the effective potential is an off-shell quantity. So it is rather natural that the effective potential possesses ambiguity.

3. EXAMPLES

3.1. The local field

Next let us see some examples. First, we take the Gross-Neveu model [6] — $O(N)$ symmetric two dimensional fermionic model — with the Lagrangian

$$L^{G.N.} = \bar{\psi}(i\slashed{\partial})\psi + \frac{1}{2}g^2(\bar{\psi}\psi)^2 . \qquad (9)$$

The source $J(x)$ is introduced in two ways:

Case I) $J(x)$ couples to the auxiliary field $\sigma(x)$: In this case we add the term $-(1/2)(\sigma - g\bar{\psi}\psi)^2 + J\sigma$ to $L^{G.N.}$ and calculate $W_I[J]$ by the Lagrangian

$$\tilde{L}^{G.N.} \equiv \bar{\psi}i\slashed{\partial}\psi - \frac{1}{2}\sigma^2 + g\sigma\bar{\psi}\psi + J\sigma . \qquad (10)$$

Case II) $J(x)$ couples to $g\bar{\psi}(x)\psi(x)$: We consider here the Lagrangian

$$L^{G.N.} - \frac{1}{2}(\sigma - g\bar{\psi}\psi - J)^2 + gJ\bar{\psi}\psi = \tilde{L}^{G.N.} - \frac{1}{2}J^2 . \qquad (11)$$

The only difference between I) and II) is the term $-(1/2)J^2$. We have to calculate the effective action Γ for two cases in order to discuss (5).

For the Case I), Γ_I is known to be obtained, for large N, by the stationary phase contribution of σ satisfying

$$-\sigma(x) - iNg\text{Tr}\{A^{-1}(x,x)\} = -J(x) = \delta\Gamma_I[\sigma]/\delta\sigma(x), \tag{12}$$

$$A(x,y) = \{i\partial\!\!\!/ + g\sigma(x)\}\delta^2(x-y).$$

We use in the following this $\sigma(x)$ for $\sigma_I(x) = \delta W_I[J]/\delta J(x)$. For the space-time independent solution $\sigma^{(0)}(x) = \sigma^{(0)}$ of $\delta\Gamma_I/\delta\sigma = 0$, after renormalization we get in Fourier space,

$$\frac{\delta\Gamma_I}{\delta\sigma_r} = \sigma_r^{(0)}\{-1 + \frac{\lambda_r}{2\pi}(\ln\frac{\mu^2}{\sigma_r^{(0)2}} + 2)\} = 0, \tag{13}$$

$$\Gamma_I^{(2)}(p^2) \equiv \frac{\delta^2\Gamma_I}{\delta\sigma_r^2(p^2)} = -1 + \frac{\lambda_r}{2\pi}\{\ln\frac{\mu^2}{\sigma_r^{(0)2}} + 2 - B\}, \tag{14}$$

$$B \equiv \sqrt{1 - M^2/p^2}\ln\frac{\sqrt{-p^2+M^2} + \sqrt{-p^2}}{\sqrt{-p^2+M^2} - \sqrt{-p^2}},$$

where $\lambda_r = g_r^2 N$, $M^2 = 4g_r^2\sigma_r^{(0)2}$ and r denotes the renormalized quantity. We have employed the renormalization condition $\partial^2 V/\partial\sigma_r^2|_{\sigma_r=\mu}=1$ with V representing the corresponding effective potential. From (13) and (14), we find that eq.(5) becomes

$$\Gamma_I^{(2)}(p^2)\Delta\sigma_I(p) = 0 \tag{15}$$

and has the following non-trivial solution $\Delta\sigma_I(p)$ for:

(i) $p^2 = M^2 > 0$ (stable) corresponding to the symmetry breaking solution $|\sigma_r| = \mu\exp(1-\pi/\lambda_r)$,

(ii) $p^2 = -g_r^2\mu^2\exp(2-2\pi/\lambda_r) < 0$ (unstable) corresponding to the symmetric solution $\sigma^{(0)}=0$. These agree with the well-known results.

Consider next the Case II). Since $W_{II} = W_I - (1/2)J^2$, $\sigma_{II}(x) \equiv \delta W_{II}/\delta J(x)$ is given by $\sigma_I(x) - J(x) = \sigma(x) - J(x)$ where $\sigma(x)$ satisfies (12). We see therefore that $W_{II}^{(2)} \equiv \delta^2 W_{II}/\delta J(x)\delta J(y)$ equals to $\delta\sigma(x)/\delta J(y) - \delta^2(x-y)$. Using this relation and the equation obtained by taking $\delta/\delta J$ of (12), we get for a space-time independent solution $\sigma^{(0)}$,

$$W_{II}^{(2)}(p^2) = -\Gamma_I^{(2)-1}(p^2) - Z_\sigma^{-1}, \tag{16}$$

where Z_σ is the renormalization constant of the σ-propagator. It is given, by using the cut off Λ, as

$$Z_\sigma = 1 + \frac{\lambda_r}{2\pi}(\ln\frac{\Lambda^2}{g_r^2\mu^2} - 2) . \tag{17}$$

Note that the solution $\sigma^{(0)}$ is the same for I) and II) since at $J=0$, $\sigma_{II}(x)=\sigma_I(x)$. Therefore eq.(5) takes the form,

$$C\Gamma_I^{(2)}(p^2)\Delta\sigma_{II}(p) = 0 , \tag{18}$$

$$C \equiv Z_\sigma(\Gamma_I^{(2)}(p^2) + Z_\sigma)^{-1} .$$

Since the factor C does not vanish for finite p^2, we get the same stability condition as in the Case I). The second derivatives of the effective potential are different for two cases and are given by

$$V_I''(\sigma_r^{(0)}) = 1 - (\lambda_r/\pi) \ln(\mu/\sigma_r^{(0)}) , \tag{19.1}$$

$$V_{II}''(\sigma_r^{(0)}) = \frac{Z_\sigma}{Z_\sigma - 1 + (\lambda_r/\pi)\ln(\mu/\sigma_r^{(0)})}\{1 - (\lambda_r/\pi)\ln(\mu/\sigma_r^{(0)})\} . \tag{19.2}$$

They are renormalization scheme dependent and also depend even on the cut off. This is because V'' is the off shell quantity.

3.2. The bilocal field

We next discuss the following action of the N-component boson field,

$$S = \frac{i}{2}\phi_i^a G_0^{-1}{}_{ij}\phi_j^a + \frac{1}{4N}\phi_i^a\phi_j^a V_{ij,kl}\phi_k^b\phi_l^b , \tag{20}$$

where $a,b=1\sim N$ and G_0 represents the free propagator. The subscripts $i,j\cdots$ represent the space-time coordinates as well as other degrees of freedom. We now introduce a bilocal auxiliary field σ_{ij} by adding the term

$$-\frac{N}{4}(\sigma_{ij}-\frac{1}{N}\phi_i^a\phi_j^a) V_{ij,kl}(\sigma_{kl}-\frac{1}{N}\phi_k^b\phi_l^b) \tag{21}$$

to (20). We thus consider

$$S = \frac{1}{2}\phi_i^a(iG_0^{-1}{}_{ij} + V_{ij,kl}\sigma_{kl})\phi_j^a - \frac{N}{4}\sigma_{ij}V_{ij,kl}\sigma_{kl} . \tag{22}$$

Case I) We add the source term $NJ_{ij}\sigma_{ij}$ to (22) and define $W_I[J]$. For large N, it is easy to obtain

$$\frac{W_I[J]}{N} = \frac{i}{2}\text{Trln}(G_0^{-1} - iV\sigma) - \frac{1}{4}\sigma_{ij}V_{ij,kl}\sigma_{kl} + J_{ij}\sigma_{ij}, \qquad (23)$$

where σ satisfies the stationary phase condition:

$$\frac{1}{2}V_{ij,kl}(G_0^{-1} - iV\sigma)_{kl}^{-1} - \frac{1}{2}V_{ij,kl}\sigma_{kl} + J_{ij} = 0, \qquad (24)$$

where $V\sigma$ represents $V_{ij,kl}\sigma_{kl}$. We use this σ_{ij} for $\sigma_{Iij}=(1/N)\delta W_I[J]/\delta J_{ij}$. The corresponding effective action is given by

$$\frac{\Gamma_I}{N} = \frac{i}{2}\text{Trln}(G_0^{-1} - iV\sigma) - \frac{1}{4}\sigma_{ij}V_{ij,kl}\sigma_{kl} \qquad (25)$$

so that we get the Schwinger-Dyson (SD) equation as a stationary equation; since we have

$$\frac{\delta\Gamma_I}{\delta\sigma_{ij}} = \frac{N}{2}V_{ij,kl}\{(G_0^{-1} - iV\sigma)_{kl}^{-1} - \sigma_{kl}\}, \qquad (26)$$

by introducing V^{-1} through $V_{ij,kl}^{-1}V_{kl,mn} = \delta_{im}\delta_{jn}$, we find

$$\sigma_{ij}^{-1} = G_0^{-1}{}_{ij} - iV_{ij,kl}\sigma_{kl}. \qquad (27)$$

Denoting one of the solutions by $\sigma^{(0)}$, eq.(5) takes the form

$$\frac{N}{2}V_{ij,mn}\sigma_{mp}^{(0)}\sigma_{qn}^{(0)}(-\sigma_{pk}^{(0)-1}\sigma_{lq}^{(0)-1} + iV_{pq,kl})\Delta\sigma_{Ikl} = 0. \qquad (28)$$

This is precisely the well-known form of the two-body bound state equation found by Y.Nambu and E.E.Salpeter-H.A.Bethe.[7,8]

Case II) We adopt the source term $J_{ij}\phi_i^a\phi_j^a$ and calculate $W_{II}[J]$. The large N limit is obtained by the similar procedure as in I). We find after some calculations

$$\frac{W_{II}[J]}{N} = \frac{W_I[J]}{N} - J_{ij}V_{ij,kl}^{-1}J_{kl}, \qquad (29)$$

where $W_I[J]$ is given by (23) and (24). The term $-JV^{-1}J$ is the source of the ambiguity of the effective potential. Now, by defining σ_{IIij} through $(1/N)\delta W_{II}[J]/\delta J_{ij}$, we calculate $\Gamma_{II}[\sigma_{II}]$. The result is

$$\frac{\Gamma_{II}[\sigma]}{N} = \frac{i}{2}\mathrm{Tr}\ln\sigma^{-1} + \frac{i}{2}\mathrm{Tr}G_0^{-1}\sigma + \frac{1}{4}\sigma_{ij}V_{ij,kl}\sigma_{kl} . \tag{30}$$

The stationary equation $0=\delta\Gamma_{II}/\delta\sigma_{II}$ is the same as for the Case I) since at $J=0$, $\sigma_{Iij}=\sigma_{IIij}$. The equation which determines the stability is

$$-i\frac{N}{2}(-\sigma_{ik}^{(0)-1}\sigma_{lj}^{(0)-1} + iV_{ij,kl})\Delta\sigma_{IIkl} = 0 . \tag{31}$$

We conclude that the stability criteria (28) and (31) are the same, since they are equivalent as the eigen value equation.

3.3. Free field case

Finally, we discuss the free field case briefly. The effective action for $G(x,y)=<\phi(x)\phi(y)>$ is known to be given by

$$\Gamma[G] = \frac{i}{2}\mathrm{Tr}\ln G^{-1} + \frac{i}{2}\mathrm{Tr}G_0^{-1}G . \tag{32}$$

So the effective potential is not bounded from below[3,4,5] as discussed in the introduction. But it does not cause any trouble. Since $\delta\Gamma/\delta G=0$ means $G=G_0$, eq.(5) takes the form

$$G_{0(+)}^{-1}(P,q)G_{0(-)}^{-1}(P,q)\Delta G(P,q) = 0 , \tag{33}$$

$$G_{0(\pm)}^{-1} \equiv (\frac{P}{2} \pm q)^2 - m^2 ,$$

where $P(q)$ or m is the total (relative) momentum or the mass of the two particle system. The non-trivial solution of ΔG exists when 1) either $G_{0(+)}^{-1}$ or $G_{0(-)}^{-1}$ vanishes, 2) both $G_{0(+)}^{-1}$ or $G_{0(-)}^{-1}$ vanish. The condition 1) is actually related to the stability in the single particle channel. Since $P^2>4m^2$, the condition 2) predicts the stability of the solution $G=G_0$ in the two particle channel if the one particle channel is stable, i.e. $m^2>0$.

4. SUMMARY

The stability of the given solution can be discussed by using the effective action without the ambiguity and the stability criterion is determined by the pole of the Green's function. This is our result. But the next statement is also valuable. For the bilocal field case, the equation which determines the stability is given in the form of the two-body bound state equation. More generally, the effective action formalism also provides for a general method of deriving exact bound state equations.[9]

REFERENCES

1. R. W. Haymaker, T. Matsuki and F. Cooper, Phys. Rev. D35, 2567 (1987).

2. M. Inoue, H. Katata, T. Muta and K. Shimizu, Prog. Theor. Phys. 79, 519 (1988).

3. T. Banks and S. Raby, Phys. Rev. D14, 2182 (1976).

4. M.Peskin, Les Houches 1982, ed. by J. B. Suber and R. Stora (North-Holland Pub. Co., Amsterdam, 1984)

5. R. Casalbouni, S. de Curtis, D. Dominici and R. Gatto, Phys. Lett. 140B, 357 (1984), 150B, 295 (1985).

6. D. J. Gross and A. Neveu, Phys. Rev. D10, 3235 (1974).

7. Y. Nambu, Prog. Theor. Phys. 5, 614 (1950).

8. E. E. Salpeter and H. A. Bethe, Phys. Rev. 84, 1232 (1951).

9. R. Fukuda, Prog. Theor. Phys. 78, 1487 (1987).

Hidden Local Symmetries

Masako Bando
Physics Division, Aichi University
Miyoshi, Aichi 470-02
JAPAN

ABSTRACT

After a short survay of nonlinear realizations the general formalism of hidden local symmetries is reviewed. Some comments are made on the applications of the formalism to such systems as strongly interacting Higgs models or technicolor models.

1. INTRODUCTION

Symmetry plays an essential role in modern physics to understand apparently complicated phenomena in a unified way. In the history of particle physics, 1960s was the time of "global" symmetry: It was found that there exists a different set of symmetry groups corresponding to each interaction. Those symmetries are all "global". For example, the strong interaction has flavor $SU(3)$ symmetry, in terms of which spectra of hadrons and their interactions were beautifully described according to neat "linear" representations. Of course now we all know well that the hadron system has a flavor symmetry as a result of the flavor-independence of color gauge interaction (QCD), although very small but flavor-dependent masses of quarks indicate weak breaking of the symmetry. The very fact of small quark masses indicated another type of symmetry (chiral symmetry), but the recognition of this symmetry needed a bit deeper insight to the structure of the symmetry realization. People came to be involved in the more exciting subjects on the strongly interacting systems, in which the symmetry principle reveals itself no more straghtforwardly and physical states are no more provided by linear representations of symmetry. However it is very important that even in such complicated physical states the symmetry principle governs the whole system, but in a *nontrivial* manner. People came to be aware of the special role of ps mesons (π mesons), which are to be identified with the NG bosons, namely, the massless particles necessarily accompanied by the dynamical symmetry breakdown, and the low energy effective lagrangian is uniquely determined with the scale parameters f_π. It is interesting to note that some physicists paid a special attention to the vector mesons (ρ mesons) which have somehow close similarity to photon, a typical gauge boson. However, for complete understanding of this we had to wait until the discovery of the notion of "hidden local symmetry".

After 1970, "local" version of symmetry, gauge symmetry, became a fundamental basis of elementary particle physics and it is now a common understanding that all existing interactions of particles are governed by the gauge principle. Gauge theories consist of two dynamical aspects, "weak" and "strong" coupling phases. Usually the gauge sector of the electroweak theory is considered to be in the weak coupling phase, while QCD is certainly in the strong coupling phase. Strong coupling gauge theories (SCGT), because of the very strength of interactions, provide us with a rich spectrum of bound states and, even more, develop dynamical symmetry breakdown, and will have a large variety of dynamical possibilities. Thus, we expect that they may play essential roles in constructing unified theories of elementary particles.

As an example of such possibilities I here would like to introduce the concept of dynamical gauge bosons of hidden local symmetries.*

2. NONLINEAR REALIZATION

Here we use the notations: the symmetry group G and its subgroup H, which are assumed, unless stated otherwise, compact groups. The set of generators T^A of G is divided into two parts, S^α of the unbroken subgroup H and X^a of the rest, and the latter spans, in general, a reducible representation of H and is further decomposed into some irreducible sectors;

$$\left\{T^A\right\} = \{S^\alpha \in \mathcal{H}, \ X^a = \{X^{a_r} \in (\mathcal{G} - \mathcal{H})^r\}\}. \tag{1}$$

They are chosen to satisfy $\text{tr}(T^A T^B) = \frac{1}{2}\delta^{AB}$, $\text{tr}(S^\alpha X^a) = 0$, where the second

---

⋆ This talk is based mainly on the series of papers.[1]

equation implies $\mathrm{tr}(S^\alpha[S^\beta, X^a]) = \mathrm{tr}([S^\alpha, S^\beta]X^a) = 0$, so that the element $[S^\alpha, X^a]$ always lies in $\mathcal{G} - \mathcal{H}$; $[\mathcal{H}, \mathcal{G} - \mathcal{H}] \subset \mathcal{G} - \mathcal{H}$.

Now we consider the cases where the lagrangian of the system is invariant under the transformation of the symmetry group G but the vacuum (or the physical states more generally) is not invariant under the group G and invariant only under the subgroup H. This kind of phenomenon is called "spontaneous symmetry breakdown". It may be that the terminology "broken symmetry" gives a bit confusion. The symmetry is never lost by the "spontaneous symmetry breakdown" but it is the Nambu-Goldstone bosons (NG bosons) that carry the original symmetry. In this sense some physicists prefer the terminology "symmetry rearrangement"; the symmetry is dynamically rearranged into the system of NG fields,[†]

It would be instructive to show first a simple example, CP^{N-1} model to see what happens.

[CP^{N-1} Model]: The lagrangian is given as

$$\mathcal{L}_\phi = \partial_\mu \phi^\dagger \partial_\mu \phi - \frac{1}{2}\lambda(\phi^\dagger \phi - v^2)^2 \tag{2}$$

in terms of an N-component complex scalar field,

$${}^t\phi = (\phi^1, \phi^2, \phi^3, \cdots, \phi^N), \qquad \phi^a \in C \tag{3}$$

which possesses $SU(N)$ symmetry; $g = e^{i\theta_A T^A} \in SU(N)$, with $N^2 - 1$ generators, T^A. The potential becomes minimal at the point $\phi^\dagger \phi = v^2$, and we can always choose as the minimal point, ${}^t\phi = (0, 0, 0, \cdots, v)$, without loss of generality. Then the field ϕ can be expressed in terms of *radial* and *angular* parts, σ and π, respectively;

$${}^t\phi = (0, 0, \cdots, v + \sigma(x))^t(\xi(\pi)), \qquad \xi(\pi) = e^{i\pi^A(x)X^A/f_\pi}, \tag{4}$$

with the $2N - 1$ broken generators X^a and f_π is a scale factor. Then in terms of

---

† The terminology was introduced by Umezawa[6] and frequently used by Takahashi,[5] whom I would like to thank for pointing out this fact.

$\pi(x)$ and $\sigma(x)$ the original lagrangian is rewritten as,

$$\mathcal{L} = f_\pi^2 \text{tr}[\partial_\mu \xi^\dagger(\pi) \partial_\mu \xi(\pi) \begin{pmatrix} 0 & 0 \\ 0 & 1 \end{pmatrix}] + (\partial_\mu \sigma(x))^2 - 2\lambda f_\pi^2 \sigma(x)^2 + \cdots. \quad (5)$$

in which we renormalized the kinetic terms of π fields by putting $f_\pi = v$. This expression clearly shows that the masses of π fields are exactly zero, leaving the masses of σ field of order f_π ($m_\sigma = \sqrt{2\lambda} f_\pi$). We see that eq.(5) determines uniquely all the low energy interactions for the NG bosons, π which are expressed in terms of the angular-like variable $\xi(\pi)$.

Now two important Theorems are in order.

[Goldstone Theorem][7,8] : The spontaneous broken symmetry necessarily leads to the existence of the massless particles (NG bosons), coupled to the broken currents, the number of which is equal to that of independent broken generators.

[Low Energy Theorems][9] : The symmetry principle determines uniquely all the interactions of NG bosons in the low energy limit ($p^2 \ll f_\pi^2$).

Although the Low Energy Theorems are universal, the proof is very complicated and process-dependent. On the other hand, once we recognize that the low energy theorem can be applied to any system realizing the symmetry G spontaneously broken down to H, and if we somehow construct a low energy effective lagrangian realizing this symmetry, as we saw already in the example, then the soft NG boson amplitudes themselves can be read off in a handy way. The simplest way to construct such effective lagrangians in terms of NG bosons is to introduce Maurer-Cartan 1 form. Here I will explain the essence of the formulation.

We have seen in the CP^{N-1} model that the important ingredients are

$$\xi(\pi) = e^{i\pi(x)}, \qquad \pi(x) \equiv \sum_{X^{a^r} \in (\mathcal{G}-\mathcal{H})^r} \frac{\pi^{a^r}(x)}{f_{\pi^r}} X^{a^r}, \quad (6)$$

with scale parameters f_{π^r}. They transform nonlinearly under G as

$$\xi(\pi') = g\xi(\pi)h(\pi,g)^\dagger, \ g \in G. \quad (7)$$

Note that this element h is dependent on $\pi(x)$ as well as on g. Now a fundamental object is the so-called Maurer-Cartan 1-form constructed from $\xi(\pi) \in G/H$:

$$\alpha_\mu(\pi) = \frac{1}{i}\xi^\dagger(\pi)\cdot\partial_\mu\xi(\pi) = \partial_\mu\pi + \frac{i}{2!}[\pi,\partial_\mu\pi] + +\cdots, \quad (8)$$

whose transformation law follows from $\xi(\pi') = g\xi(\pi)h(\pi,g)^\dagger$:

$$\alpha_\mu(\pi) \to \alpha_\mu^r(\pi') = h(\pi,g)\alpha_\mu^r(\pi)h^\dagger(\pi,g) + \frac{1}{i}h(\pi,g)\cdot\partial_\mu h^\dagger(\pi,g). \quad (9)$$

A remarkable fact is that this *algebra-valued* quantity $\alpha_\mu(\pi)$ can be divided into the sets of components which are transformed independently of each other, in sharp contrast to the *group-valued* $\xi(\pi)$,

$$\begin{aligned}\alpha_{\mu\parallel}(\pi) &= (2\mathrm{tr}S^\alpha\alpha_\mu(\pi))S^\alpha \in \mathcal{H},\\ \alpha_{\mu\perp}^r(\pi) &= (2\mathrm{tr}X^{a^r}\alpha_\mu^r(\pi))X^{a^r} \in (\mathcal{G}-\mathcal{H})^r.\end{aligned} \quad (10)$$

Since $h(\pi,g)\cdot\partial_\mu h^\dagger(\pi,g) \in \mathcal{H}$, we obtain

$$\alpha_{\mu\parallel}(\pi) \to \alpha_{\mu\parallel}(\pi') = h(\pi,g)\alpha_{\mu\parallel}(\pi)h^\dagger(\pi,g) + \frac{1}{i}\partial_\mu h(\pi,g)\cdot h^\dagger(\pi,g), \quad (11)$$

$$\alpha_{\mu\perp}^r(\pi) \to \alpha_{\mu\perp}^r(\pi') = h(\pi,g)\alpha_{\mu\perp}^r(\pi)h^\dagger(\pi,g). \quad (12)$$

We see that only the perpendicular parts $\alpha_{\mu\perp}^r(\pi)$ transform homogeneously. Thus we can construct G-invariants from $\alpha_{\mu\perp}^r$: $\mathrm{tr}(\alpha_{\mu\perp}^r(\pi))^2$. The most general lagrangian made out of $\xi(\pi)$ with the smallest number of derivatives is thus given by

$$\mathcal{L}_{CCWZ} = \sum_r f_{\pi^r}^2 \mathrm{tr}\left(\alpha_{\mu\perp}^r(\pi)\right)^2, \quad (13)$$

where the factor $f_{\pi^r}^2$ is added so as to normalize the kinetic terms of $\pi(x)$ fields. It turns out that the number of the independent invariants or the parameters of CCWZ lagrangian is equal to that of the H-irreducible representations of the coset space G/H.[10] The lagrangian (13) is precisely the one which was first derived by Callan, Coleman, Wess and Zumino.[11].

The following comments would be instructive.

Coupling to external gauge field

The system can be made coupled to external gauge field $V_\mu(x)$, the gauge group I which may be any subgroup of the global symmetry group G: $I \subset G$. When the external gauge coupling g_e is switched on, $\partial_\mu \xi(\pi)$ in the Maurer-Cartan 1-form (8) should be replaced by the covariant derivative: $D_\mu \xi(\pi) = \partial_\mu \xi(\pi) + ig_e V_\mu(x) \xi(\pi)$, where $V_\mu(x) = V_\mu^i(x) Q^i$, Q^i being the generators of I. Thus (8) should be replaced by,

$$\hat{\alpha}(\pi)_\mu = \frac{1}{i}\xi^\dagger(\pi) D_\mu \xi(\pi) = \alpha(\pi)_\mu + g_e \hat{V}_\mu(x), \qquad (14)$$

with the definition

$$\hat{V}_\mu(x) \equiv \xi^\dagger(\pi) V_\mu(x) \xi(\pi)$$
$$= V_\mu(x) - \frac{i}{f_\pi}[\pi(x), V_\mu(x)] + \frac{1}{2!}\left(\frac{i}{f_\pi}\right)^2 [\pi(x),[\pi(x), V_\mu(x)]] + \cdots. \qquad (15)$$

Thus the CCWZ nonlinear lagrangian now reads (gauged CCWZ lagrangian)

$$\mathcal{L}_{gauged\,CCWZ} = \sum_r f_\pi^2 \cdot \mathrm{tr}\left(\hat{\alpha}_{\mu\perp}^r(\pi)\right)^2 = \sum_r \left(\partial_\mu \pi^r(x) + f_{\pi^r} g_e V_{\mu\perp}(x) + \cdots\right)^2, \qquad (16)$$

If the group I is larger than the unbroken subgroup H, i.e., there exist non-zero perpendicular components $V_{\mu\perp}(x) \in \mathcal{I} - \mathcal{H}$ of the external gauge field $V_\mu(x)$, we meet two possibilities.

i) $V_{\mu\perp}(x)$ become massive by absorbing NG fields. This is the well-known Higgs phenomenon: A part of NG boson components π whose corresponding generators X^a belong to the gauge group I are absorbed into $V_{\mu\perp}(x)$. We can always make a field-redefinition by the gauge transformation (unitary gauge), namely they become unphysical if not gauged away. Especially if we consider the case where all the NG bosons are absorbed by the gauge bosons, $\xi(\pi)$ in (15) are elements of I, and we can take the unitary gauge; $V_{\mu\perp}^{(u)}(x) = V_{\mu\perp} + \alpha_{\mu\perp}(\pi)$, then $\mathcal{L}$ is reduced to just the pure mass terms ;

$$\mathcal{L} = m_V^2 \mathrm{tr}(V_{\mu\perp}^{(u)}(x))^2. \quad m_V = f_\pi^2 g_e^2. \qquad (17)$$

Note that the field redefinition is performed by gauge transformation, thus bringing no changes in the form of the kinetic terms of these massive vector bosons. Then all the interactions of the massive gauge bosons in this gauge come from the kinetic terms.

ii) The gauge bosons $V_{\mu\perp}(x)$ are already massive and they couple to the NG fields literally. There the lagrangian is expanded in terms of physical π fields as,

$$\mathcal{L}_{CCWZ} = f_\pi^2 \mathrm{tr}\left(\hat{\alpha}_{\mu\perp}(\pi)\right)^2 = \mathrm{tr}\left(\partial_\mu \pi(x) + f_\pi g_V V_{\mu\perp}(x) + \cdots\right)^2, \qquad (18)$$

from which we can read off f_π as the decay constant of π field, and 4π coupling, $\frac{1}{3f_\pi^2}$ and so on. Actually this case is realized in W and Z gauge boson sectors coupled to the hadrons, π mesons.

To these expressions we shall be back in the last section.

Matter field

The matter field $\chi(x)$ is introduced as a linear representation ρ_0 of the subgroup H; it transforms as $\chi' = \rho_0(h)\chi$ under $h \in H$. By defining the transformation of $\chi(x)$ under $g \in G$ as $\chi'(x) = \rho_0(h(\pi, g))\chi(x)$, we can convert $\chi(x)$ into what is called *a linear base* $\psi(x)$ of G with the use of any representation ρ of G whose restriction to H contains the above ρ_0: $\psi(x) = \rho(\xi)\chi(x)$, with understanding that $\chi(x)$ in the LHS is the vector embedded in the ρ-representation space. It is then easy to construct G-invariants for the matter fields also. For notational simplicity, we show here the case where $\chi(x)$ and $\psi(x)$ belong to the fundamental representation of H and G, respectively, i.e., we write $\rho_0(h) = h$ and $\rho(\xi) = \xi$. Then from the transformation law of $\alpha_{\mu\parallel}(\pi)$ and $\alpha_{\mu\perp}(\pi)$ in (11) and (12), we finally get the general fermion matter lagrangian as,

$$\mathcal{L}_{matter} = \bar{\chi} i \gamma^\mu (\partial_\mu - i\hat{\alpha}_{\mu\parallel}(\pi))\chi - m\bar{\chi}\chi + g_A^r \bar{\chi} i \gamma^\mu \hat{\alpha}_{\mu\perp}^r(\pi)\chi. \qquad (19)$$

where the first term gives the kinetic term of χ together with their multi- π interactions, and the second, corresponds to the mass term. The last term, if the external field $V_{\mu\perp}(x)$ is coupled to the system, can be expressed as

$$g_A \bar{\chi} i \gamma^\mu (\frac{1}{f_\pi}\partial_\mu \pi + g_e \hat{V}_{\mu\perp}(x) + \cdots)\chi = \frac{g_A}{f_\pi}\bar{\chi} i \gamma^\mu \chi \partial_\mu \pi + g_A g_e \bar{\chi} i \gamma_\mu \chi \hat{V}_{\mu\perp}(x) + \cdots, \qquad (20)$$

indicating that the $\chi\pi\pi$ coupling, $g_{\chi\pi\pi} = \frac{2m}{f_\pi} g_A$ and the broken charge, $G_A = g_A g_e$, with g_A being the ratio of broken charge to the unbroken charge (G_A/G_V ratio). Then we get the so-called Goldberger-Treiman relation, $2m g_A = g_{\chi\pi\pi} f_\pi$.

G-invariants out of $\alpha_{\mu\prime}(\pi)$

We could construct G-invariants out of $\alpha_{\mu\prime}(\pi)$, which transforms inhomogeneously under G: $\alpha_{\mu\prime}(\pi') = h(\pi,g)\alpha_{\mu\prime}(\pi)h^\dagger(\pi,g) + \frac{1}{i}\partial_\mu h(\pi,g)\cdot h^\dagger(\pi,g)$. To do this, it is necessary to eliminate the homogeneous part. This can be done in the following manners:

i) Introduce $F_{\mu\nu} = \partial_\mu \alpha_{\nu\prime}(\pi) - \partial_\mu \alpha_{\nu\prime}(\pi) - i[\alpha_{\mu\prime}(\pi), \alpha_{\nu\prime}(\pi)]$, which transforms homogeneously, then we have[12] G-invariants; $\text{tr}(F_{\mu\nu}(\alpha_{\mu\perp}(\pi)))^2$. This is, however, a higher derivative term, having nothing to do with low energy effective theroy.

2) As a special matter field, Weinberg[13] once considered "ρ-meson" field $\rho_\mu^W(x)$ which was assumed to transform under G as,

$$\rho_\mu^W(x) \xrightarrow{g} \rho_\mu^{W\prime}(x) = h(\pi,g)\rho_\mu^W(x)h^\dagger(\pi,g) + \frac{1}{i}h(\pi,g)\partial_\mu h^\dagger(\pi,g), \qquad (21)$$

This $\rho_\mu^W(x)$ field transforms inhomogeneously and hence does not fit to our definition of general matter field $\rho_\mu(x)$ belonging to the adjoint representation of H. However we should note that an identification; $\rho_\mu^W(x) = \rho_\mu(x) + \alpha_{\mu\prime}(\pi)$ is possible and hence this new field $\rho_\mu(x)$ transforms homogeneously under G. Then we get

$$\text{tr}(\rho_\mu(x))^2 = \text{tr}(\rho_\mu^W(x) - \alpha_{\mu\prime}(\pi))^2. \qquad (22)$$

This $\rho_\mu^W(x)$, at a glance, might be identified to the gauge bosons of another new local symmetry. However, we should note that the above *local H* symmetry is induced by the *global G* transformation, and never a *new hidden local* symmetry. Much efforts were directed to identify these matter fields with massive Yang-Mills fields, but if it is the case, then where have the NG bosons gone that must have been absorbed by massive gauge fields V_μ ? This somewhat confusing situation will be made clear logically by the introduction of hidden local symmetry.

Now we are ready to discuss *hidden local symmetry*.

3. HIDDEN LOCAL SYMMETRY AND DYNAMICAL GAUGE BOSONS

In order to introduce a new local symmetry, we need additional fields $\sigma(x)$ corresponding to this new gauge freedom, together with the gauge fields corresponding to H_{local}, $V_\mu(x) = V_\mu^\alpha(x)S^\alpha$, which transform as

$$V_\mu(x) \to V'_\mu(x) = ih(x)\partial_\mu h^\dagger(x) + h(x)V_\mu(x)h^\dagger(x), \quad h(x) \in H_{local}. \tag{23}$$

This kind of rdundant field $\sigma(x)$ is called *compensator*[14]. Let us write

$$\xi(\sigma) = e^{i\sigma(x)/f_\pi}, \quad \sigma(x) \equiv \sigma^\alpha(x)S^\alpha, \tag{24}$$

whose transformation property is defined so that the new field $\xi(x) = \xi(\pi)\xi(\sigma)$ transforms under $G_{global} \times H_{local}$ as

$$\xi'(x) = g\xi(x)h(x)^\dagger, \quad g \in G_{global}, \quad h(x) \in H_{local}, \tag{25}$$

namely,

$$\xi'(\sigma) = h(\pi, g)\xi(\sigma)h^\dagger(x). \tag{26}$$

Thus it is straightforward to construct $G_{global} \times H_{local}$ invariants out of $\xi(x)$; by defining covariant derivative $D_\mu\xi(x)$: $D_\mu\xi(x) = \partial_\mu\xi(x) - i\xi(x)V_\mu(x)$, the covariantized 1-form is given by

$$\tilde{\alpha}_\mu(x) \equiv \frac{1}{i} \cdot \xi^\dagger(x) D_\mu \xi(x) = \alpha_\mu(x) - V_\mu(x) \tag{27}$$

which transforms as

$$\tilde{\alpha}_\mu(x) \to \tilde{\alpha}'_\mu(x) = h(x)\tilde{\alpha}_\mu(x)h^\dagger(x). \tag{28}$$

By decomposing it again into its parallel and perpendicular parts, we get the following invariants:

$$\begin{aligned}\mathcal{L}_V &= f_\pi^2 \mathrm{tr}\,(\tilde{\alpha}_{\mu\parallel})^2 = f_\pi^2 \mathrm{tr}\,(V_\mu(x) - \alpha_{\mu\parallel}(x))^2, \\ \mathcal{L}_A^r &= f_{\pi^r}^2 \mathrm{tr}\,(\tilde{\alpha}_{\mu\perp}^r)^2.\end{aligned} \tag{29}$$

The most general lagrangian made out of ξ and $D_\mu\xi$ with the smallest number of derivatives is thus given by,

$$\mathcal{L} = \sum_r \mathcal{L}_A^r + a\mathcal{L}_V, \tag{30}$$

with an arbitrary parameter a. External and matter fields are introduced in the same manner as we have done in the previous section.

Note that, up to here, the hidden gauge fields are just the auxiliary fields to be eliminated by the equation of motion, and by doing so the lagrangian is reduced to the one of CCWZ. It is easy to see that, if one takes the unitary-like gauge, $\xi(\sigma) = 1$, $\xi(x) = \xi(\pi)$, the allowed transformation of the $\xi(\pi)$ under this gauge-fixed condition is expressed by eq.(7). In this way we saw that the nonlinear sigma model can be identified as the gauge-fixed version of the $G_{global} \times H_{local}$ linear model. Namely our hidden local symmetry provides merely the gauge freedom corresponding to the redundant variables $\xi(\sigma)$ with the gauge fields V_μ to be eliminated by the equation of motion.

However they become physical if the kinetic terms of these gauge fields are generated dynamically via some strong interactions.

How and when do these hidden gauge bosons become physical ? Is it theoretically possible ? or Are these gauge bosons observed in the real world ? Both we would say "yes ". I will not explain any details and just present the outline.

[Theoretical Supports]

A systematic analysis has been done by Kugo, Terao and Uehara[15] on two models explicitly; CP^{N-1} model and Nambu-Jona-Lasinio model. The former will provide us with an interesting example, that realizes the dynamical generation of *massless* gauge boson of hidden $U(1)_{local}$ symmetry and tells us the general mechanism how the gauge bosons are generated dynamically. The latter is discussed as an analogue model of quantum chromodynamics (QCD). Although it is not a nonlinear sigma model, it has an intrinsic hidden local symmetry and *massive* vector bound states appear as gauge bosons of the hidden local symmetry in spontaneously broken phase. These examples are enough to convince us that the dynamical generations of gauge bosons, massless or massive, are indeed common phenomena which can occur very easily in any systems not restricted to particular models, so far as the system provides the interaction strong enough to develop the bound-state poles in vector channels, [15,16,18]

[Phenomenological Supports]

Let us check our idea by looking for the phenomenological consequences of our hidden gauge bosons and comparing them with experiments. In fact the QCD laboratory is the only existing example of strong coupling gauge theories. Actually, it turns out[1] that the ρ meson is the dynamical gauge boson of the hidden local symmetry $H_{local} = [SU(2)_V]_{local}$ in nonlinear chiral lagrangian which is known to be a low energy effective lagrangian of the massless two-flavored QCD whose global symmetry $G = SU(2)_L \times SU(2)_R$ is expected to be spontaneously broken to the diagonal subgroup $H = SU(2)_V$. Among the direct consequences of the dynamical gauge bosons, the most remarkable relation is $g_\rho = 2g_{\rho\pi\pi}f_\pi^2$ where g_ρ and $g_{\rho\pi\pi}$ are the photon-ρ meson coupling and $\rho\pi\pi$ coupling, respectively. Both sides of the relation can be obtained from the experimental data, actually the LHS evaluated from $\Gamma^{exp}(\rho^0 \to e^+e^-) \simeq 6.62 keV$ is $g_\rho \simeq 0.12$ GeV^2, while the RHS yields $2f_\pi^2 g_{\rho\pi\pi} \simeq 0.11$ GeV^2 (corresponding to $g_{\rho\pi\pi}^2/4\pi \simeq 3.0$), a good agreement ! The old hadron phenomenology is, in this way, re-examined in the new light of dynamical structure of SCGT.

4. HIDDEN LOCAL SYMMTERY IN UNIFIED THEORIES

The standard model, QCD plus Glashow-Salam-Weinberg model, is a very successful framework for describing the elementary particles, quarks and leptons and their interactions. However the theory does not answer the questions: What is the origin of the spontaneous breaking of the electroweak gauge symmetry ? How can one explain the variety of matters; quarks and leptons ? Why do there exist at least 4-types gauge interactions in our Nature ? All those questions seem to suggest that there still exist some new types of strong coupling gauge theories (SCGT) that cause the dynamical symmetry breaking and/or provide composite Higgs, composite quarks and leptons and/or even gauge bosons, W/Z bosons, etc.., whether its energy scale is high up to 10^{19} GeV or low down to $1\,TeV$. For the scenario of $1\,TeV$ compositeness, the idea may be applied to the $J = 1$ bound states in the technicolor (TC) models or heavy (strongly interacting) Higgs model. Also the composite weak boson models may be viewed in the framework of hidden local symmetry; in fact, such a possibility in a supersymmetric model has been studied by Kugo, Uehara and Yanagida.[19] As to $10^{19}\,GeV$ compositeness, an exciting possibility would be a dynamical realization of hidden $SU(8)_{local}$ symmetry in $N = 8$ extended supergravity theory (ESGT).[20] I here shall make comments on the scenarios of the typical energy-scale (10^{19} and $10^3\,GeV$) compositeness.

[$10^{19}\,GeV$ compositeness]

As is well known,[22,21,25,23,24] in the theories having no *local* invariance, the appearance of massless particles with $j \geq 1$ is inhibited, whether elementary or composite, and hence no massless composite gauge bosons can appear. This is the reason why people had been so familiar with the view that gauge bosons are "elementary particles", until the notion of hidden local symmetry was introduced in $N = 8$ ESGT.

The unification of all particles and their interactions has become one of the main subjects in particle physics after the great success of the standard model. The $SU(5)$ model, a minimal model to unify electroweak and color gauge interactions, inevitably takes us to the GUT scale $\sim 10^{15}\,GeV$ (at least), very close to the Planck mass $10^{19}\,GeV$. Hence, if the GUT's scenario is indeed the case, there seems to be no reason to ignore the gravitational interaction anymore.

Extended supergravity theories (ESGT's) are such attempts to unify all the particles and fields including the graviton. The largest ESGT allowed theoretically is $N = 8$ in 4-dimension, or $N = 1$ in 11-dimension supergravity, which has an $SO(8)$ internal symmetry. This is, however, too small to accommodate the full standard model symmetries, $SU(3)_c \times SU(2)_L \times U(1)_Y$, and neither can it reproduce the chiral fermions, quarks and leptons.

At this stage, the very notion of hidden local symmetry came into play. Indeed $N = 8$ ESGT was found to have $SU(8)$ hidden local symmetry.

An attractive "superunification" scenario based on this $N = 8$ ESGT was first proposed by Ellis, Gaillard and Zumino.[20] The picture is the following: First the fundamental particles are the gravity multiplets. By some dynamical mechanism (via SCGT) the composite vector fields in the $SU(8)$ adjoint channel may become dynamical. Then all the gauge bosons, photon, W, Z and color gauge bosons, do appear as the bound states composed of the members of primary gravity multiplet. Accordingly, their supersymmetric partners might also become dynamical which include the matter fields appearing in the low energy world, namely, quarks and leptons together with Higgs particles.[20,26] Of course it may be that some of those gauge bosons are really elementary corresponding to external gauge group $SO(8)$, with the other gauge bosons of $SU(8)_{hidden}$ in $N = 8$ ESGT.

However, the problem is how and what kind of SCGT really generates such dynamical poles of hidden gauge bosons. Unfortunately, at the present stage we are still far from such super high energy scale. So now let us come back to the terrestrial world.

[1 TeV compositeness]

The standard Glashow-Salam-Weinberg model has left the origin of the spontaneous breaking of the electroweak gauge symmetry to the mysterious Higgs sector. From the naive understanding of the dynamical symmetry breaking, it is natural that this spontaneous breakdown is caused by some SCGT, say Technicolor(TC), which generate composite Higgs particles as NG bosons. The minimal model contains 3 NG bosons, which may be realized by the symmetry groups G/H, $SU(2) \times SU(2)/SU(2)$ or $SU(2) \times U(1)/U(1)$. In the first case NG bosons are in the 3 representation of $H = SU(2)$ and we have only one scale parameter F_Π, while in the latter case NG bosons are further divided into the charge 1- and 0-sectors of $H = U(1)$ representations, giving rise two scales, F_Π and $\rho \equiv F_\Pi/F'_\Pi$ (so-called ρ parameter). Here let me remind you the discussion of section 2, in which I explained two cases for the massive gauge bosons. Accordingly we may consider two steps A) W and Z bosons are elementary and they acquire masses of order $g_w^2 F_\Pi^2$, absorbing those 3 NG bosons. B) Then the same SCGT (or strong interacting Higgs system) may develop poles in vector channels as well, which couple to the present low energy system.

As for the second step, the dynamical vectors are just what we call hidden gauge bosons and in much the same manner we can construct the effective lagrangian as a direct application of our framework of hidden local symmetry. This will be discussed in detail by the next speaker.

Here I would like to conclude my talk by making a short comment on the story of the first step and show the essence as to how the symmetry principle is rearranged to the massive gauge fields themselves.

As we already saw in the section 1, the Low Energy Theorem ($p^2 \ll F_\Pi^2$) for NG boson system is the direct consequence of the symmetry principle. The question is how the same symmetry principle does reveal itself in the system in which NG bosons are absorbed by gauge bosons via Higgs mechanism. The answer is most explicitly demonstrated in the expression of effective lagrangian in eq.(17); the $\mathcal{L}_A$ term just reduces to pure mass terms of the gauge bosons, namely the low energy interactions of the massive gauge bosons come from the kinetic terms of gauge bosons,

$$\mathcal{L}_{kin} = \frac{1}{2}\text{tr}\,(\partial_\mu W_\nu - \partial_\nu W_\mu - ig_w[W_\mu, W_\nu])^2. \tag{31}$$

Thus we have only to take into account the 3- and 4- W vertices of this lagrangian and it is noticed that, in the relatively high energy region ($m_W^2 \ll p^2 \ll F_\Pi^2$), because of the derivative terms, the contribution of longitudinally polarized W bosons, via 3-vertices, dominates the one of direct 4-vertex in the W-W scattering processes. This would be the simplest direct derivation of what is called "Low Energy Theorem for W's and Z's" in the strongly interacting Higgs models without any use of "equivalence theorem" or current algebra technique (see the details in the separate paper[27]).

We have seen that, in the system in which Higgs mechnism occurs, symmetry principle is rearranged into NG sector and further into massive gauge boson sector. Once we recognize the fact clearly it is evident that the effective lagrangian approach provides us with the most handy method to study the low energy interactions of the system.

Acknowledgements

I would like to express my thanks to K.I.Aoki, T.Kugo and K.Yamawaki for valuable discussions and comments.

REFERENCES

1. M. Bando, T. Kugo and K. Yamawaki, *Phys. Reports* **164** (1988) 218

2. M. Bando, T. Kugo, S. Uehara, K. Yamawaki and T. Yanagida, *Phys. Rev. Lett.* **54** (1985) 1215

3. M. Bando, T. Kugo and K. Yamawaki, *Nucl. Phys.* **B259** (1985) 493

4. M. Bando,T. Kugo and K. Yamawaki, *Prog. Theor. Phys.* **73** (1985) 1541

5. Y. Takahashi,, *Phys. Rev.* **15** (1977) 1589

6. H. Umezawa,, *Nouvo Cim.* **40**(1965) 450

7. Y. Nambu and G. Jona-Lasinio, *Phys. Rev.* **122** (1961) 345

8. J. Goldstone, *Nuovo Cim.* **19** (1961) 154

9. See, e.g., S. L. Adler and R. F. Dashen, *Current Algebras* (Benjamin/ Cummings New York,1968)

10. G. Boulware and L. S. Brown, *Ann. Phys.* **138** (1982) 392

11. C. G. Callan, S. Coleman, J. Wess and B. Zumino, *Phys. Rev.* **177** (1969) 2247

12. A. P. Balachandran, A. Stern and G. Trahern, *Phys. Rev.* **D19** (1979) 2416

13. S. Weinberg, *Phys. Rev.* **166** (1968) 1568

14. S. J. Gates, Jr., M. T. Grisaru, M. Roček and W. Siegel, *Superspace* (Benjamin/Cummings, New York, 1983)

15. T. Kugo, H. Terao and S. Uehara, in *Proceedings of the Kyoto International Symposium; the Jubilee of the Meson Theory*, ed. by M. Bando, R. Kawabe and N. Nakanishi (Kyoto, 1985) [*Prog. Theor. Phys. Suppl.* **85** (1985) 122]

16. T. Kugo, *Phys. Lett.* **109B** (1982) 205

17. T. Kugo, *Soryushiron Kenkyu(Kyoto)* **71** (1985) E78 ;in *Proceedings of 1987 International Workshop on Low Energy Effective Theory of QCD*, ed. by S. Saito and K. Yamawaki(Nagoya Univ., 1987)

18. H. Terao, Master Thesis, Kyoto Univ., Jan. 1985

19. T. Kugo and S. Uehara and T. Yanagida, *Phys. Lett.* **147B** (1984) 321

20. J. Ellis, M. K. Gaillard and B. Zumino, *Phys. Lett.* **94B** (1980) 343; *Acta. Phys. Pol.* **B13** (1982) 253

21. L. Durand, *Phys. Rev.* **128** (1962) 434

22. K. M. Case and S. Gasiorowicz, *Phys. Rev.* **125** (1962) 1055

23. T. Kugo, in *Proceedings of 1981 INS Symposium on Quarks and Leptons Physics*, ed. by K. Fujikawa, H. Terazawa and A. Ukawa (INS, Univ. Tokyo, 1981)

24. J. T. Lopuszanski, *J. Math. Phys.* **25** (1984) 3503

25. S. Weinberg and E. Witten, *Phys. Lett.* **96B** (1980) 59

26. M. Bando, Y. Sato and S. Uehara, *Z. Phys.* **C22** (1984) 251

27. K-I Aoki, M. Bando and T. Kugo, in preparation

VECTOR AND AXIAL - VECTOR RESONANCES FROM A STRONGLY INTERACTING ELECTROWEAK SECTOR

Roberto Casalbuoni

Dipartimento di Fisica, Univ. di Lecce, I-73100 Lecce, Italy
and
I.N.F.N., Sezione di Bari, I-70126 Bari, Italy

ABSTRACT

The possibility that both vector and axial-vector bound states could originate from a strong interacting sector of the electroweak theory is considered. A simple lagrangian parametrization is presented where the bound states are described as gauge vector bosons of a local, non-linearly realized, $SU(2) \otimes SU(2)$ symmetry. At present the model is mostly constrained from data on W and Z masses and on neutrino-nucleon deep inelastic scattering. High-energy e^+e^- tests are suggested where visible deviations from the standard model predictions could take place. These deviations exhibit a certain pattern which allows to distinguish the model from other theoretical frameworks. We find that precise measurements of W and Z masses and asymmetries in e^+e^- collisions could put strong restrictions on the parameters of the model if no appreciable deviations occur, apart from an extreme case with the vector and axial-vector bosons degenerate in mass and coupling.

1. INTRODUCTION

Although the standard electro-weak model (SM) is extremely successful from a phenomenological point of view, it is believed to be unsatisfactory as far as its description of symmetry breaking (SB) is concerned. The main reason for this dissatisfaction has to do with the problem of fine tuning, which in turn is related to the fact that, in general, for scalar fields there is no symmetry preventing the generation of a mass term. As a consequence, the Higgs mass is expected to be close to the next nearest mass scale. Therefore, if no other mechanism takes place (like supersymmetry) keeping the Higgs mass small with respect to this scale, the theory can become rapidly nonperturbative leading to new physics in the 1 TeV region[1)2)]. These considerations can be made a little more precise if we look at the behaviour of the quartic self-coupling of the Higgs field under change of the renormalization point[3)] (neglecting for simplicity the evolution of the other couplings)

$$\frac{1}{\lambda(M)} = \frac{1}{\lambda(m_H)} + \frac{3}{4\pi^2} \log \frac{m_H}{M} \tag{1.1}$$

Clearly $\lambda(M) > \lambda(m_H)$ if $M > m_H$. It follows that there will exist a mass scale M_0 such that $\lambda(M_0) \approx 1$. Then, from (1.1) we can get upper bounds for $\lambda(m_H)$ and m_H:

$$\frac{1}{\lambda(m_H)} > \frac{3}{2\pi^2} \log \frac{M_0}{m_H}, \quad m_H^2 < \frac{4\pi^2}{3} v^2 \frac{1}{\log \frac{M_0}{m_H}} \tag{1.2}$$

where v is the v.e.v. of the Higgs field and $v^2 = (\sqrt{2}G_F)^{-1} \approx 250 \ GeV^2$. From this we see that two very different situations can arise. If we take $M_0 \approx M_{Planck}$, M_{GUTS}, then $m_H \approx 150, 170 \ GeV$, and $\lambda(m_H) \leq 1$. This means that we have to do with a relatively weakly-coupled, low-mass Higgs. However, since the Higgs mass is quadratically divergent, and sooner or later some new interaction has to show up (at least gravity), the natural value for m_H is of the order of the scale of this interaction itself. Of course, one is free to choose the renormalized parameters but, unless one is willing to fine tuning them at any order in perturbation theory, the Higgs mass can only be proportional to the scale itself. Notice that in the fermion case, the situation is quite different because, due to the chiral symmetry, the fermion self-energy is proportional to the symmetry breaking parameter, that is to the mass parameter for the fermions. The conclusion of this discussion is that if one wants to have weakly-interacting and low mass Higgs, something must happen in order to cancel the quadratic divergence in the self-energy of the scalar fields. As far as we know there is only one known answer to this question, that is supersymmetry. Here, the quadratic divergence is cancelled because the interplay of supersymmetry and chiral symmetry for the fermions gives a mechanism protecting the scalar masses as well.

The other situation we want to consider is when $M_0 \sim m_H$ which happens when both are of the order of 1 TeV. Now $\lambda(m_H) \approx 30 \div 50$, and we get a strongly-coupled, heavy mass Higgs. Here no problem arises if a new interaction sets in at a scale of order 1 TeV because this is just the right mass scale for the Higgs. Such a situation could correspond to a composite picture for the Higgs with a new interaction like technicolor. In top of all this one should also comment about the possibility that the φ^4-theory is trivial. This appears more and more plausible, but nevertheless it still makes sense to use it as an effective theory below the cutoff where new physics emerges. Then again, a lower bound can be obtained by requiring that m_H be not greater than the cutoff [4]. In this way, one gets the same kind of limitations we just discussed.

In this talk I would like to discuss in detail the possible physical consequences arising if one is going to take seriously the second option discussed before. This means that one is going to deal with a situation in which strong interactions are present, and therefore suitable tools are necessary. The best we can do, without recurring to particular approximation schemes, is to use some general principles which hold for any spontaneously broken gauge symmetry. In short these principles are: i) the equivalence theorem, ii) the low energy theorems for Goldstone bosons, iii) unitarity. The equivalence theorem[5] allows one to calculate the scattering amplitudes among longitudinal vector bosons in terms of the corresponding amplitudes for the Goldstone bosons eaten up via the Higgs mechanism. Then, the philosophy is to use low energy theorems to evaluate these last amplitudes and try to get informations about the SB scale from their high energy continuation. In fact, this continuation will break the unitarity to a scale which is presumably of the same order at which new phenomena are expected to intervene.

The use of low-energy theorems for the Goldstone bosons amplitudes can be made by copying what one does in the case of ordinary strong interactions[6]. In the present case, the global symmetry group G one has to start with must at least contain $SU(2)_L \otimes U(1)$. Then, after SB, the unbroken group H must contain at least $U(1)_{EM}$. From these requirements alone one can derive the low energy theorem[7]

$$M(w^+w^- \to zz) = \frac{1}{\rho}\frac{s}{v^2} \approx M(W_L^+ W_L^- \to Z_L Z_L) \qquad (1.3)$$

where

$$\rho = \left[\frac{M_W}{M_Z \cos\theta_W}\right]^2 \qquad (1.4)$$

is experimentally of order one. Of course, if one assumes that the breaking pattern is the same as in the SM, then $\rho = 1$. Therefore, the information we get is that, quite independently from any specific model of the symmetry breaking sector, the scale at which one can expect that new phenomena arise is of the order of [8]

$$\sqrt{s} \approx 4\sqrt{\pi}v \approx 1.8 \ TeV \qquad (1.5)$$

One must keep in mind that this low energy theorem can be valid only for $s \ll 16\pi v^2$, or for $s \ll$ SB-scale, and furthermore, in order to give to the argument any sense, we must use it for the physical amplitudes among the longitudinal vector bosons, which means $s \gg M_W^2$. It is not obvious that the window among the SB-scale and M_W exists. In fact, if the Higgs would have a small mass, lower than M_W, this argument would be empty. Given the fact that in this framework one expects that the SB-sector originates from some new strong interaction, there are various problems one can try to solve. For instance, through the equivalence theorem, strong interactions propagate into the longitudinal sector of the vector bosons, and it is possible to give some predictions about the corresponding rates.[9] Also, formation of resonances (scalar, vector, etc.) is expected. In particular, the formation of vector resonances is particularly interesting because they can mix with the ordinary vector bosons influencing a type of phenomenology that will be investigated pretty soon at LEP, and SLC.

Once one takes seriously the point of view that the SB-scale is related to new phenomena, it is natural to avoid the Higgs description but to use instead an effective description of the symmetry breaking. There are many ways of doing that, but one of the more economical possibilities consists in making use of a non-linear σ-model. In fact, this description allows to avoid the introduction of any Higgs scalar, and it is particularly useful in order to discuss vector resonances. The rest of this talk will be devoted precisely to the vector and axial-vector resonances that can originate from a strongly interacting SB sector, and to their phenomenological implications.

2. NON-LINEAR DESCRIPTION OF SB IN THE SM

Our aim here is to describe the SB within the SM context in terms of non-linear realizations instead of using the Higgs model. We know that for a breaking of a group G into a subgroup H the Goldstones can be taken as the coordinates of G/H. We know that H must contain $U(1)_{e.m.}$. We also need three Goldstones to give masses to W and Z. In addition we can guarantee $\rho = 1$ apart from weak corrections, if we have a "custodial" $SU(2)$. The minimal H in this case would have to be $SU(2)$. In the SM the breaking is realized linearly with scalars originally transforming as the $(\frac{1}{2}, \frac{1}{2})$ of $SU(2)_L \otimes SU(2)_R$ which breaks into $SU(2)_{diagonal}$, with corresponding breaking of $(\frac{1}{2}, \frac{1}{2})$ into $1 \oplus 3$, then becoming the physical Higgs and the 3 absorbed Goldstones. The non-linear realization can be seen classically as corresponding to the limit of infinite m_H. The scalars can indeed be represented as proportional to a unitary matrix U. In the formal limit $m_H \to \infty$ one is just freezing the proportionality factor to the vacuum expectation value and the scalar lagrangian is simply[10]

$$\mathcal{L} = \frac{v^2}{4} Tr[(\partial_\mu U)(\partial^\mu U^\dagger)] \qquad (2.1)$$

Such a lagrangian is obviously invariant under $SU(2)_L \otimes SU(2)_R$, namely under $U \to g_L U g_R^\dagger$ where g_L, g_R belong to $SU(2)_L, SU(2)_R$ respectively. The breaking into the diagonal $SU(2)$ is demanded by the (non-linear) unitarity condition $U^\dagger U = 1$.

A way to introduce vector resonances which has been quite successfully in ordinary strong interactions[11] is based on the concept of "hidden gauge symmetry". A procedure to discuss the "hidden gauge symmetries" in this case, that we consider as simpler and more understandable than the usually employed treatment, consists in enlarging the initial symmetry and correspondingly the scalar sector and the number of needed non-linear conditions. For the model of ref.[12] one adds to $SU(2)_L \otimes SU(2)_R$ a group $SU(2)_V$ and realizes non-linearly the breaking $SU(2)_L \otimes SU(2)_R \otimes SU(2)_V \to SU(2)_{diagonal}$. The Goldstones are six (coordinates of the quotient). They are all absorbed, giving masses to W, Z and V (the gauge particle of $SU(2)_V$). The Goldstones are described by two unitary matrices L and R, which transform as $L \to g_L L h$, $R \to g_R R h$ where g_L, g_R, h belong to $SU(2)_L, SU(2)_R, SU(2)_V$ respectively. Forgetting the unitarity conditions one would have L, R transforming as $(\frac{1}{2}, 0, \frac{1}{2})$ and $(0, \frac{1}{2}, \frac{1}{2})$ under $SU(2)_L \otimes SU(2)_R \otimes SU(2)_V$. The unitarity conditions $LL^\dagger = 1, RR^\dagger = 1$ lead to the wanted breaking. The procedure then consists in writing down the most general Lagrangian with at most two derivatives invariant under $SU(2)_L \otimes SU(2)_R \otimes SU(2)_V$ for the unitary local matrices L and R, and satisfying the symmetry $L \leftrightarrow R$. One then introduces gauge fields for the subalgebra $SU(2)_L \otimes U(1)_Y \otimes SU(2)_V$ and adds kinetic terms for them. The introduction of such kinetic terms can be justified in non-linear σ-models in two dimensions[13], but here the V-fields are introduced as just effective dynamical fields. The related gauge couplings are g, g', as usual, and g'' for $SU(2)_V$. In the formal strong coupling limit, $g'' \to \infty$, the kinetic term of the $SU(2)_V$ field vanishes, and the corresponding fields becoming auxiliary, with the original field U of lagrangian (2.1) expressed in terms of the fields L and R, as $U = LR^\dagger$. Their elimination brings back to the non-linear formulation of the SM. In this way one is able to describe massive vector fields (V-fields) mixed to W and Z-fields, with interesting phenomenological consequences that have been described in detail in ref.[14]. Here, we want to generalize this procedure, in order to describe axial-vector fields[15] in addition to the previous V-fields.

The idea is to construct a model describing vector and axial particles, such that, when these particles decouple, one obtains the non-linear σ-model lagrangian of equation (2.1), with

$$U = LM^\dagger R^\dagger \qquad (2.2)$$

where L, M, R, transform according to the following representations of

$$G = [SU(2)_L \otimes SU(2)_R]_{global} \otimes [SU(2)_L \otimes SU(2)_R]_{local}$$

as

$$L \in (\frac{1}{2}, 0, \frac{1}{2}, 0), \quad M \in (0, 0, \frac{1}{2}, \frac{1}{2}), \quad R \in (0, \frac{1}{2}, 0, \frac{1}{2}) \qquad (2.3)$$

that is:
$$L' = g_L L h_L, \quad M' = h_R^\dagger M h_L, \quad R' = g_R R h_R \qquad (2.4)$$

where
$$g_L \in (SU(2)_L)_{global}, \quad g_R \in (SU(2)_R)_{global}$$
$$h_L \in (SU(2)_L)_{local}, \quad h_R \in (SU(2)_R)_{local} \qquad (2.5)$$

In this way we have:
$$U' = g_L U g_R^\dagger \qquad (2.6)$$

that is,
$$U \in (\frac{1}{2}, \frac{1}{2}, 0, 0) \qquad (2.7)$$

and therefore, U does not see the local symmetry (hidden gauge symmetry). The lagrangian (2.1) is obviously invariant under the discrete transformation $U \to U^\dagger$, which corresponds to the parity transformation $P: L \to R$, $M \to M^\dagger$, $R \to L$.

Proceeding in a completely standard way we can build up covariant derivatives with respect to the local group:

$$D_\mu L = \partial_\mu L - L \mathbf{L}_\mu, \quad D_\mu R = \partial_\mu R - R \mathbf{R}_\mu$$
$$D_\mu M = \partial_\mu M - M \mathbf{L}_\mu + \mathbf{R}_\mu M, \quad D_\mu M^\dagger = \partial_\mu M^\dagger + \mathbf{L}_\mu M^\dagger - M^\dagger \mathbf{R}_\mu \qquad (2.8)$$

where $\mathbf{L}_\mu$ and $\mathbf{R}_\mu$ are the Lie algebra valued gauge fields with respect to the groups $(SU(2)_L)_{local}$ and $(SU(2)_R)_{local}$.

We can now construct the invariants of our original group extended by the parity operation defined before. First of all, we will construct four-vectors covariant under the adjoint representation of $(SU(2)_L)_{global}$ (of course, this can be done using any one of the various $SU(2)$ groups occurring in the original invariance group, but this does not lead to new invariants). We can construct three such objects:
$$V_0^\mu = L^\dagger D^\mu L, \quad V_1^\mu = M^\dagger D^\mu M, \quad V_2^\mu = M^\dagger R^\dagger (D^\mu R) M \qquad (2.9)$$

which transform as
$$V_i \to h_L^\dagger V_i h_L, \quad h_L \in (SU(2)_L)_{local}, \quad i = 0, 1, 2 \qquad (2.10)$$

The transformation properties under parity are
$$V_0 \to M V_2 M^\dagger, \quad V_2 \to M V_0 M^\dagger, \quad V_1 \to -M V_1 M^\dagger \qquad (2.11)$$

From the vectors (2.9) we can build up six invariants under G
$$Tr[V_i \cdot V_j], \quad i,j = 0,1,2 \qquad (2.12)$$

from which we can extract the following four invariants under the full group $G \otimes P$

$$I_1 = Tr[(V_0 - V_1 - V_2)^2], \quad I_2 = Tr[(V_0 + V_2)^2]$$
$$I_3 = Tr[(V_0 - V_2)^2], \quad I_4 = Tr[V_1^2] \qquad (2.13)$$

Using the invariants (2.13) we can write the most general lagrangian with at most two derivatives in the form:

$$\mathcal{L} = -\frac{v^2}{4}[aI_1 + bI_2 + cI_3 + dI_4] + \text{kinetic terms for } \mathbf{L}_\mu \text{ and } \mathbf{R}_\mu \qquad (2.14)$$

It is not difficult to see that this lagrangian is the same as the one one would obtain from the hidden gauge symmetry approach [16].

The $SU(2)_L \otimes U(1)_Y$ gauging can be now introduced through the substitutions

$$D_\mu L \to \Delta_\mu L = D_\mu L + W_\mu L$$
$$D_\mu R \to \Delta_\mu R = D_\mu R + Y_\mu R \qquad (2.15)$$
$$D_\mu M \to \Delta_\mu M = D_\mu M$$

By choosing the gauge $L = R = M = \mathbf{1}$ [15], we get

$$\mathcal{L} = -\frac{v^2}{4}\Big[aTr[(W-Y)^2] + bTr[(W+Y-2V)^2] + cTr[(W-Y+2A)^2]$$
$$+ 4dTr[A^2]\Big] + \text{kinetic terms for } V_\mu, A_\mu, W_\mu, Y_\mu \qquad (2.16)$$

where

$$V_\mu = \frac{1}{2}(\mathbf{R}_\mu + \mathbf{L}_\mu), \quad A_\mu = \frac{1}{2}(\mathbf{R}_\mu - \mathbf{L}_\mu) \qquad (2.17)$$

In (2.16) we have added kinetic terms for the gauge fields W_μ and Y_μ. If one neglects the kinetic terms for V_μ and A_μ, these fields can be eliminated through the equations of motion which give

$$V_\mu = \frac{1}{2}(W_\mu + Y_\mu), \quad A_\mu = -\frac{1}{2}\frac{c}{c+d}(W_\mu - Y_\mu) \qquad (2.18)$$

Substituting in (2.16) we find

$$\mathcal{L} = -\frac{v^2}{4}(a + \frac{cd}{c+d})Tr(W-Y)^2 \qquad (2.19)$$

which is nothing but the Weinberg-Salam gauging of equation (2.1), if we require the following relation among the parameters

$$a + \frac{cd}{c+d} = 1 \qquad (2.20)$$

We will assume this relation to hold in the lagrangian (2.16) in the following.

3. MASS EIGENVALUES

From the lagrangian (2.16), by introducing explicitly the gauge couplings

$$W \to gW, \quad Y \to g'Y, \quad V \to \frac{g_V}{2}V, \quad A \to \frac{g_A}{2}A \qquad (3.1)$$

and the following notations:

$$\epsilon_V = \frac{g}{g_V}, \quad \epsilon_A = \frac{g}{g_A}$$

$$r_V = \frac{\epsilon_V^2}{b}, \quad r_A = \frac{\epsilon_A^2}{c+d}, \quad z = \frac{c}{c+d} \qquad (3.2)$$

we can obtain the mass matrix. The eigenvalues and the eigenvectors of the mass matrix in the charged sector can be determined easily in the limit $g_V, g_A \to \infty$, by taking the second order in the quantities ϵ_V and ϵ_A, neglecting terms of the order $(\epsilon_V^2 \cdot \epsilon_A^2)$. The result we get is the following:

$$m_{W^\pm}^2 = \frac{v^2}{4}g^2\left(1 - \frac{\epsilon_V^2}{1-r_V} - \frac{z^2\epsilon_A^2}{1-r_A}\right)$$

$$m_{V^\pm}^2 = \frac{v^2}{4}\frac{g^2}{r_V}\left(1 + \frac{\epsilon_V^2}{1-r_V}\right), \quad m_{A^\pm}^2 = \frac{v^2}{4}\frac{g^2}{r_A}\left(1 + \frac{z^2\epsilon_A^2}{1-r_A}\right) \qquad (3.3)$$

Notice that at the zero order in ϵ_V and ϵ_A, we have

$$r_V = \frac{m_W^2}{m_V^2}, \quad r_A = \frac{m_W^2}{m_A^2} \qquad (3.4)$$

Analogously, in the neutral sector, at the same order of approximations, we get

$$m_\gamma^2 = 0, \quad m_{Z_0}^2 = \frac{v^2}{4}G^2\left(1 - \frac{E_V^2}{1-R_V} - \frac{z^2 E_A^2}{1-R_A}\right)$$

$$m_{V_0}^2 = \frac{v^2}{4}\frac{G^2}{R_V}\left(1 + \frac{E_V^2}{1-R_V}\right), \quad m_{A_0}^2 = \frac{v^2}{4}\frac{G^2}{R_A}\left(1 + \frac{z^2 E_A^2}{1-R_A}\right) \qquad (3.5)$$

where

$$E_V = \frac{\cos 2\theta}{\cos \theta}\epsilon_V[1+4\epsilon_V^2\sin^2\theta]^{-\frac{1}{2}}, \quad E_A = \frac{1}{\cos\theta}\epsilon_A$$

$$R_V = \frac{1}{\cos^2\theta}r_V[1+4\epsilon_V^2\sin^2\theta]^{-1}, \quad R_A = \frac{1}{\cos^2\theta}r_A \qquad (3.6)$$

$$\tan\theta = \frac{g'}{g}, \quad G = \frac{g}{\cos\theta}$$

4. LOW-ENERGY PHENOMENOLOGY

In order to see the effects of the vector and axial-vector resonances on the usual phenomenology of the SM, it is necessary to derive the couplings to the fermions. We will do this assuming that the only couplings that can occur among the new particles and the fermions are through the mixing to W and Z (for a more general discussion see refs.[12)14)]). The charged couplings are defined through the formula

$$(h_W W^i_\mu + h_V V^i_\mu + h_A A^i_\mu) j^{i}_{L}, \qquad i = 1, 2 \qquad (4.1)$$

and, at the same order of approximation as in the previous Section, we get

$$h_W = g\left(1 - \frac{1}{2}\frac{\epsilon_V^2}{(1-r_V)^2} - \frac{1}{2}\frac{z^2 \epsilon_A^2}{(1-r_A)^2}\right)$$
$$h_V = -g\left(\frac{\epsilon_V}{1-r_V}\right), \quad h_A = g\left(\frac{z\epsilon_A}{1-r_A}\right) \qquad (4.2)$$

The effective four-coupling fermions comes now with a Fermi constant given by

$$4\sqrt{\frac{1}{2}}G_F = \frac{1}{2}\left[\frac{h_W^2}{m_W^2} + \frac{h_V^2}{m_V^2} + \frac{h_A^2}{m_A^2}\right] \qquad (4.3)$$

leading to

$$\sqrt{2}G_F = \frac{1}{v^2} \qquad (4.4)$$

as in the standard model. It is possible to show that this is an exact result, valid at any order in ϵ_V and ϵ_A [15)]. Analogously, the neutral current written in terms of the mass eigenstates is given by

$$j_{neutral} = [Aj_{3L} + Bj_{e.m.}]Z_0 + [Cj_{3L} + Dj_{e.m.}]V_0$$
$$+ [Ej_{3L} + Fj_{e.m.}]A_0 + ej_{e.m.}\gamma \qquad (4.5)$$

where

$$A = G\cos\xi\cos\eta, \quad B = -G\sin^2\theta\cos\xi(\cos\eta - \cot\theta\tan\xi\sin\psi)$$

$$C = G\sin\xi, \quad D = -G\sin^2\theta\sin\xi(1 + \cot\theta\sin\psi\cot\xi)$$

$$E = G\sin\eta, \quad F = -G\sin^2\theta\sin\eta, \quad e = g'\cos\theta\cos\psi \qquad (4.6)$$

Here we have defined the infinitesimal angles:

$$\xi \approx -\frac{E_V}{1-R_V}, \quad \eta \approx \frac{zE_A}{1-R_A} \qquad (4.7)$$

Neglecting the photon contribution, we get the effective neutral hamiltonian:

$$\mathcal{H} = 4\sqrt{\frac{1}{2}} G_F \rho \left[(j_{3L} + X j_{e.m.})^2 + c j_{e.m.}^2 \right] \tag{4.8}$$

where

$$\begin{aligned}
4\sqrt{\frac{1}{2}} G_F \rho &= \frac{1}{2} \left[\frac{A^2}{m_{Z_0}^2} + \frac{C^2}{m_{V_0}^2} + \frac{E^2}{m_{A_0}^2} \right] \\
4\sqrt{\frac{1}{2}} G_F \rho (c + X^2) &= \frac{1}{2} \left[\frac{B^2}{m_{Z_0}^2} + \frac{D^2}{m_{V_0}^2} + \frac{F^2}{m_{A_0}^2} \right] \\
4\sqrt{\frac{1}{2}} G_F \rho X &= \frac{1}{2} \left[\frac{AB}{m_{Z_0}^2} + \frac{CD}{m_{V_0}^2} + \frac{EF}{m_{A_0}^2} \right]
\end{aligned} \tag{4.9}$$

from which

$$\begin{aligned}
\rho &= 1 \\
X &= -\sin^2\theta (1 + 2\epsilon_V^2 \cos 2\theta) \\
c &= 4 r_V \epsilon_V^2 \sin^4\theta
\end{aligned} \tag{4.10}$$

Again, it is possible to show that $\rho = 1$ holds at any order in ϵ_V and ϵ_A [15].

5. TESTING THE MODEL

In the present Section we will review the predictions of the model concerning a number of observable quantities, by comparing them with the experimental results, if they are available, or with the SM predictions for the experiments planned in future.

The low-energy tests refer to:
(i) - the measurement of the weak mixing angle characterizing the neutral current effective lagrangian, extracted from the ratio:

$$R_\nu = \frac{\sigma_{NC}(\nu_\mu N \to \nu_\mu \, any)}{\sigma_{CC}(\nu_\mu N \to \mu \, any)} \tag{5.1}$$

(ii) - the measurement of the vector boson masses at the $S p \bar{p} S$ collider, giving:

$$m_W = 80.9 \pm 1.4 \ GeV \tag{5.2}$$

$$m_Z = 92.1 \pm 1.8 \ GeV \tag{5.3}$$

In our analysis, the value of the weak mixing angle derived from R_ν, together with the Fermi constant G_F and the fine structure constant α at $q^2 = 0$, are used as input for the prediction of the m_W and m_Z values, which in turn are compared to the experimental intervals given in (5.2), (5.3). This procedure results in a limitation on the parameter space of the model, which we will discuss in a moment[17]. We will present our results by comparing three limiting situations: a vector limit (referred to as the model V) obtained decoupling completely the axial resonance, an axial limit (denoted as model A) obtained decoupling totally the vector resonance, and a degenerate case (model D) where the vector and axial resonance are present with same coupling constants and masses $m_V \approx m_A \approx m$. For what concerns the already existing tests (the knowledge of m_W and m_Z), we get a limit almost independent of the mass of the gauge particles giving $g/g_V (g/g_A) \leq .20 \,(.10)$ in the V(A) case. The D model is much less constrained.

The high-energy tests we have chosen to discuss are the e^+e^- experiments at LEP/SLC. The analysis will be based on the following observables:

a) (m_W/m_Z) at $m_Z = 92.1 \, GeV$

b) $\Gamma(Z \to l^+l^-)$ at $m_Z = 92.1 \, GeV$

c) $A_{FB}^{e^+e^- \to \mu^+\mu^-}, P_e = 0$ at $\sqrt{s} = 92.1 \, GeV$

d) $A_{FB}^{e^+e^- \to \mu^+\mu^-}, P_e = 0.5$ at $\sqrt{s} = 92.1 \, GeV$

e) $A_{LR}^{e^+e^- \to \mu^+\mu^-}, P_e = 0.5$ at $\sqrt{s} = 92.1 \, GeV$ (5.4)

where $\Gamma(Z \to l^+l^-)$ is the Z-width into charged leptons, A_{FB} and A_{LR} are the forward-backward and left-right asymmetries defined as in [14] and P_e is the degree of longitudinal polarization of the electron beam. Here we make the working assumption that these quantities will be measured with the following precision:

$$\delta\left(\frac{m_W}{m_Z}\right) = \pm 2 \cdot 10^{-3} \quad (5.5)$$

$$\frac{\delta\Gamma(Z \to l^+l^-)}{\Gamma} = 0.02 \quad (5.6)$$

$$\delta A_{FB}(m_Z) = 0.01 \quad (5.7)$$

$$\delta A_{LR}(m_Z) = 0.02 \quad (5.8)$$

If no deviations with respect to the SM predictions are observed within the given precision, (5.5) - (5.8) will give rise to limitations on the parameter space of the model. Also, in this analysis we have traded the weak mixing angle for m_Z, in fact this parameter will be measured very precisely at LEP/SLC.

In fig. 1 we show the m_W/m_Z versus m_Z plot for the SM and for the three limiting examples we have chosen. One-loop radiative corrections are included (exactly for the SM, and in a reliable approximation in the other cases). The resonance masses are taken at $250 \, GeV$ and the resonance gauge couplings $g_V =$

$g_A = g''$ are taken such that $g/g'' = 0.1$. It is interesting to note the sign of the deviation from the SM, opposite for vector and axial-vector. For such a choice of couplings the degenerate case tends to compensate the deviations and comes closer to the SM (this is a common feature visible in this m_W/m_Z plot and also in the other cases we shall discuss). Assuming that no discrepancies with the SM are observed, the assumed experimental precision given in (5.5) provides, for each of the three models considered, a permitted region in the plane of the resonance mass and coupling. The models V and A are equally limited from a measurement of m_W/m_Z. We find $g/g_V \leq 0.07$[14] and $g/g_A \leq 0.07$, respectively. The same test provides less stringent limitations for the case D, where, for instance, we find $g/g'' \leq 0.10$ at $m = 150\ GeV$ and $g/g'' \leq 0.21$ at $m = 500\ GeV$.

In Fig. 2 we show the width for $Z \to l^+l^-$, where l is a light lepton, versus the Z-mass, again for the SM and for the V, A and D models. The resonance masses are always chosen at $250\ GeV$, whereas the resonance gauge couplings are taken for $g/g'' = 0.2$. We again note, also for this measurable quantity, that the deviation goes in opposite directions for the vector and for the axial. In the case V we do not find any limitation for a resonance mass greater than $180\ GeV$ and for g/g_V up to 0.35. In the case A essentially no limitation occurs for low resonance masses (from 180 to $220\ GeV$). However for $m_A \geq 350\ GeV$ we find $g/g_A \leq 0.24$. On the contrary, in the case D, the low resonance mass region is more limited than the high one, with a g/g'' upper limit ranging from 0.15 at $m = 180\ GeV$ to 0.32 at $m = 500\ GeV$. On the whole this test is much less stringent than the previous one.

In Fig. 3 we show versus e^+e^- c.m. energy the forward-backward asymmetry in $e^+e^- \to \mu^+\mu^-$ for the three models V, A and D in a restricted region around the Z-mass (taken at $92.1\ GeV$). Here we consider unpolarized beams and an ideal detector acceptance. In all cases the resonance masses are put at $250\ GeV$ and the couplings have the same numerical value. A more detailed view of the implications of the unpolarized forward- backward asymmetry can be obtained from a study of the contour plots in a plane having the resonance mass and the resonance couplings as coordinates, obtained assuming no discrepancies from the SM predictions within the experimental precision given in (5.7). Such a plot for the case V can be found in Fig. 10 of the first ref.[14]. The axial resonance situation would thus be more severely limited from this measurement than the vector situation. Quantitatively we find $g/g_A \leq 0.1$. On the other hand, in the case D, even the assumed precision on the measurement would practically be unable to significantly restrict the parameters of the model. A substantial increase of the deviations is obtained in the forward-backward asymmetry for a longitudinally polarized electron beam, as shown in Fig. 4. As far as the limitations on the adopted parameter planes are concerned, the electron beam polarization leads to severe restrictions in the models V and A. For the assumed experimental precision (5.7) we find $g/g_V \leq 0.08$ in the V case[14], and $g/g_A \leq 0.07$ in the case A. The compensation occurring in the case D is confirmed by the insignificant constraints we find on the parameter space.

In Figs. 5, 6 and 7 we show the predicted left-right asymmetry, divided by the polarization degree, in e^+e^- for a 50% longitudinally polarized e^- beam. The parameters are again resonance mass at 250 GeV, and gauge coupling g'' always given by $g/g'' = 0.1$. The contour plot studies, not reproduced here, in the plane of coupling versus resonance mass show that, for the experimental precision given in (5.8), this measurement can provide the constraint $g/g_V \leq 0.12$ in the V case[14], $g/g_A \leq 0.11$ in the A case and again no significant limitation in the D case.

Also, very recently, possible deviations in W pair production at $e^+ e^-$ colliders, originating from the model presented here, have been studied[18].

Let us finally conclude by mentioning a possible interpretation of the two UA1 events [19], in which a W^+ recoils against an energetic dijet system whose mass is compatible with that of a W or a Z (70 $< M_{jj} <$ 100 GeV and 250 $< M_{l^+\nu jj} <$ 320 GeV). The standard model production of this kind of event gives rates which are far of one or two order of magnitude [19]. Heavy flavor production with $M_Q > M_W$ can also produce such events. Two recent calculations have however shown that for $M_Q = 90$ GeV the expected number of event is too small [20]. Since the mass of the dijet is close to M_W or M_Z, it has been conjectured that these events (if they are not statistical fluctuation) could be due to a new heavy particle, with a mass of 250 GeV which decays predominantly into W^+, W^- [21].

It has been recently shown that the model presented here can give the predicted rate, provided one makes a convenient choice of the parameters [22][14]. In [14] we have analyzed the production rate of $\bar{p}p \to V \to W^+W^-$. Asking for a total cross-section in $l^+\nu jj$ from V^0 production greater than .5 pb with $\Gamma_{V^0} \sim 25$ GeV and combining this requirement with the low energy constraints, we obtain the allowed region in the parameter space, which is presented in Fig. 8. There b parametrizes the strength of direct couplings of the new vector particle to fermions and γ parametrizes the VWW couplings (see [14] for more details). We have considered also the M_{Wjj} distribution at $S\bar{p}pS$ and at Tevatron. We predict twenty times more events at the Tevatron collider.

REFERENCES

1) M. Veltman, Acta Physica Polonica **B8** (1977) 475.

2) B.W. Lee, C. Quigg, and H.B. Thacker, Phys. Rev. **D16** (1977) 1519.

3) L. Maiani, G. Parisi and R. Petronzio, Nucl. Phys. **B136** (1978) 115;
   N. Cabibbo et al, Nucl. Phys. **B158** (1979) 295.

4) P. Hasenfratz and J. Nager, BUTP-87/18 (October 1987).

5) J.M. Cornwall, D.N. Levin, and G. Tiktopoulos, Phys. Rev. **D11** (1974) 1145;
   C.G. Vayonakis, Lett. Nuovo Cim. **17** (1976) 17;
   M.S. Chanowitz, and M.K. Gaillard, Nucl. Phys. **B261** (1985) 379;
   G.J. Gounaris, R. Kögerler and H. Neufeld, Phys. Rev. **D34** (1986) 3257.

6) S. Weinberg, Phys. Rev. Lett. **17** (1966) 616.

7) M.S. Chanowitz, M. Golden, and H. Georgi, Phys. Rev. Lett. **57** (1986) 2344 and Phys. Rev. **D36** (1987) 1490.

8) For an analysis of unitarity bounds for various models see:
   R. Casalbuoni, D. Dominici, C. Giunti and R. Gatto, Phys.Lett. **178B** (1986) 235;
   R. Casalbuoni, D. Dominici, F. Feruglio and R. Gatto, Nucl. Phys. **B299** (1988) 117.

9) M.S. Chanowitz, and M.K. Gaillard, Phys. Lett. **142B** (1984) 85;
   M.S. Chanowitz, LBL-21973 Proc. 23th Intl. Conf. on High Energy Physics (Berkeley, 1986); in Observables Standard Model Physics at SSC: Monte Carlo Simulation and Detector Capabilities, ed. H.U. Bengtsson et al. (World Scientific, Singapore, 1986) p. 183.

10) T. Appelquist, and R. Shankar, Nucl. Phys. **B158** (1979) 317;
    T. Appelquist, and C. Bernard, Phys. Rev. **D22** (1980) 200;
    A.C. Longhitano, Phys, Rev. **D22** (1980) 1166; Nucl. Phys. **B188** (1981) 119.

11) M. Bando, T. Kugo, S. Uehara, K. Yamawaki, and T. Yanagida, Phys. Rev. Lett. **54** (1985) 1215;
    M. Bando, T. Kugo and K. Yamawaki, Nucl. Phys. **B259** (1985) 493.

12) R. Casalbuoni, S. De Curtis, D. Dominici, and R. Gatto, Phys. Lett. **155B** (1985) 95; Nucl. Phys. **B282** (1987) 235.

13) A. D'Adda, P. di Vecchia, and M. Luscher, Nucl. Phys. **B146** (1978) 73, **B152** (1979) 125;
I.Ya. Aref'eva and S.I. Azakov, Nucl. Phys. **B162** (1980) 298.

14) R. Casalbuoni, P. Chiappetta, D. Dominici, F. Feruglio, and R. Gatto, CERN-TH 4876/87, October 1987; see also:
R. Casalbuoni, Proceedings of the meeting on "Search for Scalar Particles", Trieste (1987), to be published.

15) R. Casalbuoni, S. De Curtis, D. Dominici, F. Feruglio and R. Gatto, UGVA-DPT 1988/02-564 .

16) M. Bando, T. Fujiwara, and K. Yamawaki, DPNU-86-23;
K. Yamawaki, Talk presented at the 1987 International Workshop. "Low energy effective theory of QCD", Nagoya 1987.

17) R. Casalbuoni, D. Dominici, F. Feruglio, and R. Gatto, Phys. Lett. **200B** (1988) 495.

18) P. Chiappetta, and F. Feruglio, UGVA-DPT 1988/07-581.

19) C. Albajar et al., Phys. Lett. **193B** (1987) 389.

20) P. Colas, and D. Denegri, Phys. Lett. **195B** (1987) 295;
S. Geer, G. Pancheri, and Y.N. Srivastava, Phys. Lett. **192B** (1987) 223.

21) R. Kleiss and W.J. Stirling, Phys. Lett. **180B** (1986) 171.

22) P. Chiappetta and S. Narison, Phys. Lett. **198B** (1987) 421.

23) G. Altarelli, R.K. Ellis, M. Greco, and G. Martinelli, Nucl. Phys. **B246** (1984) 12.

FIGURE CAPTIONS

Fig. 1 - The ratio m_W/m_Z versus m_Z in the four cases of : the standard model (dashed line); (V) vector resonance alone; (A) axial resonance alone; (D) degenerate vector and axial resonances. The resonance masses are taken at 250 GeV and the coupling $g_V = g_A = g''$ are such that $g/g'' = 0.1$.

Fig. 2 - The leptonic width of $Z, \Gamma(Z \to l^+l^-)$, versus m_Z for the four cases of : the standard model (dashed line); (V) vector resonance alone; (A) axial resonance alone; (D) degenerate vector and axial resonances. In all cases the resonant masses are taken at 250 GeV and the figure illustrates the case $g/g'' = 0.2$.

Fig. 3 - The forward-backward asymmetry A_{FB} in $e^+e^- \to \mu^+\mu^-$ for unpolarized beam and ideal detector acceptance in a region around the Z (taken at $92.1 GeV$) for : the standard model (dashed line); (V) vector resonance alone; (A) axial resonance alone; (D) degenerate vector and axial resonances. The resonant masses are taken at 250 GeV and the couplings such that $g/g'' = 0.1$, in any case.

Fig. 4 - The forward-backward asymmetry in $e^+e^- \to \mu^+\mu^-$ for longitudinal electron beam polarization $P_e = 0.5$ and ideal detector acceptance in a region around the Z. The parameters used and the designation of the various curves are the same as in fig. 3.

Fig. 5 - Predicted left-right asymmetry divided by the polarization degree in e^+e^- versus the center-of-mass energy, for 50 % longitudinally polarized electron beam and ideal detector acceptance. The dashed line is the standard model, the full line is the model incorporating only a vector resonance with mass taken at 250 GeV and $g/g'' = 0.1$.

Fig. 6 - Same as for fig. 5, but in the case A of an axial resonance only.

Fig. 7 - Same as for figs. 5 and 6, but for the case D of degenerate vector and axial resonances.

Fig. 8 - Allowed region for $m_{V^0} = 250\ GeV$, $\Gamma_{V^0} < 25\ GeV$ and $\sigma_{l\nu jj} > .5\ pb$ in the $(b, g/g'')$ plane: a) $\gamma = .125$, b) $\gamma = .1$, c) $\gamma = .075$, d) $\gamma = .05$ The shaded area represents the region allowed by the low energy-tests.

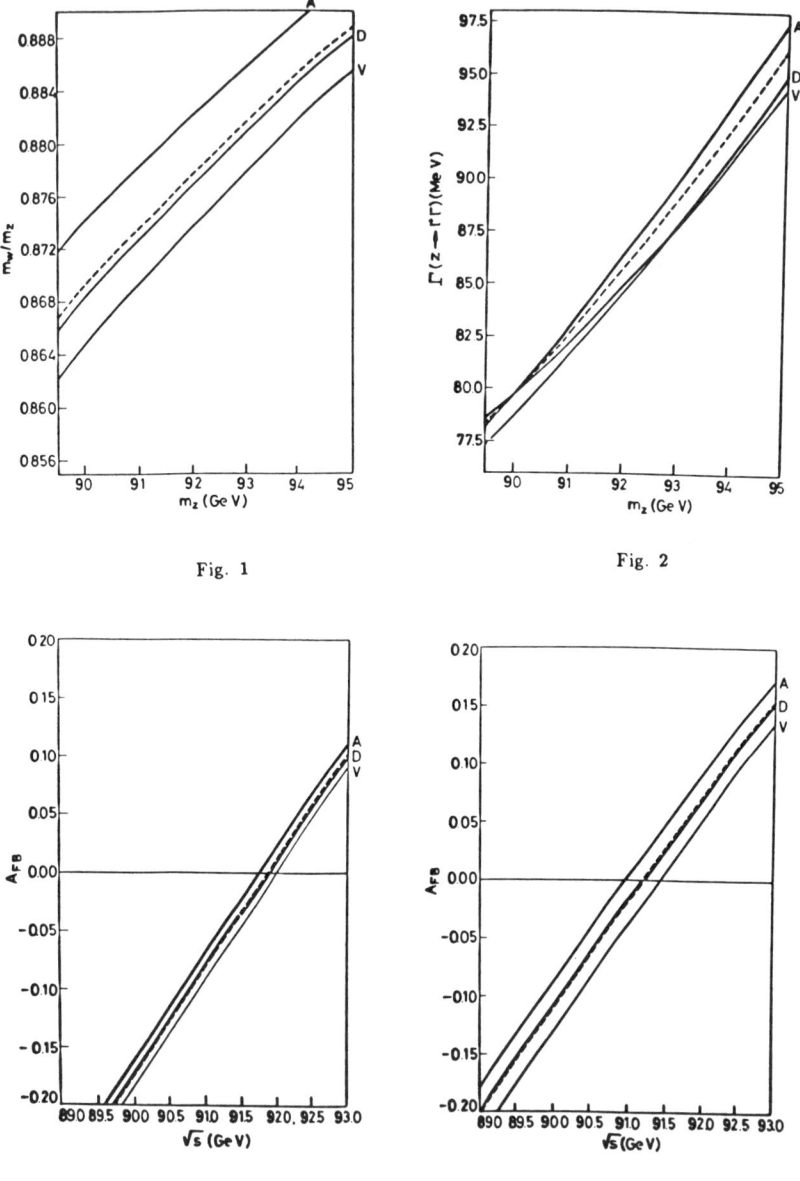

Fig. 1

Fig. 2

Fig. 3

Fig. 4

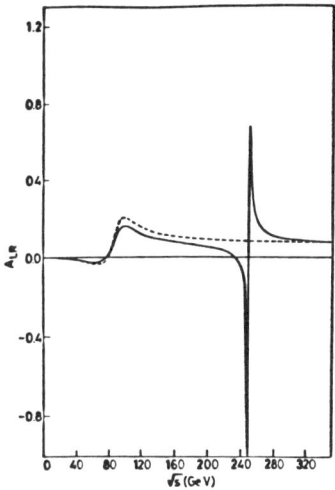

Fig. 5

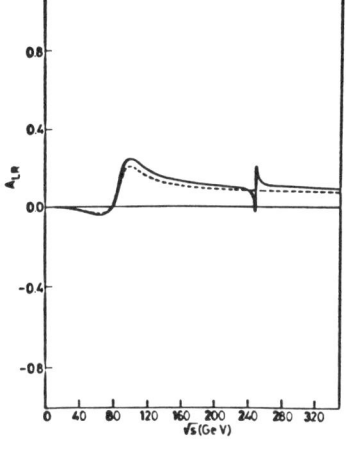

Fig. 6

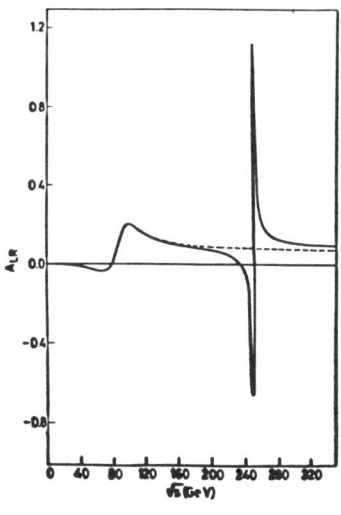

Fig. 7

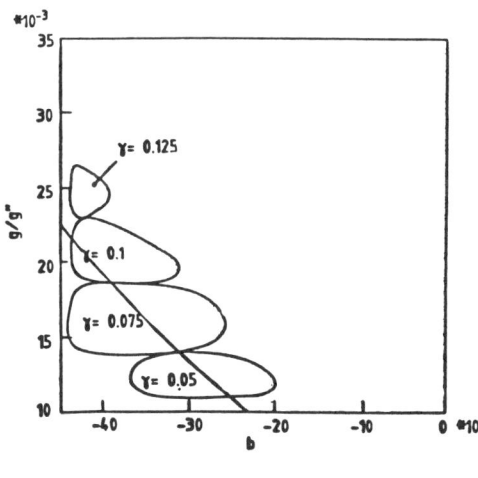

Fig. 8

THE STRONGLY COUPLED STANDARD MODEL *

Edward Farhi

Center for Theoretical Physics
Laboratory for Nuclear Science
and Department of Physics
Massachusetts Institute of Technology
Cambridge, Massachusetts 02139 U.S.A.

This talk is on the Strongly Coupled Standard Model or SCSM. The work on this model I originally did with L. Abbott[1] and then I continued working with R. Jaffe and M. Claudson.[2] The model is an alternative to the usual Weinberg–Salam theory of the weak interactions. However, I must emphasize that I know that the usual weak interaction theory works extremely well and there is no pressing need for an alternative. Yet it is interesting to explore the possibility that there may exist a phenomenologically viable alternative to the standard model. What I want to particularly emphasize are the dynamical assumptions which must be made in order for this model to work.

The model begins with the standard weak interaction Lagrangian based on $SU(2)_L \times U(1)_Y$. I call the $SU(2)_L$ coupling g_2 and the $U(1)_Y$ coupling g_1. The model has N fermions which transform as $SU(2)_L$ doublets, ψ_L^a, where $a = 1$ to N. These have hypercharges Y_a. Here I am ignoring color, it is just a subset of the labels a. For each left-handed doublet there are two right-handed singlets ψ_{R1}^a and ψ_{R2}^a with hypercharges $Y_a + 1/2$ and $Y_a - 1/2$. The theory is anomaly free. There is a scalar field φ which is an $SU(2)_L$ doublet and carries hypercharge $-1/2$. And of course there are three $SU(2)_L$ bosons and one hypercharge boson.

So far everything is exactly like in the standard model. The difference between the usual standard model and the SCSM is in the values of parameters. The $SU(2)_L$ coupling g_2 is a running coupling and can be characterized by a scale Λ_2 where it gets big. In the usual standard model the coupling never gets big at low energy because its infrared growth is cut off by the vacuum expectation value of the Higgs field which is $V \sim 250$ GeV. If V were zero in the usual standard model,

---

* This work is supported in part by funds provided by the U. S. Department of Energy (D.O.E.) under contract #DE-AC02-76ER03069.

g_2 would get large at some tiny energy and we could see that Λ_2 is small. In the SCSM we reverse Λ_2 and V. We adjust the parameters so that $\Lambda_2 \sim 250$ GeV and V is negligible. In this way the theory becomes strongly interacting and confining at the weak interaction scale.

To explore a confining gauge theory it is useful to look at its global symmetry group. For the moment, ignore g_1, all fermion masses and color. There is an exact $SU(N)$ symmetry which acts on ψ_L^a, $a = 1, N$. There is also a global $SU(2)$ which can be seen from the Higgs sector. Define a 2×2 Higgs field matrix

$$\Omega \equiv \begin{pmatrix} \varphi_1 & -\varphi_2^* \\ \varphi_2 & \varphi_1^* \end{pmatrix} \tag{1}$$

where φ_1 and φ_2 are the components of φ. The scalar potential and kinetic terms can be written as

$$\lambda \left[\operatorname{Tr} \Omega^\dagger \Omega - 2V^2 \right]^2 \tag{2}$$

and

$$\operatorname{Tr} \left[D_\mu \Omega^\dagger D^\mu \Omega \right] \tag{3}$$

where

$$D_\mu \Omega \equiv \partial_\mu \Omega + ig_2 \omega_\mu \Omega . \tag{4}$$

Left multiplication of Ω is action by the gauge group. However, right multiplication by a 2×2 unitary matrix leaves the Lagrangian invariant so the full global group is

$$G = SU(N) \times SU(2) . \tag{5}$$

Again, this $SU(2)$ is not $SU(2)_L$!

We assume that a confining gauge group produces bound states which are gauge singlets. Let's look at certain singlets. For each doublet ψ_L^a there are two ways to make a singlet with φ. Let

$$F_L^a \equiv \begin{Bmatrix} \varphi^\dagger \psi_L^a \\ \varphi \epsilon \psi_L^a \end{Bmatrix} = \Omega^\dagger \psi_L^a . \tag{6}$$

Then F_L^a is a gauge singlet which is an $(N, 2)$ under G. These F_L^a satisfy the 't Hooft anomaly matching conditions and are good candidates to be massless composite fermions.

Assumption 1: *Chiral symmetry breaking does not occur and the F_L^a are massless.*

Since the fermions satisfy the 't Hooft conditions it is reasonable that there are no other massless fermions.

We can form other $SU(2)_L$ singlet states. For example $\varphi^\dagger \varphi = \text{Tr}\,\Omega^\dagger\Omega \equiv H$ is a (1,1) under G and $\text{Tr}\left[\Omega^\dagger D_\mu \Omega \tau^i\right] \equiv W_\mu^i$ are a (1,3) under G. We can form many other $SU(2)_L$ singlet objects. For the moment focus only on F_L^a, H and W_μ^i. Turn $U(1)_Y$ back on. The $U(1)$ gauge boson a_μ remains massless since there is no spontaneous symmetry breaking. This is the photon. By adding charges of constituents we see that the two components of F_L^a have charges $Y_a \pm 1/2$. For leptons $Y = -1/2$ and the charges are 0 and 1. For quarks $Y = +1/6$ and the charges are $2/3$ and $-1/3$. Similarly the charges of W_μ^i are 1, 0, -1 and the charge of H is 0. We can therefore identify these particles as the usual left-handed fermions and the physical $W^\pm$ and Z as well as the neutral Higgs.

What about interactions? For the particles I have listed, in the absence of electromagnetism, the most general effective Lagrangian I can write down, up to dimension four is:

$$L_{\text{eff}} = \bar{F}_L^a \displaystyle{\not}\partial F_L^a + \left(\partial_\mu \vec{W}_\nu - \partial_\nu \vec{W}_\mu\right)^2 \\ + m_W^2 \vec{W}_\mu \cdot \vec{W}^\mu + \partial_\mu H \partial^\mu H + m_H^2 H^2 + \bar{g}\vec{W}_\mu \cdot \vec{j}_L^\mu \,, \quad (7)$$

where

$$\vec{j}_L^\mu = \sum_{a=1}^N \bar{F}_L^a \gamma^\mu \frac{\vec{\tau}}{2} F_L^a \,. \quad (8)$$

The exchange of the W_μ^i leads to an effective four Fermi interaction

$$\frac{\bar{g}^2}{2m_W^2} = \frac{4G_F}{\sqrt{2}} \,. \quad (9)$$

Now M_W has been measured and G_F is known so we infer $\bar{g} = 0.63$. Here $\bar{g}$ is the effective W-fermion–anti-fermion interaction. It is not the underlying gauge coupling. For the model to make sense we must make:

Assumption 2: *The SCSM has an effective W-fermion–anti-fermion interaction which is small, $\simeq 0.63$.*

You may argue that this number is just what it is observed to be and we are experimentally determining it to be 0.63. However, the analogous $\rho - N - N$ coupling is much larger. The assumption is that a strongly coupled gauge theory can have a small effective interaction.

If we include electromagnetism, we see that $g_1 = e$ since the $U(1)_Y$ is actually $U(1)_{\text{QED}}$. If we include electromagnetism by minimal covariance, then L_{eff} from (7) is modified to be

$$L_{\text{eff with }EM} = L_{\text{eff}}(\text{gauged } U(1)) \\ + \frac{k}{2}\mathcal{F}_{\mu\nu}\left(\partial^\mu W^{\nu 3} - \partial^\nu W^{\mu 3}\right) + ie\lambda \mathcal{F}_{\mu\nu} W^{\mu +} W^{\nu -} \quad (10)$$

The first term is a gauge invariant photon-W^3 mixing. The resulting mass eigenstates are the photon and Z. To diagonalize, define

$$A_\mu = a_\mu + kW_\mu^3$$
$$Z_\mu = \sqrt{1-k^2}\,W_\mu^3 \ . \tag{11}$$

Then we discover

$$\frac{m_W}{m_Z} = \sqrt{1-k^2} \tag{12}$$

and the $W^\pm, Z$ fermion interaction is

$$\frac{\bar{g}}{\sqrt{2}}\left(W_\mu^+ \cdot j_L^{(-)\mu} + W_\mu^- j_L^{(+)\mu}\right) + \frac{\bar{g}}{\sqrt{1-k^2}} Z^\mu \left(j_{3L}^\mu - k^2 J_{em}^\mu\right) \ . \tag{13}$$

This part of the model looks exactly like the usual perturbative theory if $k = \sin\theta$ where $\sin\theta$ is the experimentally observed weak mixing angle.

In the standard model there is a relation between the weak mixing angle and the other couplings. We can show that under certain circumstances such a relationship also exists in SCSM. To see this, consider the electromagnetic form factor of the left-handed electron. There are two obvious contributions to this form factor. The first comes from the direct photon electron coupling and the second from a diagram where the photon first mixes with W^3 which then interacts with the electron. These two add to give

$$F_V(q^2) = e + \frac{\bar{g}kq^2}{m_W^2 - q^2} + \text{higher order terms} \ . \tag{14}$$

Now $F_V(0) = e$ as is required. If we also require $F_V(q^2 \to \infty) = 0$, so that the electron looks pointlike at ultra-high energies, then we get $k = e/\bar{g}$. Taking the numerical values of e and $\bar{g}$ gives $k^2 = 0.22$ which is the same as $(\sin^2\theta)$ exp. Thus the SCSM has the same relations amongst observed couplings as the usual model.

Can we believe this? In hadronic physics the analogous relationship is $k_{\rho\gamma} = e/g_{\rho N \bar{N}}$ which works at the 10% level. In order for the relation $k = e/\bar{g}$ to be exact we must assume that the contributions to the electron form factor, other than from the photon-W^3 mixing, can be neglected. In other words the form factor is dominated by a single pole. The excitation spectrum should be high in mass and weakly coupled which leads to assumption 3.

Assumption 3: *The SCSM has a weakly coupled excitation spectrum much above the W and Z.*

An alternative formulation is that the W and Z are unusually light relative to the natural scale of the theory which is, say, above 500 GeV.

If you buy assumptions 1, 2 and 3 then you have a viable alternative to the standard weak interaction theory. This model has a host of excitations not to be found in the standard weakly coupled model. The physics to be discovered at high energies would be much richer than we now expect! In fact, if deviations from the standard model are discovered, physicists should look seriously at the SCSM.

REFERENCES

1. L. F. Abbott and E. Farhi, *Phys. Lett.* **101B**, 69 (1981); *Nucl. Phys.* **B189**, 547 (1981).

2. M. Claudson, E. Farhi and R. L. Jaffe, *Phys. Rev.* **D34**, 873 (1986).

Chiral Gauge Theories on a Lattice

Koichi Funakubo and Taro Kashiwa

Department of Physics, Kyushu University 33
Fukuoka 812, Japan

ABSTRACT
We formulate a chiral gauge invariant theory of lattice fermions by introducing extra degrees of freedom. It is applied to the chiral U(1) gauge theories in two and four dimensions and the effective actions of the gauge fields are calculated which indicate the mass generation of the gauge bosons. However the difficulty is pointed out to execute the perturbation with a finite gauge boson mass in four dimensions.

§1. INTRODUCTION

Chiral gauge theories have been playing an important role in elementary particle physics after the proposal of the electroweak theory. Upon quantization, they suffer from a so-called gauge anomaly which violates the gauge invariance, unitarity and renormalizability.[1] The cancellation, then, of the anomalies among the fermion contents has been a guiding principle of unification models. However, Jackiw and Rajaraman[2] showed that the chiral Schwinger model can be consistently quantized to yield the massive gauge boson. While Faddeev[3] suggested that the introduction of extra bosons through the Wess-Zumino (WZ) term[4] maintains the gauge invariance, without adjusting the fermion contents, to give a consistent quantum theory.

Faddeev's strategy becomes promising since the result of Jackiw-Rajaraman can be understood as a special gauge case of the gauge invariant theory with an extra scalar.[5]

Following this spirit, one could have a chiral gauge theory on a lattice: the famous no-go theorem prevents chiral fermions from being put on a lattice without spoiling a correct chiral anomaly.[6] For example, the Wilson's fermion action inevitably contains both chiralities due to the Wilson term introduced to avoid the species doubling. However this is not a problem since we see fermions only through the (gauge) interaction so that we could have a prototype model with left-handed fermions interacting but with right-handed ones being free. Now if one adopts the prescription of quantizing a chiral gauge theory one can have chiral gauge invariant theory on a lattice and study chiral gauge theories, the electroweak theory and the grand unified theories, by various methods as has been done with the lattice QCD.

§2. CHIRAL GAUGE INVARIANT ACTION

Here we construct a lattice fermion action which is invariant under $G_L \otimes G_R$ gauge transformation (G_L and G_R are any compact Lie groups.);

$$\psi(n) \rightarrow (h_L(n)P_L + h_R(n)P_R)\psi(n),$$

$$\bar{\psi}(n) \rightarrow \bar{\psi}(n)(h_L^\dagger(n)P_R + h_R^\dagger(n)P_L),$$
(2.1)

starting from the Wilson's lattice fermion action.[7] The result is

$$S_f[\psi,\bar{\psi},A,\theta] = S_D[\psi,\bar{\psi},A] + S_W[\psi,\bar{\psi},\theta] + S_M[\psi,\bar{\psi},\theta]. \quad (2.2)$$

where

$$S_D[\psi,\bar{\psi},A] = -\frac{1}{2}\sum_{n,\mu}[\bar{\psi}(n)\gamma_\mu(U_{L\mu}(n)P_L + U_{R\mu}(n)P_R)\psi(n+\mu)$$
$$- \bar{\psi}(n+\mu)\gamma_\mu(U_{L\mu}^\dagger(n)P_L + U_{R\mu}^\dagger(n)P_R)\psi(n)]. \quad (2.3)$$

$$S_W[\psi,\bar\psi,\theta] = \frac{r}{2}\sum_{n,\mu}[\bar\psi(n)(g_L^+(n)P_R + g_R^+(n)P_L)(g_L(n+\mu)P_L + g_R(n+\mu)P_R)\psi(n+\mu)$$

$$+ \bar\psi(n+\mu)(g_L^+(n+\mu)P_R + g_R^+(n+\mu)P_L)(g_L(n)P_L + g_R(n)P_R)\psi(n)$$

$$- 2\bar\psi(n)(g_L^+(n)P_R + g_R^+(n)P_L)(g_L(n)P_L + g_R(n)P_R)\psi(n)],$$
(2.4)
$$S_M[\psi,\bar\psi,\theta] = - M\sum_n \bar\psi(n)(g_L^+(n)P_R + g_R^+(n)P_L)(g_L(n)P_L + g_R(n)P_R)\psi(n).$$

$$P_{L,R} = \frac{1 \pm \gamma_5}{2},$$

and $\gamma_5 = i\gamma_1\gamma_2$ ($\gamma_1\gamma_2\gamma_3\gamma_4$) in two(four) dimensions. Here $h_L(n) \in G_L$, $h_R(n) \in G_R$ and ψ belongs to some representation of $G_L \otimes G_R$, and the link variables and the scalars are understood to transform as

$$U_{L\mu}(n) \to h_L(n)U_{L\mu}(n)h_L^+(n+\mu), \quad U_{R\mu}(n) \to h_R(n)U_{R\mu}(n)h_R^+(n+\mu),$$
(2.5)
$$g_L(n) \to g_L(n)h_L^+(n), \quad g_R(n) \to g_R(n)h_R^+(n).$$

S_W is needed to avoid the species doubling (in perturbative sense) and its coefficient, r, is some nonzero constant assumed r>0 for convention. S_M is introduced to control the infrared singularity even in the case of massless fermions. Our conventions of defining fields and parameters on a lattice are summarized in Ref.8.

§3. EFFECTIVE ACTION OF GAUGE FIELDS

Regarding the gauge fields as external ones, we evaluate the effective action. For simplicity we consider only the case of $U(1)_L$ gauge theory in this section. The effective action is defined by

$$e^{W[A]} = \int[d\theta]e^{S_g[A] + \Gamma[A,\theta]} = \int[d\theta]e^{S_g[A]}\int[d\psi d\bar\psi]e^{S_f[\psi,\bar\psi,A,\theta]} \quad (3.1)$$

with S_f being (2.7) and $S_g[A]$ being the Wilson action of the gauge fields written by the plaquette variables.

(i) Chiral Schwinger Model:[9,10]

In the continuum limit we have

$$\Gamma[\tilde{A},\theta] = -\frac{\tilde{e}^2}{8\pi}\int d^2x (\tilde{A}_\mu - \frac{1}{\tilde{e}}\partial_\mu\theta)[\alpha(r)\delta_{\mu\nu} - (\delta_{\mu\alpha}+i\varepsilon_{\mu\alpha})\frac{\partial_\alpha\partial_\beta}{\Box}(\delta_{\nu\beta}+i\varepsilon_{\nu\beta})](\tilde{A}_\nu - \frac{1}{\tilde{e}}\partial_\nu\theta)$$
(3.2)

where $\alpha(r)$ is a numerical constant such that $0<\alpha<4$ for $r>0$. After performing the θ-integral, provided $\alpha(r)>1$, we obtain

$$W[\tilde{A}] = \frac{1}{2}\int d^2x\, \tilde{A}_\mu(x)[\Box - \frac{\tilde{e}^2}{4\pi}\frac{\alpha(r)^2}{\alpha(r)-1}](\delta_{\mu\nu} - \frac{\partial_\mu\partial_\nu}{\Box})\tilde{A}_\nu(x).$$
(3.3)

This shows that if $\alpha(r)>1$ the resulting theory is unitary and then the transverse component of the gauge boson has a mass such that

$$m(r)^2 = \frac{\tilde{e}^2}{4\pi}\frac{\alpha(r)^2}{\alpha(r)-1},$$
(3.4)

in agreement with the previous works in the continuum theory.[5]

The condition, $\alpha(r) \geq 1$, imposes restriction on r such that $0<r\leq r_c$ contrary to the free theory where $r>0$. Here r_c is defined through $\alpha(r_c)=1$, whose numerical value is about 1.5184313. $m(r)^2$ takes a value between $\tilde{e}^2/\pi$ and infinity for $0<r\leq r_c$ in accordance with the continuum theory.

(ii) Four-Dimensional $U(1)_L$ Theory:

In contrast with the two-dimensional theory, the expansion of $\Gamma[A,\theta]$ does not end up with finite number of terms.

$$\Gamma[\tilde{A},\theta] = \int d^4x \{-\frac{e^2 c(r)}{2a^2}(\tilde{A}_\mu - \frac{1}{e}\partial_\mu\theta)^2 + \frac{e^2}{48\pi^2}\tilde{A}_\mu(\Box\delta_{\mu\nu} - \partial_\mu\partial_\nu)(-\ln\tilde{m}a)\tilde{A}_\nu$$
$$+ \frac{ie^2}{24\pi^2}\theta\varepsilon_{\alpha\beta\mu\nu}\partial_\alpha\tilde{A}_\beta\partial_\mu\tilde{A}_\nu\} + \sum_{N=3}^{\infty}\Gamma^{(N)}[\tilde{A}],$$
(3.5)

where $c(r)$ is positive definite for $r>0$ and we preserve only the divergent terms and throw away the finite terms which however contains Lorentz noninvariant pieces in the A^2 term. The third term in the braces is nothing but the WZ term and it compensates the gauge variation of $\Gamma^{(3)}[\tilde{A}]$. Here we have a quadratically divergent term in addition to the usual logarithmically divergent one which is absorbed by the wave function renormalization of the gauge field.

§4. RENORMALIZABILITY

So far we have treated the gauge fields as external ones, here we develop the weak coupling perturbation including their quantum effects. We consider a chiral G_L gauge theory in two and four dimensions with G_L being any compact Lie group and exploit the power counting argument.

We start from the action

$$S_0 = S_f[\psi,\bar{\psi},A,\theta] + S_g[A] + S_\theta[A,\theta], \qquad (4.1)$$

where

$$S_\theta[A,\theta] = \frac{\mu^2}{2e^2} \sum_{n,\mu} \text{Tr}[\, g(n)U_\mu(n)g^\dagger(n+\mu) - 1 + \text{h.c.}\,]. \qquad (4.2)$$

S_θ is needed to perform the perturbation expansion. For our purpose it is convenient to fix the gauge and extend the compact region of A_μ to infinity. In the unitary gauge ($\theta=0$), where the theory reduces to the massive Yang-Mills theory, it is apparently nonrenormalizable in four dimensions. However this difficulty may be circumvented by choosing the so-called R_ξ-gauge.(As for the gauge fixed action, see ref.8.)

In two dimensions the power-counting argument indicates that the anomalous gauge theory in two dimensions is renormalizable just as the nonanomalous ones.[11]

In four dimensions we find that, in the same way as the massive vector theory, the more WZ internal lines increase, the more insuperable divergence occurs: the nonrenormalizability comes from the massive vector propagator in the unitary gauge but from the θ propagator in the R_ξ-gauge.

§5. DISCUSSIONS

We have constructed a chiral gauge invariant action of lattice fermions by introducing extra degrees of freedom, WZ scalars without conflicting with the no-go theorem. Based on this action we have calculated the fermion determinant of the chiral U(1) gauge theories in two and four dimensions. They are shown to have not only the WZ term, which is also generated in the continuum theory with a Yukawa coupling of fermions and the WZ scalar[12], but also the A^2-term whose coefficient depends on the regularization parameter, r. The latter causes the mass generation of the gauge boson. Next we have examined the renormalizability of chiral nonabelian gauge theories in terms of the power counting rule, adding to the theory the kinetic term of the WZ scalar which is also the mass counterterm of the gauge boson. The theory is found to be renormalizable in two dimensions but not in four dimensions.

Now we add some comments on these results.

1) The A^2-term in the fermion determinant exists also in the continuum theory depending on the regularization scheme, but it has been considered to be eliminated by a local counterterm until noted that this term is responsible for the mass generation of the gauge boson to make the chiral Schwinger model consistent. Its existence is due to the chiral coupling nature of the gauge fields so that it has the same origin as the gauge anomaly. Indeed if we use the Dirac fermion action instead of the Wilson, both the anomaly and the A^2-term do not appear.

2) The renormalizability in two dimensions guarantees that one can safely take $a \to 0$ limit under the weak coupling expansion. On the contrary the nonrenormalizability in four dimensions means that it is impossible to take $a \to 0$ limit in the weak coupling expansion with a finite gauge boson mass. Therefore the theory should be regarded as an effective low energy theory up to the energy level of the cut-off$\sim 1/a$, the gauge boson mass scale.[13] Or the theory may yield some meaningful results after a full analysis. It is, in principle, possible since the theory is defined on a lattice and the continuum limit will be achieved if some ultraviolet fixed point exists at nonzero coupling

constant. The resultant theory might effectively be described by a theory which contains topological configurations such as vortices.[14]

Finally we remark on other attempts to study the chiral gauge theories on a lattice. One of them is the lattice Weinberg-Salam model[15] in which the Wilson term is generated by the Higgs mechanism.[16] This is intriguing in that it provides an origin of the Wilson term while our theory is phenomenologically more favorable in that it does not contain unobserved particle. This theory might be power-counting renormalizable but has a serious problem pointed out by J.Smit[17]: demanding the decoupling of the species doublers with the gauge boson mass kept finite, one must treat the fermion-Higgs coupling nonperturbatively. Another is to integrate out the WZ scalars at first as done by S.Aoki.[10] The theory then reduces to a chiral gauge invariant one with higher order interactions of fermions, which shows the renormalizability by the power counting. However in this approach the mass generation of the gauge bosons may be unclear. Moreover, in a perturbative sense, one has additional fermions because of the disappearance of the Wilson term by the θ-integral: this clearly changes the initial motive; we do not want to introduce additional fermions but scalars to cancel the anomaly.

REFERENCES

1. D.J.Gross and R.Jackiw, Phys.Rev. D6, 477 (1972).
2. R.Jackiw and R.Rajaraman, Phys.Rev.Lett. 54 1219, 2060(E) (1985).
3. L.Faddeev, in "Supersymmetry and its applications", edited by G.W.Gibbons, S.W.Hawking and P.K.Townsend (Cambridge University Press, Cambridge, 1986);
   L.Faddeev and S.Shatashvili, Phys.Lett. 167B, 225 (1986).
4. J.Wess and B.Zumino, Phys.Lett. 37B, 95 (1971).
5. R.Banerjee, Phys.Rev.Lett. 56, 1889 (1986); K.Harada, H.Kubota and I.Tsutsui, Phys.Lett. B173, 77 (1986); R.D.Ball, Phys.Lett. B183, 315(1987).
6. H.B.Nielsen and M.Ninomiya, Nucl.Phys. B185, 20 (1981);B193, 173

(1981); L.H.Karsten, Phys.Lett. 104B, 315 (1981).
7. K.G.Wilson, in "New Phenomena in Subnuclear Physics", edited by A.Zichichi (Plenum Press, New York, 1977).
8. K.Funakubo and T.Kashiwa, Phys.Rev.D.(to be published)
9. K.Funakubo and T.Kashiwa, Phys.Rev.Lett. 60, 2113 (1988).
10. S.Aoki, Phys.Rev.Lett. 60, 2109 (1988) and ref.11; T.D.Kieu, D.Sen and S.-S.Xue, Edinburgh preprint 88/432 (1988).
11. K.Shizuya, Nucl.Phys. B121, 125 (1977).
12. E.D'Hoker and E.Farhi, Nucl.Phys. B248, 59 (1984).
13. R.Jackiw and K.Johnson, Phys.Rev. D8, 2386 (1973).
14. D.Foerster, Phys.Lett. 77B, 211 (1978); M.Stone and P.R.Thomas, Phys.Rev.Lett. 41, 351 (1978); J.M.Cornwall, Nucl.Phys. B157, 392 (1979); Phys.Rev. D26, 1453 (1982).
15. P.V.D.Swift, Phys.Lett. 145B, 256 (1984); J.Smit, Acta Phys. Pol. B17, 531 (1986).
16. L.H.Karsten, in "Field Theoretical Methods in Particle Physics", edited by W.Ruhl (Plenum Press, New York, 1980).
17. J.Smit, talk at Seillac LATTICE 87 meeting, preprint ITFA-87-21 (October, 1987).

CONCLUDING REMARKS

Y. Nambu[†]

Enrico Fermi Institute and Department of Physics
University of Chicago, Chicago, IL 60637 USA

I am very happy to have been invited to this workshop on strong coupling gauge theories. I am happy because, being a non-expert, I have learned what the latest topics and problems are in many lively sessions which covered subjects ranging from technical controversies around technicolor theories, solutions of the mass gap equation (Schwinger-Dyson Industry), progress in lattice gauge theories (sheer computing power as well as cleverness), speculations about a new phase in the good old QED (any relation to the e^+e^- peaks at GSI?), hidden symmetries in hadron dynamics and beyond (are some gauge fields composite?), etc. Nevertheless it seems fair to say that no really significant results or firm conclusions have emerged. Of course one does not expect such things to happen often or as planned. Most of the time physics is hard work, frustration, and some wishful thinking. So instead of summarizing the conference, which I am not qualified to do any way, I would rather like to talk about another reason why I am happy to be here. Nagoya has a special meaning to me: it is the place where the great physicist Shoichi Sakata (1911-1970) spent his most fruitful years. To me, and to my generation of physicists in the immediate postwar period in Japan, Sakata gave a lasting impact. He had a penetrating view of the way particle physics has developed, especially since Yukawa, and how it should and will develop in the future. Not only did he articulate it forcefully in many writings and lectures, but also he practised it with some notable successes. Among his contributions are the two-meson

---

[†] Work supported in part by the National Science Foundation: grant no. PHY-85-21588.

hypothesis long before the discovery of the pion, the C-meson theory of self-energy, and the advocacy of a composite model of hadrons. I will come back to them in a moment, but I also recall what his close friend M. Taketani said about the course of particle physics. The development of particle physics proceeds through various stages, insisted Taketani following a dialectic philosophy. There are times when we are faced phenomena which we do not understand. So we analyze the data, trying to find some rules or regularities behind them. In other words we do phenomenology. There are times when we have found some regularities, and begin to build models to explain them, then further test the models by more experiments. This is of course not yet the ultimate goal. Eventually there will come a stage where we have found a true theory, in a precise mathematical form, which underlies the phenomena and the models, or at least we get convinced of its basic correctness. But the happy situation may not last very long. Eventually there will come another time when we are faced with an unexpected phenomenon, so the process will have to repeat itself. You will agree that the development of particle physics since the Thirties up to the recent years nicely fits this characterization. The important thing for a physicist, then, is to be properly aware of what stage he or she is in at a given moment, and recognize the larger picture of physics as a human endeavor. On the other hand, it is dangerous to take anything for a dogma. History does not repeat the past exactly. Research is a creative process.

With this in mind, a few years ago I tried to characterize the trends in particle theory in terms of various operating modes.[1] Making a slight revision, I would like now to identify three different modes:

Yukawa-Sakata Mode

Einstein Mode

Dirac Mode

In the Yukawa-Sakata Mode, one tries to explain phenomena in terms of particle models. In other words, one freely assumes, or invents, new particles as necessity arises. By nature it is ad hoc and heuristic. Nowadays one takes this mode of operation for granted, but it was a radical approach in the Thirties, and has proved to be enormously successful in the subsequent development of particle physics.

In the Einstein Mode, on the other hand, certain physical principles reign supreme. Theories are constructed according to the proclaimed principles, and then tested against reality. In particle physics, the symmetry principle is a prime example. It was absent in Yukawa's approach, as has been pointed out by Maki, but it has also proved to be extremely potent and indispensable when combined with the Yukawa-Sakata Mode. It has further led to the principle of gauge fields and to the unified field theories as the basis for model building.

The Dirac Mode is somewhat different from the Einstein Mode. Theories are created for the sake of their inner mathematical beauty as if they were works of art. Paraphrasing Dirac when he invented the theory of magnetic monopoles (and at the same time the theory of fiber bundles), we might say, "It looks so beautiful that nature must have adopted it." But experimental confirmation, although essential in order to qualify as science, becomes a somewhat distant goal: "Perhaps some day somehow it will find relevance in some form or other. For the moment, we will pursue it for its own inner logic and beauty." You will realize that this is the prevailing mode that is operating at present: supersymmetry, superstring theory, Kaluza-Klein paradigm, even a p-adic numbers game. (My colleagues at Chicago seem to like my characterization of the present situation as "postmodern physics".) I suppose the trend has come about mainly because of an impasse we seem to have reached along the other modes. In the past ten years or so there has not been any real progress on the experimental side, which used to go, until then, hand in hand with theoretical progress in the other modes. Proton decays and monopoles have not been found. The top quark and the Higgs are yet to be produced or detected. On the other hand, our mathematical power and imagination have soared to the realms of the Planck mass without, however, being able to say much about the currently accessible energy regime.

What kind of real benefits will come out of the current stage? It is not likely that a convincing "Theory of Everything" will have been found in the near future. How long will this stage last? That will depend on what the coming generation of accelerators will reveal to us. What will be the next successful theories? Instead of an answer which I do not have, I would like to look back again into the past, and pick some key events in which there were competing theories but one of them came out the winner. Viewed in this way, history is a sequence of turning points. By contrasting opposing theories, one can learn something which one would not learn from the winners alone. Moreover, an idea which was not right in one instant often re-emerges in later occasions and may turn out to be the right one. Interesting ideas never really die. Below are some notable examples.

1. The cosmic ray meson puzzle: the two meson theory versus the strong coupling theory.

This refers to the problem that arose when the new particle discovered in the cosmic rays by Anderson and others was identified with one predicted by Yukawa as the quantum of nuclear forces. The cosmic "mesons" were created with large cross sections in the upper atmosphere, yet they came down to the ground with little interaction thereafter. Many theorists (Heitler, Wentzel, Tomonaga, Pauli, Oppenheimer, Schwinger, etc.) being psychologically resistant to the proliferation of new particles, tried to explain it away as dynamical effects of strongly coupled meson fields. The puzzle, however, was simply resolved by the two-meson hypothesis of Sakata: The cosmic ray particle (muon) is different from Yukawa's particle (pion); the latter decays into the former after production. (The later proposal of Marshak and Bethe was similar in spirit but not correct regarding spin assignment.) Of course the puzzle still remains, in a different sense: "Who ordered the muon?" Or who needs the generations?

The strong coupling theory of Wentzel and others is not a total loser either. It predicted the existence of nucleon isobars, which was at least partially vindicated later by the delta (3-3) resonance. Here is a back-of-an-envelope explanation of the relevance of the strong coupling theory: As a simplified model, consider the scattering of a scalar meson by a nucleon. The two Compton-like Feynman diagrams (crossed and uncrossed) that exist for a neutral meson cancel each other if nucleon recoil is neglected, so the cross section tends to be suppressed, as would be desirable to explain the puzzle. For a charged meson, however, only one of the diagrams is allowed, so there is no suppression. But if excited nucleons carrying higher charges exist, there will again be both types of diagrams which tend to cancel.

The strong coupling theories with which we are grappling now, as in this conference, are much more sophisticated. But I note, for one thing, that Wentzel had already used a lattice theory, which was later reinvented. For another thing, the Skyrmion model shares some properties of the old-fashioned strong coupling theory.

2. The infinite self-energy problem: renormalization versus cohesive fields.

This age-old problem is now commonly disposed of by invoking the idea of renormalization. But at the time of the birth of QED, Tomonaga was helped in developing his renormalization theory by the competing theory of Sakata because he could compare them and sharpen his arguments. Sakata had proposed to render the self-energy of electron finite by a compensating massive neutral scalar field. An identical proposal was also made independently by Pais. (The scalar coupling f must be related to the electric charge e by $f^2 = 2e^2$.) But this realistic version of the more formal regulator method introduced by Pauli later, did not work for the vacuum polarization.

Since then, renormalizability has become a general principle on which to build quantum field theories, and its deep physical meaning has been clarified by Wilson. But superstring has again raised the possibility that there might be finite and realistic theories.

3. Hadron dynamics: quark model and QCD versus bootstrap.

Before the quark model established itself, there was a competing view that the hadrons form a self-consistent set of states in the sense that exchange of hadrons between two hadrons is responsible for their existence as composites of each other. It was a dynamical theory which, however, failed to make clear predictions. This was in sharp contrast to the quark model, which did not have any dynamics but succeeded brilliantly in terms of phenomenology. In the above mentioned three stages scheme, one was still at the first two stages; the third stage would arrive later in the form of QCD.

But Chew's bootstrap idea did not die. It led to the concept of Regge trajectories, then to duality, which in turn led to the Veneziano model, then to the string model. Now with superstring theory we have come full circle; it claims to have subsumed QCD in a Hegelian synthesis.

A corollary to this development is the problem of invisible quarks: fractionally charged confined quarks versus integrally charged nonconfined quarks. Fractional charges and confinement being a revolutionary concept, it required an equally unprecedented concept of asymptotic freedom for the whole scheme to gain acceptance. Note that QCD and string model stand together on the same side in this confrontation. Both can explain confinement.

One theme that emerges from these observations is that some concepts have a long life. They may not always be right, but they can undergo reincarnations and become relevant later ("old wine in a new bottle", a remark we have heard). The Kaluza-Klein theory has made a comeback after 60 years. The hadron dynamics characterized by composite bosons and chiral symmetry breaking may be repeated at the electroweak scale, as many have suggested. I hope the historical review I have presented here may be of some help in anticipating the next stage of particle physics that may be just over the horizon.

1). Y. Nambu, Prog. Theor. Phys. Suppl. **85** (1985) 104.

CLOSING ADDRESS

Yoshio Ohnuki
Department of Physics, Nagoya University
Nagoya 464-01, Japan

Before closing the Workshop, where more than 30 papers have been presented, on behalf of the Organizing Committee I would like to extend my sincere thanks to those speakers and those session chairpersons who elaborately prepared their direct contributions to the Workshop. Though it has been our first experience to have an interantional workshop of this size, I believe that the Workshop has been very successful and fruitful. This is of course due solely to the dedicated and ardent discussions by all participants, for which I wish to express my hearty appreciation. I am also very grateful to all members of the secretary group including wakatés (young stars) who have continually collaborated to carry out elaborate tasks needed for the Workshop. Further, Prof. Matsuura and Prof. Nakano of Aichi University, though they are not physicists, have kindly made so valuable contributions that are indispensable to the office management of the Workshop and also to the ladies' program.

Moreover, this Workshop has been financially supported by many institutions and foundations, that are Aichi University, the respective Laboratories of Particle Physics in Nagoya and Hiroshima Universities, Miyoshi-cho, Casio Foundation, Daiko Foundation, Inoue Foundation for Science, Ishida Foundation, Kajima Foundation, Kato Foundation of Nagoya University, and Shimazu Foundation. In addition, Aichi University has generously offered the confatable meeting places for our Workshop. I would like to express my deep gratitude to all of them.

Finally in closing the Workshop I truely hope that based on vigorous discussions and valuable outcomes of our Workshop new developments of particle physics with deeper understanding of substructure of the microscopic world will be attained in near future.

Thank you.

LIST OF PARTICIPANTS

ABE, Y.	Mie Univ.
AKIBA, T.	Tohoku Univ.
AOKI, K-I.	RIFP, Kyoto Univ.
APPELQUIST, T.W.	Yale Univ. – U.S.A.
ARAMAKI, S.	Nagoya Univ.
ASANO, H.	Niigata Univ.
ATKINSON, D.	Univ. of Groningen – the Netherlands
BANDO, M.	Aichi Univ.
BARDEEN, W.A.	Fermilab–U.S.A.
BERNARD, v.	MIT–U.S.A.
CALDI, D.G.	Univ. of Connecticut – U.S.A.
CASALBUONI, R.	Univ. of Lecce–Italy
DAGOTTO, E.R.	Univ. of Illinois – U.S.A.
FARHI, E.	MIT – U.S.A.
FUJIKAWA, K.	RITP, Hiroshima Univ.
FUKUDA, R.	Keio Univ.
FURUI, S.	Ibaraki Univ.
FUSAOKA, H.	Aichi Medical Univ.
HASEBE, K.	Aichi Univ.
HATTORI, T.	Tokushima Univ.
HIGASHIJIMA, K.	KEK
HIRATA, Y.S.	Kitazato Univ.
HOLDOM, B.	RIFP, Kyoto Univ. and Univ. of Toronto – Canada
HOSHINO, Y.	Hokkaido Univ.
IGI, K.	Univ. of Tokyo
IMACHI, M.	Kyushu Univ.
IMAI, H.	Niigata Univ.
INAZAWA, H.	Shoin Women's Univ.
INOUE, M.	Hiroshima National Coll. of Maritime Tech.
KANADA, H.	Niigata Univ.
KASHIWA, T.	Kyushu Univ.
KATSUBE, K.	Kanazawa Univ.
KIKKAWA, K.	Osaka Univ.
KIKUCHI, H.	Tohoku Univ.
KIKUGAWA, M.	Hiroshima Univ.
KIKUKAWA, Y.	Nagoya Univ.
KIMURA, T.	Chiba Univ.
KIMURA, M.	Saitama Univ.
KITAMURA, I.	Chubu Univ.
KOBAYASHI, A.	Niigata Univ.
KOBAYASHI, T.	Tokyo Metropolitan Univ.
KODAIRA, J.	Hiroshima Univ.
KOGUT, J.B.	Univ. of Illinois – U.S.A.
KOIDE, Y.	Univ. of Shizuoka
KOMACHIYA, M.	Keio Univ.
KONDO, K.-I.	Nagoya Univ.
KORETUNE, S.	Fukui Medical School

KUGO, T.	Kyoto Univ.
KUNIHIRO, T.	Ryukoku Univ.
KUNITOMO, H.	Kyoto Univ.
LOW, M.F.	Waseda Univ. and Univ. of Sidney – Australia
MAEDAN, S.	Kanazawa Univ.
MAKI, Z.	RIFP, Kyoto Univ.
MASKAWA, T.	RIFP, Kyoto Univ.
MATSUBARA, Y.	Nanao Junior College
MATSUDA, M.	Aichi Univ. of Education
MATSUNAGA, M.	Mie Univ.
MATUMOTO, K.	Toyama Univ.
MEISSNER, U.-G.	MIT – U.S.A.
MIDORIKAWA, S.	INS, Univ. of Tokyo
MINAKATA, H.	Tokyo Metropolitan Univ.
MINO, H.	Nagoya Univ.
MIRANSKY, V.A.	Nagoya Univ. and Inst. for Theor. Phys., Kiev – U.S.S.R
MITANI, N.	Hiroshima Univ.
MORII, T.	Kobe Univ.
MORITA, K.	Nagoya Univ.
MOROZUMI, T.	KEK and Kyoto
MOSHE, M.	Technion – Israel
MUTA, T.	Hiroshima Univ.
NAKAMURA, A.	Freie Univ., Berlin – F.R.G.
NAKANISHI, K.	Kyoto Univ.
NAKATANI, H.	Nagoya Univ.
NAMBU, Y.	Univ. of Chicago – U.S.A.
NG, Y.J.	Univ. of North Carolina – U.S.A.
NINOMIYA, K.	Nihon Fukushi Univ.
NISHIJIMA, K.	RIFP, Kyoto Univ.
NISHITANI, T.	Kikuchi College of Optometry
NOJIRI, M.	Kyoto Univ.
NOJIRI, S.	Kyoto Univ.
NONOYAMA, T.	Nagoya Univ.
OGAWA, S.	Nagoya Univ.
OHNO, S.	Kanazawa Univ.
OHNUKI, Y.	Nagoya Univ.
OKAWA, M.	KEK
OTOFUJI, T.	Akita Univ.
SAITO, J.	Hiroshima Univ.
SAITO, S.	Nagoya Univ.
SAKAKIBARA, K.	Kanazawa Univ.
SAKAMOTO, M.	Kobe Univ.
SAWADA, S.	Nagoya Univ.
SEO, K.	Gifu City Womens's Junior Coll.
SHIOZAKI, T.	Nagoya Inst. of Tech.
SHIOKAWA, K.	Kanazawa Univ.
SO, H.	Niigata Univ.
SUEHIRO, K.	Kyoto Univ.
SUEMATSU, D.	Kanazawa Univ.
SUGIYAMA, T.	Nagoya Univ.
SUWA, M.	Niigata Univ.
SUZUKI, C.	Nagoya Univ.

SUZUKI, T.B.	Nagoya Univ.
TAKEUCHI, T.	Yale Univ. – U.S.A.
TANABASHI, M.	Nagoya Univ.
TERASAKI, K.	RITP, Hiroshima Univ.
TERAZAWA, H.	INS, Univ. of Tokyo
UKITA, M.	Keio Univ.
WAKAIZUMI, S.	Tokushima Univ.
YAMADA, M.	Ibaraki Tech. Junior Coll.
YAMAWAKI, K.	Nagoya Univ.
YANAGIDA, T.	Tohoku Univ.
YASUE, M.	INS, Univ. of Tokyo
YASUNO, M.	Nagoya Univ.
YOKOTA, T.	Japan Atomic Energy Res. Inst.
YOTSUYANAGI, I.	Kanazawa Medical Univ.
YU, H.-L.	Academia Sinica – Taiwan